Automated Data Analysis Using Excel

Automated Data Analysis Using Excel

Brian D. Bissett

Pfizer Global Research and Development
Groton, CT, U.S.A.

Chapman & Hall/CRC
Taylor & Francis Group

Boca Raton   London   New York

Chapman & Hall/CRC is an imprint of the
Taylor & Francis Group, an **informa** business

Chapman & Hall/CRC
Taylor & Francis Group
6000 Broken Sound Parkway NW, Suite 300
Boca Raton, FL 33487-2742

© 2007 by Taylor & Francis Group, LLC
Chapman & Hall/CRC is an imprint of Taylor & Francis Group, an Informa business

No claim to original U.S. Government works
Printed in the United States of America on acid-free paper
10 9 8 7 6 5 4 3 2 1

International Standard Book Number-13: 978-1-58488-885-7 (Softcover)

Visit the Taylor & Francis Web site at
http://www.taylorandfrancis.com

and the CRC Press Web site at
http://www.crcpress.com

Dedication

For Crocetta, my baby ...

Table of Contents

Preface

"First it was, "Don't worry about the manufacturing jobs, electronics is where our future is."
Then it was "Don't worry about the electronics jobs, software is where our future business will
be." Did it ever occur to these genius MBAs that once they had the manufacturing and the
electronics industries it was only a matter of time before the software went, too?"

Thomas A. Freehill, 1952–2004

Thoughts on World Trade and Outsourcing

Software development has to be one of the more frustrating jobs in the world, primarily because developers spend a great deal of time analyzing why things that should work do not work. Often times a bug in the application itself prevents a property or method from functioning properly or even at all. The more routine the application, the less likely the developer is to run into problems as most of the bugs will already have been discovered and either fixed or have had some kind of work around readily published and available for use. Sadly, when someone tries to take on the task of doing something radically different that might yield a real time savings and competitive advantage, they can often run into a problem not of their own doing because they are in essence operating "off the grid" and in untested water. In such instances the developer is in effect doing the beta testing for the more sophisticated processes in the product. The really frustrating thing for a developer is that when they find such a bug or oversight in a product that is readily repeatable, but it falls within the space of a product that is seldom utilized, the software company will usually not be willing to fix it. The reason for this is understandable in that often times when software is fixed, something somewhere else in the application gets broken as a result of the fix. Why take a chance on fixing a rarely used functionality if it might end up compromising a more widely utilized area of the application? Further, if such a small percentage of the user base utilizes this area of the application, why bother to invest resources to fix it when they could be enhancing the areas of the application utilized by the masses. For an MBA, this school of thought would seem reasonable but in reality it is extremely self-limiting. As a result of this thinking, the complexity of many applications quickly reaches an asymptotic limit as companies are not willing to invest the resources to fix and upgrade applications in areas which will not reap significant financial gain.

I think for this reason open source code will become the way of the future. When the source code is available for anyone to review, there are thousands of eyes looking at the code and proposing fixes and upgrades that are then looked at by a governance committee. For many applications today, the only eyes who can look at the source code are those developers who work at the software company that produces the application. Many of those developers are simply at the company to earn a living as opposed to the precious few who live and breathe to code. These are the people who are primarily responsible for making the application robust and accurate. The problem is no company can ever get enough of these people.

We have all seen the headlines of things that can result from an undiagnosed software glitch, the parachute fails to deploy on a billion-dollar spacecraft, the airplane crashes, the cancer patient gets a lethal dose of radiation or chemotherapy, or the structure some poor, unfortunate person happens to be depending on collapses. These are unnecessary costs to society and really hinder our ability to develop intellectually as a society. Simply put, too many people are duplicating the same work, and most of them are doing what human beings do too well — making mistakes.

TARGET AUDIENCE

I have operated "off the grid," in terms of the software I have developed, for years. In my opinion, it is the only interesting place to be in software development. When I end up doing routine things out of necessity I feel more like a file clerk as opposed to an engineer or scientist. I started writing algorithms for experimental property determinations in Origin in the early 1990s. From that point forward, things just got out of control, and I started using any software tool or combination of software tools to create custom applications to get the results required for a particular task or drug-screening method.

When I have presented at conferences the overwhelming response from people seems to be "Wow, I didn't know you could do that with tool X." Most jobs that I have started *I* didn't know could be done using the tool I was using. I just started designing the application to the specifications of the job requirement and tackled the problems as I encountered them. When a tool could not perform a task the job required, I simply looked for another tool that could and tried to integrate it into the system somehow. Some of the projects turned out very elegant, others worked but were nothing to brag about, but in the end the task got done and that was all that mattered.

I think the type of person who will get the most out of this text is someone who has an engineering or scientific background and has some experience coding in a higher level language. If the reader does not know VBA, do not despair; there are many excellent books on VBA (my favorites are from QUE), and this book would make an ideal companion to an introductory type text. Where most texts tend to leave off, this one picks up.

A programming purist may not get a lot out of this text because in their mind it may break some rules. It reminds me of the constant arguments I see between mathematicians and engineers. An engineer will use any equation provided it can be proven to work over a specified interval. Mathematicians, however, want a proof for everything. No proof equals no good in their minds. This differential in thinking can lead to some really interesting (and sometimes heated) arguments.

Another person who stands to gain a great deal from this book is a person who works at a small company and does not have the resources to buy a lot of software to do different things. This book will show them how to stretch Excel's capabilities to the extreme limits and to do the most with the one package nearly every business has — Microsoft Office.

Oddly enough, people at the other extreme who work at big companies also stand to get a lot out of this text. Large corporations tend to queue up those who request technical resources. The person who has the task deemed most important, or who is a political animal, tends to get their resources while others can be left hanging for quite some time. So if you happen to be a newcomer to an organization and would like to make an impact but have no clout, this book can help you bypass the system and do things yourself. Be advised, however, that such behavior can earn you labels such as "difficult," "a nonconformer," or "not a team player." Such labels are extremely popular with the IT people you may bypass or other more senior associates if your projects should show more momentum. Remember, sticks and stones may break your bones but whiners will never hurt you.

The Author

Brian Bissett holds graduate degrees in Business (MBA) and Electrical Engineering (MSEE) from Rensselaer Polytechnic Institute in Troy, NY and is employed as a Staff Scientist in the Molecular Properties Group of the Exploratory Medicinal Sciences department at Pfizer Global Research and Development in Groton, CT. He started working at Pfizer nearly thirteen years ago for Dr. Christopher Lipinski who is best known for his discovery of the Rule of 5, and continued to work for Chris until his retirement from Pfizer in 2002. He then reported to Dr. Franco Lombardo, who is well known for his development of an outstanding experimental LogD method (ElogD), until

Franco's recent departure to work at Novartis Pharmaceuticals in Cambridge, MA. Presently he reports to Dr. Alan Mathiowetz, Director of Computational Chemistry and Molecular Properties, constructing assays and automating measurements and data analysis routines for parameters such as ElogD/P, pKa, Solubility, Permeability, and Stability. He is a member of both the national chapter of The Institute of Electrical and Electronics Engineers (IEEE) and the IEEE's Consultants Network of Connecticut. He is the author of the textbook "Practical Pharmaceutical Laboratory Automation" (May 2003), numerous technical articles, and co-author of several book chapters. He has also been credited in many papers for creating automated data analysis routines to automate the analysis of pharmaceutical drug discovery parameters such as ElogD/P, pKa, Solubility, and Stability. Many of these routines are utilized world wide across numerous Pfizer sites.

His undergraduate work was done at The University of Rhode Island, where he received a degree in Electrical Engineering and worked as a co-op student at the Naval Undersea Warfare Center (NUWC). Prior to working at Pfizer, he worked on site at NUWC on projects ranging from digital design on the New Sonar Intercept System (NSIS) for Robert J. White, to analog preamplifier testing and fabrication for towed arrays for Thomas A. Freehill, technical director of the Hybrid Microelectronics Laboratory.

He is a product of Endicott, NY known as "Home of the Square Deal" to locals. The idea of the "Square Deal" came from George F. Johnson's philosophy of "a fair day's work for a fair day's wage." It was George F. Johnson who founded Endicott Johnson Corporation (a shoe manufacturing company) in the 1890's. In 1906 a new company, Bundy Manufacturing (a.k.a. International Time Recording Company), broke ground in Endicott. At the time many locals considered those who chose to work at Bundy Manufacturing fools if they had the opportunity to work at Endicott Johnson. Today however the world knows Bundy as International Business Machines, but few are aware that IBM started in Endicott.

Among his favorite pastimes are rollerblading (his favorite trail is from South Beach to Oak Bluffs in Martha's Vineyard, MA) and mountain biking. In the winter he is a competitive shooter on the Pfizer Pistol Team and is a Benefactor member of the National Rifle Association (NRA). In the summer he can usually be found on a beach somewhere eliminating any vitamin D deficiency he might have.

NOTICE ON SOFTWARE COMPATIBILITY

The software contained on the CD-ROM was developed under both the Windows 2000 and Windows XP Operating Systems using Microsoft Office versions 2000 and 2003. At the time of this writing, Windows XP had two major service packages, Windows 2000 had four major service packages, Office 2000 had three major service packages, Office XP had three major service packages, and Office 2003 had two major service packages. It is beyond the ability of the author to test the software on every possible configurable permutation of Microsoft Office and Windows, not to mention all the minor upgrades that a system may have which support other Microsoft products or security updates which may be system specific. It would not be unexpected therefore for some of the examples to fail to work properly on some systems while working flawlessly on others. The author found some code in the Chapter 4 module which runs fine under MS Office 2000 in Windows 2000 SP4, but fails under MS Office 2003 in Windows XP SP2. (The code is that which automatically generates linear and polynomial curve fits and reports.) A file that contains revised code for Chapter 4 which will function correctly with MS Office 2003 in Windows XP SP2 has been included on the CD-ROM. At the time of publication this was the only known failure to the author, but there will undoubtedly be others given the vast number of configurable systems the code will be run on. Failures often occur because of what is referred to as stronger typecasting, meaning Excel requires VBA macros to be more specific as to what exactly it is they are referring to. Stronger typecasting occurs as a product matures because more features are added and often a single feature may be replaced with many more options requiring the user to specify exactly which feature (or the corresponding property or method) will be utilized. In newer versions of software, a prior default value may no longer be valid when a plurality of new options are now available. It is also possible that the original option may no longer be referred to using the same syntax. This is a common source of headaches when updating software.

Acknowledgments

I have had the pleasure of working with some really excellent people in the Molecular Properties Group at Pfizer Global Research and Development. Both Marina Shalaeva and Gus Campos have been exceptional colleagues and friends. My former supervisor, Dr. Franco Lombardo, was a man who could teach a people something or help them correct a mistake without making them feel 3 inches tall. He was also very quick to promote the work of his colleagues, and reported and always gave credit and recognition where it was due. These are rare qualities to find in a supervisor, and while he will be missed we wish him the best in his new position.

I would also like to thank three former Pfizer employees, including my former supervisor Dr. Chris Lipinski, and Dr. Beryl Dominy, both of whom retired from Pfizer, Inc. Also, I would like to thank Dr. James F. Blake who now works at Array Biopharma Inc. Chris, Beryl, and Jim always made themselves available and shared their expertise with their colleagues. In doing so, they helped to build the strength of the organizations that they served.

From my former job at the Naval Undersea Warfare Center in New London, Connecticut, I would like to recognize some engineers who helped "teach me the ropes." They are Robert J. White, Thomas A. Freehill, Joseph J. Maher, Gerald Messina, George Battista, Edward G. Marsh, David Cain, Thomas A. Bouchard, and Al Jagaczewski.

From a personal standpoint I would like to thank my fiancée, Crocetta Argento, who puts up with my wanting to be left alone some nights to work on projects such as this, as opposed to doing things I am supposed to be doing like fixing the kitchen disposal, which still doesn't work, or finishing the tile floor in the entryway.

Introduction

"He who turns the other cheek shares responsibility for the second blow."
— *The Talmud*

PURPOSE OF THIS TEXT

"When is the answer the answer." At one time I worked for someone who asked this question all the time. In today's world, a lot of people devote a lot of time to finding "the answers" to a lot of different problems. Answers, much like people, can be elusive things to determine. So many tools exist to analyze data, and so many transforms and algorithms can be applied to data that a clever person can shape any set of data to reflect the properties he or she wishes to see in the results. Hence the question, "When is the answer the answer." This book is about helping people find answers. It doesn't matter if you work at a pharmaceutical company trying to find the cure for cancer or if you work for a bank that needs to prepare financial reports, this book will help you find the answers to your problems. Which brings about the question, "What kind of solutions is this text best suited for in helping me determine?"

Chances are no matter what industry you work in, you are flooded with information from many sources. It is a common requirement in many industries to have the ability to take in a great deal of information and condense that information into some sort of readable report about which an intelligent person can make some reasonable conjectures. That is the focal point of this text. The goal of this text is to enable the reader to be able to construct an application that can import information (data) from a variety of data sources (both manually and automatically), apply some sort of algorithms upon the data which has been acquired (or imported), and then construct a meaningful report based on the results of the algorithms applied to the imported information (or data).

An example of such a process might be something like this: Suppose a machine takes substances out of containers and does some kind of measurements on them. Such a process from start to finish might utilize three data files. One file might list the substances and which container they reside in. Another file might have information relating to the substances, such as what color they are or when they were manufactured. A third file might be generated by a machine performing the measurements on the substances and would contain raw data related to each substance measurement. It might be desirable to take these three files (which might be different types of files such as: comma delimited, text, or Excel files), and extract information from each of the files to create a common report. In this example, it would probably be most useful to have a report that listed each substance and extracted certain measurements from the raw data file. Perhaps only certain "windows" of data should be extracted, depending upon what color the substance is. It might also be desirable to add a comment if the substance was tested X number of days after it was manufactured (perhaps degradation can occur rapidly after these substances are manufactured). It is easy to see how such reports could prove very useful in an industrial, scientific, and academic type of setting. If such reports are routinely generated, their results can be stored in legal notebooks. Creating such reports on a routine basis also provides the flexibility of control charting the results for quality control purposes.

It is this type of perspective that makes this book so much different from other types of Excel VBA programming books on the market today. (It is not meant to imply that this makes this book better than other books on the market, but its format does, perhaps, make the book more appealing to certain segments of the marketplace.) The fundamental goal of this book is to teach the reader, step by step, how to construct a useful automated data analysis application that would prove useful in both an industrial and academic type of setting.

"You have done so much with so little for so long that I'd like you to move on to doing everything with nothing."

FIGURE I.1 Doing more with less.

Below are listed some common tasks that could be encountered in any industrial, scientific, or academic setting. If the reader is interested in learning how to accomplish any or all of the tasks listed below, this text can help them.

Files in a Worksheet-readable format [comma delimited (*.csv), text (*.txt), or -(*.xls)] are utilized in my workplace that need to be reformatted or otherwise adjusted to be utilized on a machine or other type of computational tool.

Our workplace runs a machine that stores data in multiple files to multiple folders. In order to be really productive, it is imperative to have the ability to automatically open all these files and "cherry-pick" the proper data, placing it into a report Worksheet.

Our group has a proprietary method that we have begun to utilize in a high-throughput fashion. We need the ability to do some basic analysis on the data (simple regression, linear interpolation, finding the minimums and maximums of a dataset, etc.) and create a consistent custom report for our legal notebooks.

Process X is done time and time again in our workplace. We really need a method of automating the data analysis portion of this process and creating a consistent report for legal and archival purposes.

The capability to look through large amounts of data is desired. If the results meet certain criteria they should be reported without question. Any results which fall outside of the specified criteria (i.e., multiple solutions converged on, outside of reasonable range, etc.) will be flagged and sent to a scientist or engineer for interpretation.

It would be desirable to create a data report in which certain options could be chosen, and the results of the entire report would be automatically recalculated based on those chosen options.

The software that came with machine X creates a fine report but it is really unsuitable for our purposes. We need the ability to automatically take the report generated by machine X's software and create a new report more appropriate for our purposes.

When the data analysis on a process is complete, a mechanism is required for automatically creating a file which can be loaded into a relational database.

When the data analysis on a process is complete, a mechanism is required for automatically mailing out the custom report we have generated to our customers.

Chances are if you work in an industrial, scientific, or academic setting with any degree of sophistication, a need exists to accomplish some or all of the processes listed above. As workers' time is one of the greatest significant costs any organization can bear, especially that of highly trained workers such as scientists and engineers, it would be most productive and economically beneficial to automate as much of the routine data analysis tasks as possible.

WHAT DIFFERENTIATES THIS TEXT FROM ITS COMPETITORS

This text is different from others in two respects, the approach it takes toward teaching the reader to solve problems, and the comprehensive suite of tools and working examples it provides to the reader.

Most texts on the market have concepts arranged in categories that are grouped within the table of contents by their order of relative difficulty (from easiest to hardest). While this approach is useful for an advanced programmer, it leaves the newcomer and intermediate developer left scratching their head. Often people know what they want to do, but they lack the knowledge to know where an item of interest to them at any particular moment will reside in a text. The user may wish to know how to write a formula to a cell, but may be completely ignorant of the property and method naming conventions and hierarchy required that will allow them to do so.

Rather than grouping topics in the order of relative difficulty, the text takes the user forward through the topics in the order that they will most likely be needed when an automated data analysis application is being constructed. Thus, the first topics covered are how to utilize the basic elements in Excel from VBA code, starting with controlling Workbooks, Worksheets, and Cells. Next would be creating custom analysis functions in Excel VBA, and utilizing built-in worksheet functions to analyze data of interest that was segregated or identified through robust control of Excel's basic elements. Using more advanced VBA functions from third-party libraries is covered next, along with how to format and create reports to store that which has been analyzed. Extraction of supplemental information from databases and writing information to databases is also covered, along with control of third-party applications in Excel using COM, DDE, and OLE.

The working examples in the text are probably one of the most important elements in it. It is the little nuances in programming that kill the average programmer. Not knowing the proper syntax of how to code exactly what is wanted can cost programmers hours, especially should they be unaware that what they want to do isn't even possible to invoke using code. Many times what a

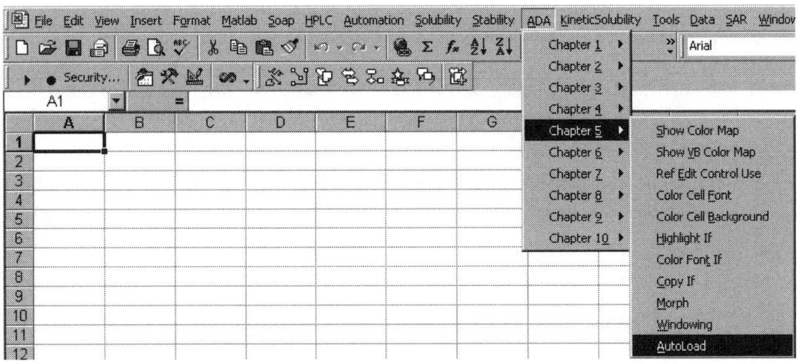

FIGURE I.2 A menu of the examples included with the text.

programmer wishes to accomplish *is* possible but is either very poorly or not at all documented. In purchasing this text the reader is provided working syntax for most of the common tasks that one would be expected to perform using ADA.

The text also discusses limitations the author encountered when writing some automated data analysis applications of his own. Knowing limitations of common tasks will prevent the reader from searching for answers to questions that at the current time lack a solution.

EVOLUTION OF THE SPREADSHEET

Spreadsheet tools have grown exponentially in popularity since their inception at the advent of personal computers. Initially the tool of choice for bean counters, they were utilized primarily for accounting purposes. They were great at showing how much money was made during period X or creating a bar chart of widgets produced each month. Like most products, people started finding out how to utilize them in ways never imagined by their creators.

In the beginning, the amount of creativity one could apply toward a spreadsheet application was limited, as the products available had very little flexibility. As time went on, however, most products developed macro languages which allowed greater customization of the application toward meeting the end users' needs. The first macro languages (notably Excel's) were hard to learn and difficult to use. Then along came Visual Basic (VB) and its adaptation to Visual Basic for Applications (VBA). Unlike its predecessors, VBA is easy to learn and intuitive to even the nonprogrammer. Many programs now license VBA as their macro language, and VBA is the current macro language for both Excel and its distant contender Quattro Pro.

VBA has changed the whole scheme of spreadsheet applications. It is now possible for the user not only to manipulate data within the spreadsheet application itself but to access functionality built into the Windows operating system. By utilizing technologies such as DDE and OLE, VBA applications can control, exchange, and manipulate data with other programs found on the resident operating system. More impressive is that Excel VBA is nearly an all-encompassing macro language. For many manipulations that can be performed manually in the Excel environment, there is a property or method built within Excel VBA to automate them.

Strangely enough, even with all the newfound functionality built into the modern-day spreadsheet, many times spreadsheets are not utilized for data analysis. A number of such applications are used as "transporter" vehicles for information to be shared across a corporate division, department, or group. Although this may seem wasteful, in many ways it makes sense. There is a great advantage in being able to drop information into a known format and being able to send it to anyone in the organization, knowing that they will be able to open and utilize the information because everyone in the organization utilizes the same suite of office tools.

SHOULD I BE USING EXCEL? WHAT OTHER OPTIONS EXIST?

This text focuses on utilizing Microsoft Excel for data analysis purposes. Is Excel the best spreadsheet program on the market? No. Excel is by far, however, the dominant player in terms of market share for spreadsheet programs. Which brings us to the question "What is the best spreadsheet tool on the market?" Without a doubt it is Origin Labs' Origin. Origin's macro language is more comprehensive than Excel's, and Origin can handle more math intensive problems than Excel. The graphics are far superior to those found in Excel — publication quality, in fact. There are only two downsides to Origin. First, the macro language is C-based and is about three times more difficult for a novice to learn how to use. Second, the package costs three to four times as much as Excel, depending upon whether the standard version or professional version is purchased. If the work to be done is very scientific in nature, as opposed to financial or industrial, Origin is worth a hard look.

Excel, like any package, has its good points and its bad points. I always like to start with the good in things. With the exception of Origin, Excel has the most comprehensive scripting language

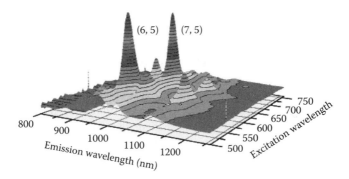

FIGURE I.3 Example graph in Origin.

in terms of properties and methods available to the user through the use of its scripting language (Excel VBA). Although I like WordPerfect's Office suite much better than MSOffice (and so do most attorneys), neither Quattro Pro in the WordPerfect suit or the spreadsheet program within Star Office suite has as sophisticated macro capabilities as Excel does. WordPerfect is in the process of building a VBA-based macro language into its Quattro Pro product, but at the time of this publication this was still in the early stages.

Excel has a really nice feature that lets the user performing a task record it via the macro recorder. The user can then actually look at the code generated by the macro recorder to complete the task. This is a very good tool for giving a novice programmer an idea of how to accomplish a task, but the code generated by the macro recorder often does not lend itself to generic use. The reason for this is that most subroutines generated by the macro recorder are "hard coded," meaning that the parameters within the subroutines generated by the macro recorder are fixed to performing the one task recorded by the macro recorder. Anyone with a reasonable amount of intelligence, however, can take the code generated by the macro recorder and modify it to utilize variables to enable variations on the recorded subroutine (i.e., perform the formatting on five rows instead of three).

Until the last few years I was a big proponent of Excel. What has changed my opinion about the product is the number of difficulties I have encountered in using it. That is really the bottom line for me: How much grief is a product causing me in terms of use? Until about two years ago I could honestly say that Excel had its quirks, but I was happy with the product. I am at the point now however where I must reboot Windows and close Excel three to four times a day to get the product to behave when I am doing intensive development work. The problems that I encounter routinely requiring me to exit out of Excel and/or reboot the machine are detailed at the end of this section under "Odd Behavior in Excel to Watch Out For."

Needless to say, having to reboot a machine three to four times a day is an annoyance. Excel was a decent product when it started, but nothing has really been done to upgrade the product in 5 years. It is true that new features are added with each release, but the vast majority of users will never utilize most of the new features that are incorporated into each subsequent release of the product. I would be happy if Microsoft would make the next upgrade a "code review" upgrade, meaning Microsoft would not add any new functionality to the product, just clean up the internals and make the product more stable and reliable. That would be worth a great deal to most users. Having the ability to strike through a superscript font and color it 40 shades of purple really isn't necessary for most users.

I would say my second biggest disappointment with the product other than reliability is the graphics. The plots and charts have not improved much in over 5 years. It is almost embarrassing to submit a manuscript using Excel's graphics. The lines in many cases look ragged, and the plots and charts are just not well thought out with regard to presentation. It's almost hard to believe that they continue to release the product the way it is.

However, I must be fair to Excel in the respect that in my work I utilize over 20 macros on my machine. I have about 20 hidden Workbooks with code in them that load every time Excel starts. I also have about 5 additional menus added to the Excel environment every time Excel starts, so I may be pushing the product to its absolute limits. However, if the code were clean, the product might run slowly because of this, but it wouldn't cause the product to crash as often as it does.

Other areas Excel does not excel at are higher math and curve fitting. One of the nice things about dealing with Excel, however, is the variety of aftermarket tools available for purchase which will integrate right into the Excel environment. With a lot of persistence, some hard work, and the right aftermarket tools, Excel can be made to do nearly anything. The question is "How persistent can you be?" vs. "How frustrated can you get?"

SCOPE OF COVERAGE

The text can be divided into three functional sections or categories. Chapter 1 to Chapter 6 are dedicated toward manipulation and calculation of data within the Excel environment. Here the reader will learn the basic syntax of accessing fundamental elements in Excel, such as Workbooks, Worksheets, and Cells. The methods of importing and accessing data within Excel are covered, as well as conceptuals such as Ranges, Referencing, and the use of object variables to access Excel elements. Functional calculations are covered utilizing both worksheet and custom VBA functions, as well as basic data mining techniques. The last and final chapter in this functional section, Chapter 6, is dedicated toward showing the developer how to create a report to store custom analyses.

The second functional unit of the text relates to utilizing Excel to obtain data from nonnative sources such as databases and third-party calculation tools. Chapter 7 and Chapter 8 focus on how to extract information from Microsoft Access Databases using SQL, DAO, DOA, and the XLODBC, Microsoft's ODBC add-in. In this section the reader will learn how to access database files from Excel, with clear working examples given using the Microsoft Access Database Program. Connectivity to any ODBC compliant data source such as Oracle is also covered. Here, basic SQL queries are reviewed, which allows users to return specified subsets of stored data. The basic elements of a relational database are also covered, such as tables, records, and fields, as well as methods for creating, accessing, and modifying them.

The last chapter in this functional section, Chapter 9, focuses on allowing the developer to utilize third-party tools to carry out more complex calculations. As Excel has limited built-in functionality, it is often necessary to employ a third-party tool such as Matlab, Origin, Mathematica, SAS, Mathcad, or similar tool to execute more robust functionality. Chapter 9 explores using DDE, OLE, and COM to allow Excel to control other applications, pass data to those applications, utilize their computational engines, and return any calculated results to Excel. The Excel environment is not the extent of the universe, however, and it is here that the reader will learn how to interface Excel with the rest of the known universe.

The final chapter (10) and functional section shows how to put the concepts covered in two functional areas together to create a cohesive and robust application. It is important to have a complete sample Automated Data Analysis sample application available as a resource for the reader. Without one, the text is relegated to being a collection of useful VBA routines. In this final chapter the user learns the methodology involved in constructing a "real" ADA application using the methods discussed within the text. For example purposes, the reader is shown how to construct a project that analyzes the manufacturing process of strain gages on a substrate and produces a report of the results. The sum of the parts is not necessarily the whole, thus the whole is presented so as not to leave the reader with just the sum of the parts!

Excel is often utilized as a "carrier vehicle" for data, simply utilized as a standard for moving information around from one person to the next. Often, analyses done on the data contained in

Excel Worksheets are done using third-party tools outside Excel. One goal of this text is to enable the reader to have the skills to complete all the tasks desired within the Excel environment, although for some calculations this may require harnessing an additional computational engine, but still within the Excel environment.

Like any other software package on the market, Excel has its strengths and its limitations. Tasks for Excel fall into three basic categories, the first being those tasks that Excel was designed to accomplish and can accomplish quite well. The second group includes those tasks that Excel is capable of accomplishing but was never designed to accomplish. Although such tasks can be completed using Excel, typically the solution is not elegant nor is the calculation process as robust as can be found in other products. The last category is comprised of those tasks that Excel is simply either incapable of doing or incapable of doing reliably. This text focuses primarily on the first category and touches on the second category.

In the beginning, the text introduces the reader to the Excel environment from a macro writer's perspective, as opposed to a passive user's perspective. Here, the reader learns the elements found within an Excel Workbook and how to access, manipulate, and change them utilizing VBA. With the knowledge of how to control data within the Excel environment, the user is then shown how to load and save data within the Excel environment. Once data is loaded into Excel, it will be necessary to perform some type of customized analysis on it, and the next section discusses creating custom functions that can be utilized within the Excel environment.

The next sections of the text involve working with Worksheets in Excel. Every Workbook in Excel must have at least one Worksheet, and here the user will learn how to determine how many Worksheets a particular Workbook contains, what Worksheets contain data, how to create new Workbooks and Worksheets, and methods for extracting data from Worksheets in an efficient manner. In any type of scientific or industrial setting, reports are critical, and utilizing Worksheets to create custom data reports are discussed at length. The reader will learn how to accomplish such tasks as utilizing templates, setting up single and multiple section reports, creating report headers, and formatting the contents of the report to the user's liking.

As readers learn more, it will be natural for them to want to tackle more complex projects such as those in which the coding requirements are not at all intuitive and, in fact, in some cases are obscure. Readers will learn how to utilize the processes in place within Excel to assist them in finding the syntax necessary to complete the kinds of specialized tasks their processes require. By knowing how to utilize the macro recorder, the user can determine what code is necessary to automate any process that is capable of being done by hand. The macro recorder also has the tendency to add code that is unnecessary. However, if the user knows what to look for, the code generated by the macro recorder can be modified to perform generic-type processes. In addition to the macro recorder, the VBA editor has that built-in process called Auto Complete, which can help users determine if a property or method is available to perform the task they had in mind. The Auto Complete function has its limitations, and it does not list properties and methods for every object within Excel. Knowing the limitations of this function will help the user make the best use of this tool.

As the reader develops projects which become more and more complex, the mechanisms for coping with them must improve as well. Readers will be taught how to incorporate error-handling mechanisms into their projects. Errors really fall into two categories: those with local consequences and those with global consequences. An error of local consequence would be defined as an error that may corrupt or inhibit part of the preparation of one section within a report. An error of global consequence would be defined as an error that halts the entire process, thus keeping the data following it from being processed. Obviously, it is the errors with global consequences that are most destructive, as they inhibit the processing of further data. Therefore, to have an effective automated data analysis (ADA) routine, an error-handling mechanism must be built in to correct any errors with global consequences as they occur. Errors which have local consequences can be handled a little bit more casually than those with global consequences. In some instances, simply highlighting the values of Cells in which something went awry will be sufficient. Other times, more elaborate mechanisms must be put in place.

Any application with any degree of sophistication will require that the user choose some options and add some information. The standard procedure for doing this is with the Graphical User Interface (GUI). A little bit of attention to detail can make or break the professional impact of a potential application. The text will talk about the best way to convey the information and options to the potential user. Small details such as utilizing the Windows registry to save the last options selected when a program was run can save users the drudgery of having to select the same information time and time again. Equally as critical is to have a way of conveying to the user how the program is progressing. It can be very nerve-racking to run a program that may do many intense calculations requiring a lot of time (in terms of hours, not minutes, here) and have no idea if the program is progressing through its tasks properly or if the machine has locked up. Utilizing visual cues such as status bars, check-off lists, and progress indicators can be invaluable both for the end users' confidence and for the programmer as well when it comes time to debug a stalled process.

Of course, it does little good to spend a lot of time developing a pristine Macro application if it is difficult for the user to execute. Methods of macro execution are covered with special emphasis on running macros from custom menu items, which is the most intuitive and easiest method for most users. If one is creating only a few macros, setting up menu items does not merit an exhaustive thought process. However, for the user that sets up more than 10 or more macros on a machine, some thought really should be given as to how to manage all the different menu items. The author will give some valuable hints as to the do's and don'ts for anyone faced with this situation.

Linear and nonlinear regression is a fact of life for those who work within the sciences. Typically, an instrument will acquire a great many points around the area of interest for a particular parameter. In order to extract the information for the region or point to be identified, it is necessary to construct some sort of model (equation) that fits the data points acquired reasonably well. Excel is capable of both linear and nonlinear regression, although I like using other tools (most notably Origin Lab's Origin) for such purposes. Setting up Excel to do some rudimentary curve fitting is shown, along with some algorithms which can aid in analyses.

A picture is worth a thousand words, and a graph is a picture. Excel graphs leave a lot to be desired, so some mechanisms of creating graphs utilizing other applications will be covered. A graph can add a great deal of professionalism to an automatically generated report.

The Excel environment does not constitute the whole universe, and the reader will explore how to interact with applications outside the Excel environment. The user is also shown how to set up variables to be recognized between different Workbooks and how to wait for user input from another workbook or application. Running code from a "nonresident" VBA module is covered, along with utilizing the SendKeys function to send data to other applications.

Information is only good if the people who need to see it have prompt access to it. When an analysis macro has completed, it will probably be desirable to mail out the report to certain users or groups of users. The text covers setting up distribution lists and how to mail out a workbook or individual worksheet. The reader is also shown how to send notification slips to clients and how to utilize routing slips in the mail environment.

Finally, a "real world" sample application is put together utilizing many of the tasks covered in the text. First, a problem is defined for the analysis of data from substrates on which strain gages are fabricated. The task is then divided up into manageable modules or functional units. A GUI is designed for the application, as are file loading, data extraction, data analysis, quality control, report generation, and file-saving routines. Complete code for the sample applications as well as all the code contained within the various chapters is included on the accompanying CD-ROM.

PROJECTS THAT LED TO THIS TEXT

My first involvement in Excel Macro writing began back in 1998 when I was working in the Molecular Properties Group at Pfizer Global Research and Development in Groton, Connecticut. The group was running a number of drug screens including solubility, stability, and Experimental LogD (ElogD). We had only one person running the ElogD assay, and she was doing all the data analysis by hand, constructing the same Worksheets over and over again. The popularity of the assay grew, and soon submissions were overwhelming her ability to analyze the data by hand. It soon became apparent that the scientist's job should really focus on (1) generating quality data (attention to detail when running the screen) and (2) deciding when the answer is truly the answer and not just a number. Instead what was happening was the scientist ended up dedicating a great deal of time to mundane tasks like constructing these Excel Worksheets. What was needed was a way to automatically generate these report sheets in the same fashion time after time. In addition to just generating the report sheets automatically, some flags were set up to draw the attention of the scientist when the results obtained were questionable. The scientist could then take a quick look at the data and make some basic decisions about the quality of the data.

The program had a methodology that worked like this (See Figure I.4). A comma-delimited file which was originally generated by an Agilent HPLC was imported into Excel. The number of rows in the file were counted and stored. The data was sorted by compound number, and any duplicate compounds were deleted. The largest area and corresponding retention time were then extracted for each method under which the compound was run and written to the appropriate worksheet. Using the INTERCEPT function, the point at which the best fit regression line will cross the Y-Axis was determined from the extracted data points. The "goodness" of fit, or R^2 value, was then determined for the regression line utilized in the INTERCEPT function by means of the RSQ function. The value returned by the RSQ function served as a flag to indicate if the data generated correlateD well or correlateD poorly to the ElogD model. The ElogD can then be calculated using the value returned by the INTERCEPT function. A complete report is subsequently generated, which can be entered in the scientist's laboratory notebook. (See Figure I.5.) The macro doesn't end with just spitting out the answers. When an improper correlation is found, the macro recommends rerunning that compound under a different method. Additionally, when questionable data is found, the macro generates a list of compounds for which the chromatograms should be printed out and adds comments to the worksheet for that compound such as "exceeds reasonable range of measurements."

	MOLECULAR PROPERTIES GROUP																
	Lipophilicity Study																
pH = 7.40		Prepared By: bissettbd															
Flow = 2.0 mL/min		Formula: ElogD = 1.1073*logk'w +0.2023															Print Report
Columns Shown		MeOH (t₀) 040203D_D															
Expr:	0.460	Column: Supelcosil LC-ABZ#: abc-123															
Expected:	0.285	C:\HPCHEM\1\DATA\092299\															

| Sample Name | Position | MeOH Content in eluent - t,xx | | | | | | log k.60 | log k.65 | log k.70 | log k'w | RSQ | Elog D | Print | ReRun | Comments: |
		t.60	t.65	t.70	k.60	k.65	k.70	60.00	65.00	70.00						
MeOH 2nd Inject		0.329	0.712	0.339												
Bifonazole	12	3.68	2.04	1.23	11.92	6.15	3.30	1.08	0.79	0.52	4.42	1.00	5.10			
Chlorpromazine	62	0.98	0.75	0.61	2.45	1.64	1.12	0.39	0.21	0.05	2.41	1.00	2.88		M	
Clotrimazole	2	3.68	2.09	1.27	11.92	6.33	3.47	1.08	0.80	0.54	4.29	1.00	4.95			
Clozapine	45	1.10	0.82	0.64	2.86	1.87	1.24	0.46	0.27	0.09	2.63	1.00	3.11			
Deprenyl	46	0.90	0.72	0.60	2.17	1.53	1.09	0.34	0.19	0.04	2.13	1.00	2.56		M	
Diazepam	17	1.01	0.77	0.62	2.55	1.71	1.18	0.41	0.23	0.07	2.41	1.00	2.87		M	
Dichlorophen	4	1.45	1.03	0.78	4.08	2.61	1.72	0.61	0.42	0.24	2.86	1.00	3.37			
Diethylstilbestr	13	1.22	0.83	0.62	3.26	1.90	1.17	0.51	0.28	0.07	3.18	1.00	3.73			
Estradiol	3	1.00	0.75	0.59	2.49	1.61	1.08	0.40	0.21	0.03	2.57	1.00	3.05			
Finasteride	16	1.19	0.85	0.65	3.16	1.99	1.28	0.50	0.30	0.11	2.85	1.00	3.35			
Lorazepam	18	1.20	0.88	0.68	3.21	2.07	1.37	0.51	0.32	0.14	2.72	1.00	3.22			
Lormetazepam	19	0.80	0.63	0.53	1.80	1.23	0.86	0.25	0.09	-0.06	2.17	1.00	2.60		M	
Naphthalene	11	1.36	0.52	0.76	3.77	0.83	1.68	0.58	-0.08	0.23	2.52	0.28	2.99		M	
Testosterone	14	1.19	0.88	0.68	3.17	2.08	1.39	0.50	0.32	0.14	2.66	1.00	3.14			
Thioridazine	89	0.99	0.74	0.59	2.49	1.61	1.07	0.40	0.21	0.03	2.58	1.00	3.06			
Tolnaftate	1	4.29	2.38	1.41	14.06	7.35	3.94	1.15	0.87	0.60	4.46	1.00	5.14			
Trazodone	49	0.96	0.74	0.60	2.36	1.59	1.10	0.37	0.20	0.04	2.36	1.00	2.82		M	
Trifluromazine	90	1.30	0.92	0.69	3.55	2.23	1.43	0.55	0.35	0.15	2.92	1.00	3.44			

FIGURE I.4 ElogD macro report with all parameters shown.

MOLECULAR PROPERTIES GROUP

Lipophilicity Study

Columns Hidden

Print Report

pH = 7.40		Prepared By: bissettbd	
Flow = 2.0 mL/min		Formula: ElogD = 1.1073*logk'w +0.2023	
MeOH (t₀)	040203D_D		
Expr:	0.460	Column: Supelcosil LC-ABZ#: abc-123	
Expected:	0.285	C:\HPCHEM\1\DATA\092299\	

MeOH (t$_0$): Expr: 0.460, Expected: 0.285

Sample Name	Position	MeOH Content in eluent - t.xx			log k'w	RSQ	Elog D	Print	ReRun	Comments:
		t.60	t.65	t.70						
MeOH 2nd Inject		0.329	0.712	0.339						
Bifonazole	12	3.68	2.04	1.23	4.42	1.00	5.10			
Chlorpromazine	62	0.98	0.75	0.61	2.41	1.00	2.88		M	
Clotrimazole	2	3.68	2.09	1.27	4.29	1.00	4.95			
Clozapine	45	1.10	0.82	0.64	2.63	1.00	3.11			
Deprenyl	46	0.90	0.72	0.60	2.13	1.00	2.56		M	
Diazepam	17	1.01	0.77	0.62	2.41	1.00	2.87		M	
Dichlorophen	4	1.45	1.03	0.78	2.86	1.00	3.37			
Diethylstilbestr	13	1.22	0.83	0.62	3.18	1.00	3.73			
Estradiol	3	1.00	0.75	0.59	2.57	1.00	3.05			
Finasteride	16	1.19	0.85	0.65	2.85	1.00	3.35			
Lorazepam	18	1.20	0.88	0.68	2.72	1.00	3.22			
Lormetazepam	19	0.80	0.63	0.53	2.17	1.00	2.60		M	
Naphthalene	11	1.36	0.52	0.76	2.52	0.28	2.99		M	
Testosterone	14	1.19	0.88	0.68	2.66	1.00	3.14			
Thioridazine	89	0.99	0.74	0.59	2.58	1.00	3.06			
Tolnaftate	1	4.29	2.38	1.41	4.46	1.00	5.14			
Trazodone	49	0.96	0.74	0.60	2.36	1.00	2.82		M	
Triflupromazine	90	1.30	0.92	0.69	2.92	1.00	3.44			

FIGURE I.5 ElogD sample report with hidden columns for clarity.

Notice in Figure I.4 and Figure I.5 that there are two images of an ElogD Sample Report. In one instance, some columns are hidden in the report, in the other all of the columns are shown. The user can toggle between hiding some of the columns or showing all of the columns simply by pressing a button. This is a nice feature because when the report is ready to be printed it really should not be necessary to show any columns which contain intermediate calculations. The intermediate calculations can prove valuable, however, when a result is generated that is incorrect, and the user wishes to track down just what went wrong. The report can also be printed by pressing the "print" button on the right-hand side of the worksheet.

Another problem in which the use of macros made the method more manageable for the scientist to run occurred when setting up a process to run an oxidative stability assay. The instrument that was chosen for this assay had two problems. The first is almost universal when any instrument is purchased. The software did not prepare a report of the parameters in the format that the user needs. (In this case the software was also incapable of extracting the parameter we were interested in.) The second problem stemmed from the fact that the software that ran the instrument had to be manually triggered to export data about a compound in a comma-delimited format. If users wanted a row of comma-delimited parameters about a particular compound, they had to click on that compound in a listbox (to select it), and then press a button. Although this may not seem like too much trouble for a few compounds, more often than not a plate of 96 compounds was being run. This, of course, meant the scientist had to scroll and click 96 times, taking great care not to omit or repeat any of the compounds on the list. In addition to leading to boredom and unhappiness on the part of the scientist, it was also a terrible waste of a resource (the scientist certainly could be doing more productive things).

The solution to these problems was implemented using two macros. The first macro automatically extracted the comma-delimited data from the application. The macro would start by opening the instrument's software package and then prompting the user for the dataset to be loaded. Once the data set was loaded, the application used the SendKeys function to send the appropriate keystrokes to the instrument's software to automatically export the comma-delimited information for each compound. The second macro took in the comma-delimited data file generated by the first macro using the instrument's software and performed some automated data analysis routines on the data to create a report. Three derivatives of the report were created: one for the laboratory notebook, one amenable for upload to a corporate database (a flat type of file), and a final report that was to be mailed out to those who had requested the results of the assay.

BEFORE BEGINNING: SETTING UP THE EXAMPLES

This text is loaded with examples. This is not an accident. Most people tend to learn processes efficiently through the use of examples. There simply is no substitute for a working example of how to carry out a process. When such an example is present, a reasonably intelligent person can then look at such an example, reverse-engineer it, and modify it as needed. Some extremely bright people are capable of learning a process by simply having someone *tell* them how it should work rather than *demonstrate* the actual process. This is not true for most people, however.

It would be very impractical to try to teach someone how to program in Excel without the utilization of examples. This is because there are so many peculiarities within Excel VBA that the reader would waste a lot of time trying to accomplish things that should work but, for a variety of reasons, will not work due to bugs in Excel, incomplete object models, and other known and unknown system problems. In many instances the reader will be shown a routine that works and, if appropriate, a routine that should work to accomplish the same task but will not. This is not done for esoteric purposes. It is to prevent the reader from looking at a routine and thinking "It could be done much more efficiently like this … ," only to find upon trying a different method that ultimately it will not work for one reason or another. The author has spent years programming in Excel VBA and has tried numerous routines to accomplish many tasks. With less than complete documentation for many Excel processes, a lot of trial and error was utilized to find the processes that comprise the bulk of this text. It is the author's hope to save the reader some of the same miseries of going through endless iterations of trial and error trying to make something work that should work — but won't work — because of a nuance in Excel.

All of the examples for this text can be run directly from a menu in Excel. The menu is titled "ADA," and below this menu item are submenus for nearly every chapter in the book, each of which contain the example programs pertinent to that particular chapter. All of the code snippets and examples utilized in this text are contained within the workbook "ADAUsingExcel.xls." If this workbook is placed in the "XLStart" directory for Excel, the ADA menu item will be created in Excel upon start-up. This is because any workbook placed in Excel's XLStart directory is loaded into Excel (or opened) upon startup. Contained within the "ADAUsingExcel.xls" is a shareware program called Menu Maker, developed by John Walkenbach, who has published several excellent books on Excel. John also runs a Web page that has a plethora of valuable information on Excel, which can be found at this address: http://j-walk.com/ss/. Specific information on the menu maker application can be found at this address: http://j-walk.com/ss/excel/tips/tip53.htm.

To utilize the example programs from Excel, the Workbook "ADAUsingExcel.xls" must be placed in Excel's XLStart directory. Excel's XLStart directory is *usually* found at this location:

C:\Program Files\Microsoft Office\Office\XLStart

or

C:\Program Files\Microsoft Office\OFFICE11\XLSTART

However, there are also instances where this folder is located under the user's user name in documents and settings folder, in a location like this:

C:\Documents and Settings\bissettbd\Application Data\Microsoft\XLStart

Or:

C:\Documents and Settings\bissettbd\Application Data\Microsoft\Excel\XLSTART\

Once the workbook "ADAUsingExcel.xls" has been placed in Excel's XLStart directory, the following menu item will appear when Excel is started as shown in Figure I.6.

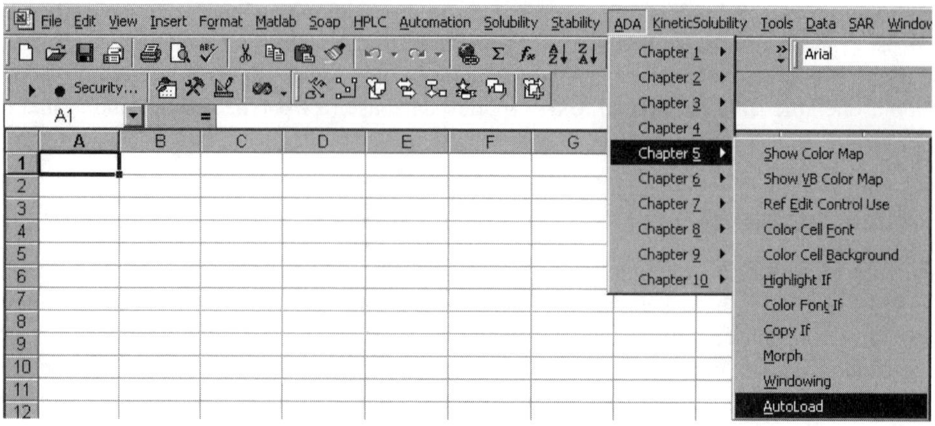

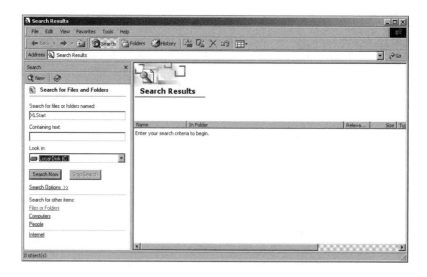

FIGURE I.6 The ADA menu utilized to run the examples in the text.

If difficulty is encountered in trying to locate the XLStart folder, it can easily be searched for by right clicking on My Computer and selecting "Search …" from the menu items. Doing so will produce a dialog box like that shown in Figure I.7. Then simply search for the folder XLStart on just the local hard drive (C:\).

Some of the examples require specific data files to be run. These files are stored in a hierarchy type structure on the CD-ROM depicted in Figure I.8.

Once the "ADAUsingExcel.xls" Workbook is loaded into the Excel environment, all of the associated code with the Workbook can be viewed easily by entering the Visual Basic Editor. This can easily be done in Excel by pushing the button shown in Figure I.9.

If the button shown in Figure I.9 is not present in Excel, it can easily be shown by going to the View Menu Item and selecting Toolbars->Visual Basic as depicted in Figure I.10.

Once inside the Visual Basic Editor, it is easy to look through all the code associated with each workbook. This is done within the "Project Explorer" window in the Visual Basic Editor, which is shown in Figure I.11. In the Project Explorer window, all the forms within a Workbook

FIGURE I.7 Searching for the XLStart Folder.

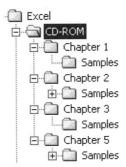

FIGURE I.8 Location of test files to run examples in the text.

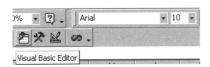

FIGURE I.9 How to enter the Visual Basic Editor.

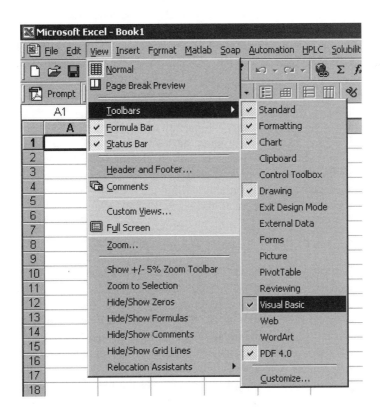

FIGURE I.10 Showing the Visual Basic toolbar.

FIGURE I.11 The Project Explorer window.

are contained in the folder "Forms," and all the code modules contained within a Workbook are contained in the folder "Modules." Objects within the workbook that may contain code, such as the Workbook itself or any of the Worksheets, have their code stored in the "Microsoft Excel Objects" folder. In the "ADAUsingExcel.xls" workbook, there is a unique module associated with every chapter in the text. Within each module is the code associated with that chapter. Similarly, forms are notated "frm#*Name*," where # is the chapter number the form is utilized in, and *Name* is the name of the sample application the form is utilized with. The methodology utilized to store the sample code associated with this text in the "ADAUsingExcel.xls" workbook is shown in Figure I.11.

Installing the "ADAUsingExcel.xls" workbook in the readers XLStart folder allows them to access all the properties, methods, functions, and subroutines present in the "ADAUsingExcel.xls" workbook. The reader can view a comprehensive list of all the above parameters simply by selecting View->Object Browser within the VBA programming environment. If the user has the ADA project selected in the Project Explorer window (see Figure I.11), they can then select the library titled "ADA" within the Object Browser window. Doing so will expose all of the properties, methods, functions, subroutines, constants, and variables within the

"ADAUsingExcel.xls" Workbook, as illustrated in Figure I.12.

ODD BEHAVIOR IN EXCEL TO WATCH OUT FOR

When intensive development work is done in Excel, it does not take long before the user will begin to notice odd behavior, which will usually manifest itself after a 2- to 3-h period of running and rerunning macros. When developing new programs, it is very common to run a chunk of code, debug that section of code, add more code, and then start the process all over again. What happens in the Excel environment (or any programming environment for that matter) is that each time the macro (or program) is run, a chunk of memory in the operating system is allocated for exclusive use by the program. When the program has finished, the operating system is supposed to deallocate (or release) that chunk of memory so that it is again available for use by any program within the operating system. What happens in Excel (and this is true with many products, not just Excel) is that not all of the memory allocated when the program is executed is released when the program is terminated. Such behavior is known as a "memory leak." What then happens over time is more

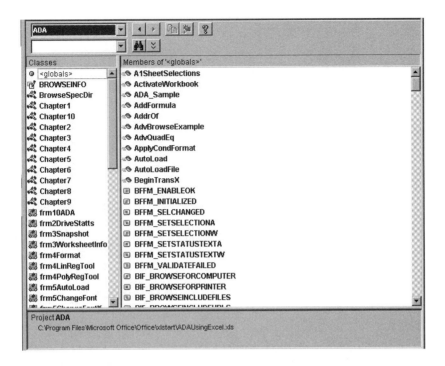

FIGURE I.12 Exposing functionality of the ADA using Excel.xls Workbook.

and more system memory is lost as macros are executed over and over again. At some point the amount of memory lost reaches a critical mass where there is no longer enough memory for basic system procedures to be performed reliably. When this point is reached things will begin to happen like: macros will no longer execute from menu items properly, the auto complete functions will not work when writing code, Excel will momentarily seize up, etc. These actions are the precursors to Excel reaching a state where it will no longer be able to function at all. When this happens a dialog box like that shown in Figure I.13 will appear.

This lovely dialog box is basically telling the user that Excel has dribbled away so much memory that it can no longer reliably function. Notice all the hexadecimal numbers (they have "0x" in them), which are pointing to the areas of memory within Excel that are either discontinuous or are no longer accessible due to the memory leak. These hexadecimal numbers mean nothing to the average user although they might mean something to a software engineer at Microsoft (but probably not, judging by how quickly these memory leaks have not been fixed.)

The only cure for this type of scenario — and it will happen a few times a day during intensive development work — is to reboot Windows. It is not recommended to just restart Excel. Although this may work for a time, the problem lies just as much within the operating system as it does within Excel. Excel probably will not behave in a robust manner until Windows is rebooted.

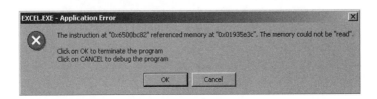

FIGURE I.13 The result of a memory leak.

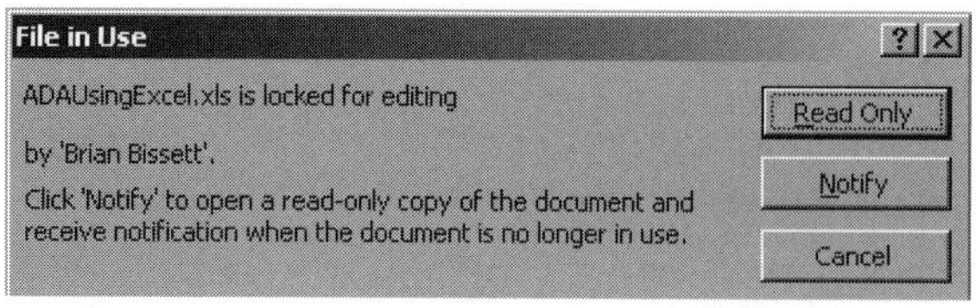

FIGURE I.14 The Locked for Editing Upon Startup dialog.

Another common memory leak problem is depicted in Figure I.14. Unlike the earlier memory leak problem discussed, this problem manifests itself when Excel is first opened. The dialog box in I.14 will immediately appear, telling the user that certain files are locked for editing when this should not be the case. A file should only be locked for editing if it is already opened in another application. This problem is caused by memory leaks in other applications which drain the system memory down so low that when Excel starts, it malfunctions right from the beginning, telling the user they will not be able to edit files they should be able to edit. The files locked for editing will always be those present in the XLStart directory that are loaded upon the start of Excel. Like its predecessor, the only cure for this problem is to shut down Excel and reboot Windows.

The last problem to be discussed is the Program Error dialog box, which tells the user that Excel has generated errors and will be closed by Windows, and that the user will have to restart the program. This dialog box is shown in Figure I.15.

Fatal errors such as this are sometimes the result of a VBA programming error, but more often than not, they are the result of some internal bug within Excel that generates the error and forces its closure. This type of fatal error is often encountered if the user utilizes a RefEdit control, and an invalid range is inadvertently assigned to the control. A number of other mistakes made within VBA that may not be readily apparent can also cause such an error to occur. The most frustrating case is when this error occurs on one machine with one macro, and does not occur on another machine with the identical macro. If a macro causes a crash on every machine it is installed on, chances are it is the user's macro that is at fault. If the macro functions fine on one machine and crashes on another, chances are something is wrong with Excel or Windows on one of the machines. It would then be prudent to check the version number of Excel and Windows on both machines and make sure that the machine that is crashing has the latest service release or service pack. This information is readily available by selecting Help->About Microsoft Excel. The version number and Service Release number (SR-#) are located in the top right-hand side of the about form.

THE TOP PRODUCTIVITY HINDRANCES IN EXCEL

No product in today's marketplace is perfect, and Excel is no exception. One particularly irritating aspect of Microsoft's development policy toward Excel is to continue to add new features to Excel and not fix the bugs present in the existing features of the product unless the use of these features,

FIGURE I.15 Fatal errors generated by Excel force closure.

and thus the frequency rate at which these bugs are encountered, rises above some predetermined threshold. The absurdity of this policy is that many of the new features added to Excel are not utilized by the majority of users, or they are utilized quite infrequently. Therefore, when bugs are present in the new features, they are rarely fixed, as the majority of users are never utilizing them and therefore not encountering the bugs. Excel has continued to grow, with more and more new features added, which, ironically enough, will never be debugged. As my grandfather used to always say, "A house is only as good as its foundation." Excel has many bugs present in today's product that existed three versions ago. This is somewhat pathetic.

I have asked numerous Microsoft premier support personnel why many of the old established bugs in the program have never been fixed. Most of the Microsoft engineers said they have been hearing the same complaint from users for years, which is that most Excel users wish the existing features in the product would be debugged before adding any new features.

As explained to me by numerous Microsoft personnel, the current bug fixing mechanism works as follows. Bugs are logged into a database with a notation indicating if they were encountered by a premier support user or just a regular customer. The determining factors that decide if a bug gets fixed are as follows: (1) How many users encounter the bug? (2) How much will the bug cost to fix? (3) How much is the bug costing Microsoft's clients?

On its face, this policy seems reasonable. Microsoft, like any company, has a limited amount of resources, and it must make decisions on where to allocate them. If the model you operate under is closed, and the only considerations are the current resources vs. the immediate users, the model is perfect. The real world market, however, is open, which is why this policy is failing. The flaw of this policy lies with the parameters it fails to consider, the most important of which are: (1) Where is the future market space for this product located? (2) Can people enter this market space with the functionality the current product has? (3) What amount of current vs. future market share is being lost by continuing under the existing policies?

It is safe to say that the future market share for Excel does not lie with users who simply want to utilize a spreadsheet for basic calculation purposes. Although this category clearly represents the majority of current users, it is saturated and, worse yet, there now exist many competing products (most notably Star Office and the WordPerfect Office suite), which continue to erode Microsoft's market share. I believe the future market for Excel is in higher-end data analysis, data mining, and scientific analysis and presentation.

The problem is that Excel cannot compete in this market space with the functionality the current product has. The limiting factors here are (1) poor data modeling abilities, (2) limited higher-end mathematical functions, (3) incomplete VBA object models that hinder users trying to build their own tools, (4) poor graphics, and most importantly (5) bugs present in the VBA Macro Language, which have remained unfixed for years.

The consequence of this is that Excel is relegated into competing in a saturated market with an eroding market share. Worse yet, when users in the scientific community try to write custom routines to utilize the product for more sophisticated analyses, so many bugs and holes in the object model are encountered that many simply give up in disgust and start utilizing another product. It is the author's intention to try to provide a resource for people wishing to utilize Excel for higher-end analyses, which will help them navigate around the limitations in the Excel VBA framework and bridge the gap between the existing documentation and what is required as a practical matter to create a successful ADA application.

The author has written numerous automated data analysis algorithms in Excel VBA and would like to bring to the attention of the reader his views on the top 10 productivity hindrances (or terrible 10) currently present in the product:

1. Holes in the Autocomplete Feature. Ex. ActiveSheet.Name.
2. Macro Recorder does not record all VBA functions when done by hand.
3. Help Index incomplete: Try looking up InStr function.

4. Bugs in the RefEdit Control make it nearly unusable.
5. Bugs in Conditional Formatting Features make many features unusable from VBA.
6. Many built-in dialogs cannot be utilized from VBA. Perfect Example: xlDialogColor-Palette.
7. When a macro jumps to an error handling routine, the subroutine causing the error is not identified to the user.
8. No documentation on how to use the Analysis Toolpack VBA functions.
9. The graphs are substandard for a package at this price. They look like something recycled from MS-DOS. They hardly constitute presentation quality graphics, and the default gray, pink, and purple color scheme is horrible.
10. A lack of documentation is present across the board for the product. Not just the parent product, but add-ins and toolpacks authored by Microsoft. Even when such tools have basic instruction on their use, rarely is documentation provided on how to utilize them from within the VBA framework.

In many ways Microsoft is not aware of what it has with its Excel product. In many organizations this product is utilized simply as a transporter mechanism because nearly everyone has Excel on their desktop. Often, however, the information is taken out of Excel and utilized in another application only because Excel cannot perform the tasks people require of it. If the application were made robust, it would have far greater potential than that which it is currently being utilized for now. This should be self-evident by looking at all the aftermarket add-ins that people are creating for Excel to accomplish tasks that the package itself is incapable of performing. However, if the bugs in the Excel VBA programming language are not fixed, the utility of this package will continue to be held at its present asymptotic limit. It is the author's sincere hope that the product will be improved and would look forward to the opportunity to work with a much improved version of the product.

FINAL THOUGHTS

I was working on a patent some years back with a patent attorney named Stephen Gilbert at Bryan Cave LLC who said, "Patents are never really finished. You just run out of time." I think the same thing applies to books. You have an idea where you want to go and what you want to cover, but resource constraints and time only allow so much before the deadline is up. If the reader feels some topic should have been covered but was omitted, please let me know. Also, please report any bugs found within the text or CD-ROM, with instructions on how to duplicate them. The author will compile the comments on the text and keep them in mind for a revised edition. Anyone whose ideas were utilized will be acknowledged in the next version of the text. Thank you in advance for helping bring a better product to the market. The author can be reached at excel. errors@yahoo.com.

1 Accessing Data in Excel: A Macro Writer's Perspective

1.1 INTRODUCTION

A macro writer must have an understanding of Excel that is different from that of someone who utilizes it as a spreadsheet application. The programmer must have a fundamental understanding of how Excel interacts with its environment, processes information, and accesses information.

Excel utilizes Visual Basic for Applications (VBA) as its macro language. VBA has broader functionality than just programming macros in Excel. It has much of the same functionality as Visual Basic (VB), the language upon which it was derived. VBA also has the ability to interact with applications outside of Excel that provide a great deal of flexibility. Examples of such functionality are covered in this text in Chapters 7–10. In Chapter 2, examples are given showing how to utilize VBA to interact with the Windows API to accomplish some handy tasks such as browsing for a directory or obtaining the username of the person currently logged into the windows session. At the top of the Excel hierarchy is the Object. Loosely speaking, objects can be thought of as "things," meaning, if it can be seen in the Excel environment, it is an object. Workbooks, Worksheets, and Charts are all examples of objects in the Excel VBA environment. VBA utilizes objects to access information in Excel and perform manipulations on elements in the Excel environment. Table 1.1 lists some of the most common objects in VBA.

The most common objects utilized when working with Excel VBA are the Workbook, the Worksheet, and the Range. A fourth object, which is slightly more abstract, is the Application. An application is simply an instance of Excel that is running. Several application objects may be running on the same machine, but when the application object is called in VBA, it can refer only to objects that are contained in the application that the code was called from. These four objects are illustrated in Figure 1.1.

In Figure 1.1, the application can be thought of as the running program, which encompasses all elements in the figure. The Workbook is named "Book2," and it contains three Worksheets named "Sheet1," "Sheet2," and "Sheet3," respectively. Sheet1 is the active sheet in this Workbook because it has the focus of the program at this time. Within Sheet1, a Range of Cells has been selected. In this case, the range is from column B row 2 to column D row 4 (B2:D4). To successfully manipulate data using VBA, it is necessary to have a mastery of the Application, Workbook, Worksheet, and Range objects.

1.2 THE WORKBOOK

The Workbook is the topmost object in the Excel paradigm. Typically when gathering information, it is necessary to specify, first, which Workbook has the information; next, which sheet in the Workbook has the information; and, finally, what cell or range of cells has the information that is sought or to be acted upon.

Although several Excel Workbooks may be open in an application, only one will the active Workbook, or the Workbook that has the focus of the program. It is easy to determine from the VBA code which Workbook is active by means of the following code snippet that can be executed in the immediate (debug) window.

```
Debug.Print ActiveWorkbook.Name
```

TABLE 1.1
Table of Common Excel Objects

Object	Description
Workbook	An Excel File that can contain a collection of Worksheets
Worksheet	A Single Worksheet contained within a Workbook
Range	A Cell or Group of Cells contained within a Worksheet
Application	An Instance of Excel that is running
Chart	A Chart that can stand alone or be embedded in a Worksheet

If, as in the case of Figure 1.1, the active Workbook is named "Book2," then "Book2" will be printed in the immediate (debug) window. If the programmer wishes to make a different Workbook, the active Workbook, it can be done with the following command:

```
Workbooks("Book3").Activate
```

Here, the active Workbook is changed from "Book2" to "Book3" by means of activating Book3. If this command is typed in the immediate window, Excel will change the Workbook focus to "Book3" provided a Workbook named "Book3" exists!

There is one major oversight to this seemingly simple command, however, which is termed "The Hide File Extensions for known file types" bug. Here is how the bug occurs. Double-click on My Computer, under the View Menu item select Options, and choose the view tab. There will be a check box that says "Hide File Extensions for known file types." If this box is checked, then Workbooks in Microsoft Excel will have a caption at the top — "filename." If the box is unchecked, then the Workbooks will have a caption — "filename.xls."

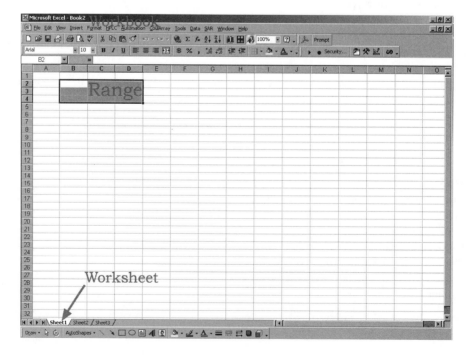

FIGURE 1.1 Illustration of common Excel objects.

The problem typically manifests itself when a macro is deployed across a variety of machines, and one of the machines has a different setting for the "Hide File Extensions for known file types" option. What then happens is that Excel will only recognize Workbooks if they are referenced with or without the *.xls extension, depending upon the setting on the machine. If the extension types are hidden on a machine, and you try to activate a Workbook with the command

```
Workbooks("Book3.xls").Activate
```

you will end up with an error message like the following:

FIGURE 1.2 Subscript out of range error.

"Run-time error '9' " "Subscript out of range"
What this means is that Excel cannot find a Workbook named "Book3.xls." This is because the extensions are hidden on this machine, and Excel only recognizes the Workbook as "Book3." (The reverse condition can happen as well.) To avoid this situation, a subroutine can be created such as the one below to activate Workbooks. If it proves to be necessary to add the extension to the Workbook name, the subroutine will automatically do so.

```
Sub ActivateWorkbook(wbname$)
'Activate the desired Workbook
'Compensates for Hide Extensions of Known File Type Error 'in Excel
On Error GoTo ChangeStatts

Workbooks(wbname$).Activate
Exit Sub
ChangeStatts:
Select Case Err
  Case 9
    wbname$ = wbname$ & ".xls"
    Workbooks(wbname$).Activate
    Debug.Print "New Name: "; wbname$
  Case Else
    dummy = MsgBox("An Unanticipated Error No." & Str(Err) &
    " Occurred !", vbOKOnly, "Error")
End Select
End Sub
```

Use this subroutine whenever a different Workbook must be activated, and pass the Workbook name using wbname$ without the ".xls" suffix. If the ".xls" suffix is required, an error will be triggered, and the error handling mechanism within the subroutine will automatically add the suffix ".xls" to the Workbook name. This will allow Worksheets to be activated trouble-free from any computer regardless of the "Hide File Extensions for known file types" setting on a particular machine.

This is also a good opportunity to mention that the Workbook name in the Workbooks Activate sequence need not be a literal string. A variable string can also be utilized. This provides great flexibility in programming. For instance:

```
Workbooks("Book3.xls").Activate
```

and

```
WorkbookName$ ="Book3.xls"
Workbooks(WorkbookName$).Activate
```

will accomplish the same thing.

It is also possible to structure a loop to iterate through the number of existing Workbooks. This is an extremely handy feature because, oftentimes, at the end of an analysis, it is desirable to close Workbooks that meet certain criteria. Perhaps, some Workbooks may be utilized for intermediate calculations so as not to clutter the "official" report Workbook. Others may contain no data for one reason or another and should be removed from the workspace. The next subroutine will print out the names of all the Workbooks open in Excel to the debug window. Users can select if they wish to have the hidden (or invisible) Workbooks in Excel displayed by means of the includeInvisible parameter.

```
Sub CountWorkbooks(ByVal includeInvisible As Boolean)
Dim bookCounter As Workbook
For Each bookCounter In Application.Workbooks
  If Windows(bookCounter.Name).Visible = True Then Debug.Print
  bookCounter.Name
  If Windows(bookCounter.Name).Visible = False And
  includeInvisible = True Then
      Debug.Print bookCounter.Name
  End If
Next bookCounter
End Sub
```

One last comment on Workbooks. It is not possible to change the name of the Workbook by setting a string to equal the (.Name) property of the Workbook. This will result in an error. This is because the (.Name) property is used by Excel to facilitate loading and saving of Workbooks in the windows file system. Therefore, a statement like:

```
ActiveWorkbook.Name = "SpecialReport"
```

will result in an error. The only way to change the name of a Workbook is to save the Workbook under a different filename. This process is covered in Chapter 2.

1.3 THE WORKSHEET

Every Workbook must contain at least one visible Worksheet. When a new Workbook is created, three Worksheets are included within it by default. The user can set the number of Worksheets to be created at startup by going to the menu and selecting Tools->Options, and under the "General" tab there is a spin button that allows the user to set the number of default sheets in a new Workbook. (see Figure 1.3.)

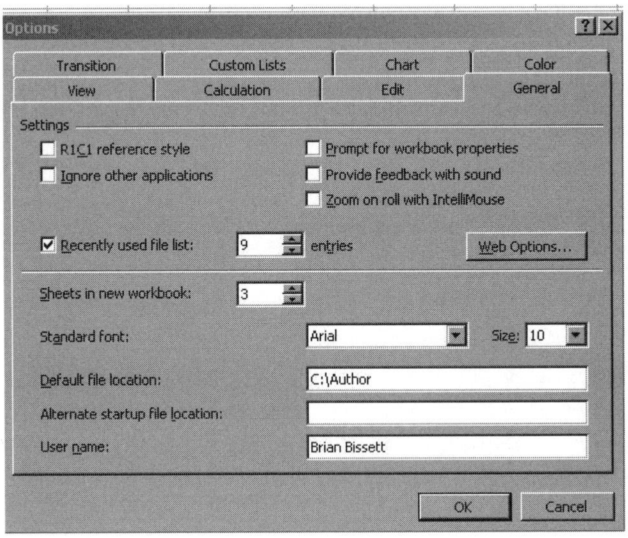

FIGURE 1.3 Changing the number of sheets in a new Workbook.

No matter how many sheets are within a current Workbook, it is simple enough to add another one. The command

```
Sheets.Add
```

will add another sheet to the active Workbook. The sheet will appear immediately preceding the active sheet. Before anything on a sheet can be referenced or changed, it must be either implicitly referenced or made active, or explicitly referenced using the Workbook and Worksheet names in a defining statement. A sheet can be made into the active sheet by utilizing the commands

```
Sheets(1).Activate
```

or

```
Sheets("Sheet1").Activate
```

Notice that using the Sheets method, a specific sheet can be referenced in one of two ways: either by the sheet name shown on the tab within the Workbook (by default names are Sheet1, Sheet2, …, Sheetn), or by number. The numbers refer to the order in which the sheets were created in the Workbook. For a new Workbook with three sheets created at startup, the one-to-one correspondence between references is shown in Table 1.2.

A word of caution is in order when utilizing the activate method with sheets. If a Workbook contains only one *visible* sheet, and the activate method is called on that sheet, an error will result.

TABLE 1.2
Numerical and String Correspondence
of Sheet Names

Worksheet Name	Numerical Reference
Sheet1	1
Sheet2	2
Sheet3	3

Why? Probably because if only one visible sheet exists in a Workbook, it *must* be the active sheet. (This really should not happen and is a poor design on the part of Microsoft. If the sheet is already activated, and a command is issued to activate it, then nothing should happen.) Because this error can occur, it is handy to have a function that can determine if a Workbook contains only a single sheet. The function OneVisibleWorksheet provides this capability.

```
Function OneVisibleWorksheet(wkbook) As Boolean
'Determines if Only One Visible Worksheet in Workbook (wkbook)
'Useful when calling Sheets().Activate
Workbooks(wkbook).Activate
For Each wks In Worksheets
  If wks.Visible = 1 Then
     TotalVisWkshts = TotalVisWkshts + 1
     If TotalVisWkshts > 1 Then
        OneVisibleWorksheet = False
        Exit Function
     End If
  End If
Next wks
OneVisibleWorksheet = True
End Function
```

If the possibility exists that a Workbook will only have one Worksheet, this function can be called before the activate method is utilized on a sheet. If the function returns True, then the activate method should not be executed. The following line of VBA code illustrates such a check:

```
If OneVisibleWorksheet(ActiveWorkbook.Name) = False then
Sheets(2).activate
```

Now suppose Sheet2 has its name changed to RawData. To activate the sheet now named "RawData," the following commands would have to be typed:

```
Sheets(2).Activate
```

or

```
Sheets("RawData").Activate
```

What has happened here is that even though Sheet2 had its string name changed from "Sheet2" to "RawData," its numerical reference number does not change. The numerical reference number is indicative of the order in which a sheet was created, and will never change even when a sheet is renamed.

While on the subject of renaming sheets, a sheet can be renamed at will by using the (.Name) property. Here are two examples.

```
ActiveSheet.Name = "pKa Data"
Sheets(1).Name = "ELogD Data"
```

To explicitly reference a sheet allows a user to change values on a sheet *without that sheet being the active sheet*. To do this, both the Workbook and the Worksheet, and possibly the range, must be referenced in the command. This form of referencing will be covered in this chapter in more detail when Ranges and Cells are discussed. However, an example of using an explicit reference to change the name of a Worksheet in any Workbook (not just the active Workbook) is as follows:

```
Workbooks("Book2.xls").Sheets("RawData").Name = "Analyzed Data"
```

What is interesting about the foregoing statement is that it is acting upon the Worksheet "RawData" without the Worksheet "RawData" having to be the active sheet in the Workbook. In fact, the Workbook "Book2.xls" need not even be the active Workbook for the above statement to change the name of the Worksheet "RawData" to "Analyzed Data." The only requirement is that a Workbook named Book2.xls be open in the Excel environment and that the Workbook named Book2.xls contain a sheet named "RawData." (If such a Workbook does not exist, or does not contain the Worksheet specified using Sheets, a subscript out of range error will occur like that shown in Figure 1.2.)

While explicitly referencing a sheet requires a bit more coding than merely acting upon an active Worksheet, it is really the superior method to use when coding a program. One obvious benefit of explicit referencing is that anyone looking at the code knows what Workbook and Worksheet are being acted upon for any particular statement. A less obvious benefit of using explicit referencing is that the programmer does not have to change the active sheet or Workbook to perform an operation on that sheet or Workbook. Why is this a benefit? When the active Workbook or Worksheet is changed in Excel, the screen will flicker and be redrawn with the active Workbook and Worksheet now at the forefront. If the user is passing information between several Workbooks and/or Worksheets, this constant screen updating takes a great deal of time and slows the program down. By explicit referencing each Worksheet, one Worksheet can remain the active Worksheet of focus, yet any and every Worksheet present in the Excel environment can be acted upon. This results in faster program execution, and while the macro is executing, the screen will not be flashing to give the end user a headache.

Just as is the case with Workbooks, it is possible to loop through Worksheets one at a time and look at their attributes. The following subroutine loops through each Worksheet in the active Workbook. Each Worksheet name is then displayed in a message box.

```
Sub LoopThruSheets()
'This Subroutine Loops through sheets in the ACTIVE Workbook
Dim ws As Worksheet
For Each ws In Worksheets
  'Display the Name of each Worksheet in a message Box
  MsgBox ws.Name
Next
End Sub
```

This subroutine works fine but only if the sheets to be looped through are contained within the active Workbook. To loop through the sheets in any Workbook, the following subroutine can be used. The Workbook in which the sheets to be looped through are located in must be passed to the subroutine, in this case by utilizing the variable BookX$.

```
Sub LoopThruSheetsBookX(BookX$)
Dim sh As Sheet
For Each sh In Workbooks(BookX$)
  Debug.Print sh.Name
Next sh
End Sub
```

In addition to adding and renaming sheets, it is also possible to delete them. This is often particularly useful if a Worksheet is used as a place where intermediate values are stored (a scratch sheet) during the process of complex calculations. To delete a sheet, the following syntax is utilized:

```
Sheets(2).Delete
```

or

```
Sheets("Sheet2").Delete
```

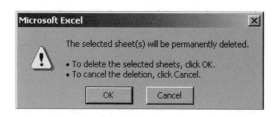

FIGURE 1.4 Typical Excel warning dialog box.

When the delete method is executed, ordinarily the user will be prompted by a dialog box (see Figure 1.4) which asks the user if in fact the sheet should be deleted. This makes sense as a precaution to keep a user from inadvertently deleting some important information with the mere click of a mouse button.

Unfortunately, however, this system of warning notifications carries over when such methods are executed from VBA code. The ramification of this is that if the developer is trying to create a process that is totally automated, when some methods are executed, dialog boxes will pop up, which require a user to click on them to allow the program to continue. (Such dialog boxes are referred to as modal forms.)

There is a means of suppressing all warning messages in Excel. By setting the Application DisplayAlerts property equal to false, all of Excel's Dialog messages to the user will be suppressed. This means the user will not be prompted when tasks are performed such as deleting items, saving files to an existing filename, or closing files without having saved the changes. This property can be set by using this syntax at the beginning of any VBA program.

```
'Suppress all Excel Warning Dialog Boxes
Application.DisplayAlerts = False
```

Setting this property to false is a must for creating a macro that does not require user intervention. However, care must be taken when running and testing code with this property set to false. The code will execute without any regard to what it is doing. It will delete and overwrite files, Worksheets, and Workbooks with impunity. Although this is probably one of the properties that makes VBA so useful, it is also one of the properties that can make VBA so dangerous. (Danger and usefulness oddly enough seem to travel in pairs in the programming world.)

Another useful tool is the Copy method of the Sheets Object. The following code will create a new Workbook with a clone of all the existing sheets in the active Workbook.

```
Sheets.Copy
```

To copy a single sheet within the same Workbook, utilize this variation of the method.

```
ActiveSheet.Copy Before:=ActiveSheet
```

Or

```
Sheets(1).Copy Before:=ActiveSheet
```

This will create a copy of the active sheet and place it immediately before the active sheet in the Workbook. The copied sheet will have the same name as the original sheet but will have a number in parentheses after indicating what number copy it is. (Example: "Sheet2(2)).")

To copy a single sheet and place that sheet within a new Workbook, truncate the above syntax to just

```
ActiveSheet.Copy
```

Or

```
Sheets(1).Copy
```

The ability to copy a sheet at will is very lucrative, especially when creating a variety of reports. The initial program can create a report sheet with every conceivable parameter included. Copies can then be made of this report, and it can be customized for distribution to different people or groups. If certain parameters are stored in different columns or rows, those columns or rows can be retained or deleted depending upon who the report is to be distributed to.

1.4 RANGES IN WORKSHEETS

Whenever an operation is performed on a Worksheet, the programmer must tell the program where in the Worksheet the operation is to take place. A Range is essentially a selected region within a Worksheet upon which actions will take place. When the value of a cell within a Worksheet is to be changed, the program must know the location of the cell for the value to be replaced. When an item is to be pasted within a Worksheet, the program must know a starting position (an upper left cell) that will serve as an anchor to the object to be pasted. Similarly, if a group of Cells is to be pasted from one Worksheet to another, that location must be specified by either a single cell starting point, or a range of Cells which matches *the exact size and shape* of the data to be pasted. It is the Range object that allows all of these tasks to be accomplished.

Four types of ranges predominate in the Excel VBA programming environment. They are a rectangular selection of Cells, a single cell, a column, or a row. Although an infinite configuration of ranges is possible, it is these four that lend themselves useful time and time again. The rectangular section of Cells is often utilized to copy and paste a section of information from one Worksheet to another Worksheet. The single cell is often utilized to write a single discreet value to a particular cell. In a flat file format, the same parameter is stored down an entire column with each row representing a unique data point. By performing operations on columns or rows, parameters from a unique data set can be added, removed, or altered at will.

Ranges within Worksheets are identified by Cells. Cells are identified by rows and columns, in a format somewhat like latitude and longitude on a map. Columns run left to right across a Worksheet page, and columns can be identified using either a number or a letter, which can lead to some confusion. Table 1.3 shows the column letter references and their numerical equivalents. Rows run top to bottom across a Worksheet and are only identified using a number.

It is the Cells form of the Range method that is the most useful in VBA programming because it is the only method that allows the programmer to directly define the limits of the range using numerical variables. For instance, the data in Table 1.3 can be readily duplicated by means of a macro that utilizes the Cells form of the Range method as follows.

```
Sub CreateColEqTable()
Dim ii As Integer
  For ii = 1 To 26
    Workbooks(ActiveWorkbook.Name).Worksheets(ActiveSheet.Name).
    Cells(1, ii).Value = Chr(64 + ii)
    Workbooks(ActiveWorkbook.Name).Worksheets(ActiveSheet.Name)
    .Range(Cells(1, ii), Cells(1, ii)).Value = Chr(64 + ii)
Workbooks(ActiveWorkbook.Name).Worksheets(ActiveSheet.Name).Cells
(2, ii).Value = ii

'Workbooks(ActiveWorkbook.Name).Worksheets(ActiveSheet.Name).R
ange(Cells(2, ii), Cells(2, ii)).Value = ii

  Next ii
End Sub
```

TABLE 1.3
Column Letter and Number Reference Equivalents

A	B	C	D	E	F	G	H	I	J	K	L	M	N	O	P	Q	R	S	T	U	V	W	X	Y	Z
1	2	3	4	5	6	7	8	9	10	11	12	13	14	15	16	17	18	19	20	21	22	23	24	25	26

This short macro brings several techniques to the forefront that will be utilized throughout this text. First, notice that the For loop is utilized to transition from column to column in this example. Also notice how the Cells method references individual Cells. The Cells method is followed by two numbers separated by commas in parentheses. The first number is the row (which, in this example, is fixed at one or two), and the second number is the column. When taking a linear algebra class, a professor said it was good to memorize the following statement to remember how to multiply a matrix: "Rows by Columns, write in Rows." If this statement is committed to memory, the first statement is how the Cells form of the Range method references a cell: Cells(row, col). Notice also that there are two ways of utilizing the Cells method to point to a single cell in a Worksheet. The first utilizes just the Cells method, and the second utilizes the Cells method as part of a range.

Using the Cells method is the simplest and easiest way to reference an individual cell. Here the Cells method directly follows the Worksheet reference, and only the coordinates for a single row and column must follow it as in this statement:

```
Workbooks(ActiveWorkbook.Name).Worksheets(ActiveSheet.Name).Cells
(1, ii).Value = Chr(64 + ii)
```

In this example, it is worth pointing out that both the column and the row references could be replaced by variables that could be altered by means of a loop or another method. Here the row is fixed (at 1), and the column varies as the loop `ii` increments.

The Cells method can also be incorporated as part of the Range method as shown in this statement (which was commented out in the CreateColEqTable subroutine):

```
'Workbooks(ActiveWorkbook.Name).Worksheets(ActiveSheet.Name).
Range(Cells(2, ii), Cells(2, ii)).Value = ii
```

Notice here that the Cells method is nested inside the Range method. Also notice that the Cells method occurs twice inside the Range method. This must always be the case. The first Cells statement defines the upper left corner of a rectangular Range. The second Cells statement defines the lower right-hand corner of a rectangular Range. To reference a single cell, both Cells statements must point to the same cell, resulting in a single-cell selection bounded by itself. Many VBA programmers run into confusion (and rightly so) when they try to implement a statement such as this:

```
Workbooks(ActiveWorkbook.Name).Worksheets(ActiveSheet.Name).
Range(Cells(2,2)).Value = 100
```

On its face this statement seems completely reasonable. Clearly, here the programmer wishes to assign the value 100 to the cell in the second row of column B. Trying to execute this statement will result in an error every time. For whatever reason, Excel was designed so that when the Cells form of the Range method is utilized, two instances of the Cells method must appear within the Range method even if the Range is to reference a single cell. The only valid way to write such a statement would be

```
Workbooks(ActiveWorkbook.Name).Worksheets(ActiveSheet.Name).
Range(Cells(2,2), Cells(2,2)).Value = 100
```

Although such a statement is counterintuitive (as opposed to the previous statement), it is the only correct way to assign a value to a cell using the Cells form of the Range method.

For example, the following subroutine would select a square range from row 2 col B to row 4 col D (B2:D4).

```
Sub SelectSquareRange()
'Example to Select a square range from B2:D4
Workbooks(ActiveWorkbook.Name).Worksheets(ActiveSheet.Name).Ra
nge(Cells(2, 2), Cells(4, 4)).Select
End Sub
```

Executing this subroutine will result in a Square from (B2:D4) being highlighted in the active Worksheet as shown in Figure 1.5.

Notice that in the `SelectSquareRange` subroutine that the Cells method does not reference columns by the letter designator shown on the Worksheet. In this case, column B is referenced using 2, and column D is referenced using 4, as shown in Table 1.3. Further notice that the single statement in this subroutine does not alter the contents of the Cells referenced using the Range method; it merely selects them, which is indicated on the Worksheet using the purple highlighting . What purpose does it serve to select a grouping of Cells? If a grouping of Cells of the same size and shape had been copied to the clipboard and the paste command were executed, the values would be neatly pasted into the selected region. The programmer could also select a region of Cells and format them in a particular way to suit the needs of the report by setting the number of decimal places, or perhaps changing the color of the font in those selected Cells.

Although the Cells form of the Range method is the only method that allows the programmer to directly define the limits of the range using numerical variables, there are methods of defining Ranges using a form of indirect addressing. Recall that columns are capable of being referenced by both letter and number formats. The following command will select the upper left-hand cell in any Worksheet:

```
Range("A1").Select
```

A 3 × 3 rectangular-shaped area of Cells with corners at row 2 col 2 and row 4 col 4 could be selected with this command:

```
Range("B2:D4").Select
```

FIGURE 1.5 Use of the Range method to select an area of a Worksheet.

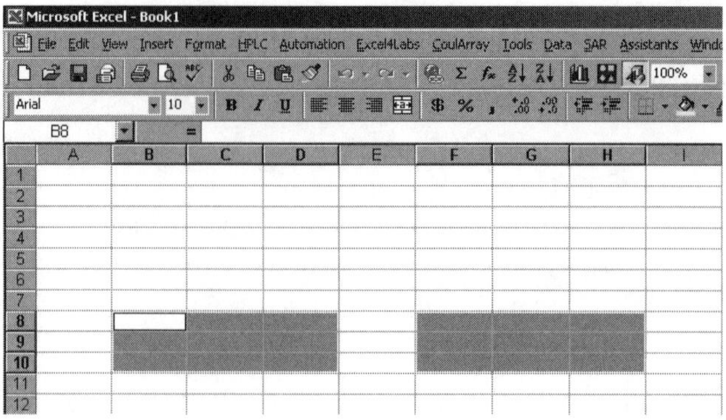

FIGURE 1.6 Simultaneous selection of two ranges in Excel.

It is also possible to select multiple areas on a Workbook simultaneously. This is done by grouping the addresses in a single Range statement and separating them with commas. This statement

```
Range("B8:D10, F8:H10").Select
```

will select two rectangular ranges on a Worksheet — one with corners at row 8 col 2 × row 10 col 4, and the other at row 8 col 6 × row 10 col 8, as shown in Figure 1.6.

Although it is possible to reference ranges in Worksheets using the (column letter)(row number) as shown in the preceding text, the method, on its face, is severely limited in that ranges cannot be dynamically defined using variables as with the Cells form of the Range method. Taking this method of selecting Ranges in the examples presented, its only real use in programming would be to select areas in Worksheets where parameters would be fixed in the same location on every Worksheet. This is seldom the case because reports will often contain varying numbers of samples run, and even the report parameters themselves can be altered depending on the selections made at the beginning of a macro program. However, there is a trick that can be employed that makes the A1 style [(column letter)(row number)] of referencing more flexible to use.

Notice that in all the examples presented using the A1 style of the Range method, the actual Range reference is passed as a string. It is possible to replace the Range reference with a string variable that represents a valid Range. Therefore, in the statement

```
Range("B2:D4").Select
```

the Range reference of "B2:D4" can be represented with a string variable. The statement could be written with the same functionality using the following two lines:

```
StrRange$ = "B2:D4"
Range(StrRange$).Select
```

Here, a valid range is replaced with the string variable `StrRange$`. This is more forcefully illustrated using the following subroutine:

```
Sub ReplaceStrInRange(ByVal Index As Integer)
Dim StrRange$
Select Case Index
  Case 1
    StrRange$ = "A1"
  Case 2
    StrRange$ = "B2:D4"
```

```
  Case 3
    StrRange$ = "B8:D10, F8:H10"
End Select
Range(StrRange$).Select
End Sub
```

In the `ReplaceStrInRange` subroutine, an integer value is passed to the subroutine when it is called. The Range to be selected will vary depending upon the value of the integer passed. However, this could be streamlined even further by simply passing a string variable to the subroutine that contains a valid range. The next subroutine demonstrates this:

```
Sub SelectPassedRange(ByVal CellRange$)
Range(CellRange$).Select
End Sub
```

which brings the discussion to the true point of focus, which is the flexibility of utilizing a string variable to represent a range. The point here is that the programmer can "build" the string variable to represent any Range that is desired. The following is an example to illustrate the point.

Suppose it was desired to print out the letters of the alphabet to a Worksheet in a specific manner. What if the specification called for a letter to be added to each cell columnwise, and after the third column (C) was filled, the letters should be added restarting at the first column but one row down (so as not to overwrite the previous additions)? So, row 1 would have A,B,C, and row 2 would have D,E,F, until row 9 col B is reached whereby Z is written into this cell and the macro terminates.

Obviously, to accomplish this task, the range to which the Cells are written is changing for each subsequent letter. Although this could be done utilizing the Cells form of the Range method, it is done in the following subroutine by indirectly defining the limits of the range by using a string variable. This importance of knowing how to do this is that not every property and method in Excel VBA is capable of recognizing a cell address based on a variable parameter. Some only recognize the A1 style [(column letter)(row number)] form of addressing. In many instances, building a string to represent the address to be acted upon is the only way of coding a particular property of method. The Build-StringRange subroutine shows how to build a string range to satisfy the previously defined specification.

```
Sub BuildStringRange()
Dim ii As Integer, CellRange$, ColPos As Integer
Dim RowPos As Integer
RowPos = 1
For ii = 0 To 25
  If ii Mod 3 = 0 And ii <> 0 Then
     ColPos = 0
     RowPos = RowPos + 1
  End If
  ColPos = ColPos + 1
  CellRange$ = Chr(64 + ColPos) & Trim(Str(RowPos))
  'Debug.Print CellRange$
Workbooks(ActiveWorkbook.Name).Worksheets(ActiveSheet.Name).Range
(CellRange$).Value = Chr(65 + ii)
Next ii
End Sub
```

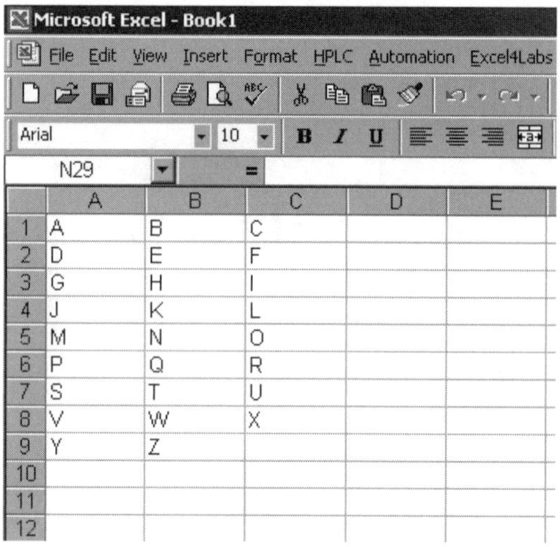

FIGURE 1.7 The BuildStringRange subroutine uses string-defined ranges.

The result of running the BuildStringRange subroutine can be viewed in Figure 1.7. It is important to understand the methodology employed within this macro to manipulate the cell range by utilizing a string variable. Notice that the row and column positions are "built" from the numeric variables `ColPos` and `RowPos`. Further notice that the value of these variables is controlled by decisions rendered in the `ii` looping structure. The range is then built from the numeric variables with the statement

```
CellRange$ = Chr(64 + ColPos) & Trim(Str(RowPos))
```

Notice that the column position is built using the `Chr()` function. Looking at Table 1.4, notice that the decimal equivalent of the character A is the value 65. The statement: `Debug.Print Chr(65)` will print "A" to the immediate (debug) window. In this example, the `ColPos` variable increments from 1 to 3. To return the corresponding character representative of the column to be included in the Range function, the `ColPos` variable is added to 64 in the `Chr()` function. This will result in Chr(65), Chr(66), or Chr(67) being called, which will return an "A," "B," or "C," respectively.

Also notice the snippet of code `Trim(Str(RowPos))`, which builds the row position of the `CellRange$` variable. Here the numeric variable `RowPos` is converted to a string by means of the `Str()` function. Most importantly, note the use of the `Trim()` function on the string returned by the `Str()` function. When the `Str()` function returns a string form of a number, a leading space is always present for a positive number. (This space is reserved for a negative sign (–) should the string returned be from a negative number.) If this leading space is not removed when the cell range string is being built, the result will be an invalid cell range because an unintended space will be present in the range.

Since using the A1 form [(column letter)(row number)] of addressing can be the only way of coding a particular property or method, it is useful to have a subroutine that will take a column number and return the equivalent alphabetical column reference. The function `ColNoToColLet` accomplishes this task. This function takes into account the fact that column numbers greater than 26 are referenced by adding an incrementing second letter. For example, if 27 is passed to the function, it will return "AA," 28 <-> "AB," 29 <-> "AC," etc. ...

```
Function ColNoToColLet(ByVal ColumnNo As Integer) As String
'This function will convert any column # to the letter Equivalent
If ColumnNo < 27 Then
```

TABLE 1.4
ASCII Table

Dec	Hex	Char	Dec	Hex	Char	Dec	Hex	Char	Dec	Hex	Char
0	0	NUL	32	20		64	40	@	96	60	`
1	1	SOH	33	21	!	65	41	A	97	61	a
2	2	STX	34	22	"	66	42	B	98	62	b
3	3	ETX	35	23	#	67	43	C	99	63	c
4	4	EOT	36	24	$	68	44	D	100	64	d
5	5	ENQ	37	25	%	69	45	E	101	65	e
6	6	ACK	38	26	&	70	46	F	102	66	f
7	7	BEL	39	27	'	71	47	G	103	67	g
8	8	BS	40	28	(	72	48	H	104	68	h
9	9	TAB	41	29	)	73	49	I	105	69	i
10	0A	LF	42	2A	*	74	4A	J	106	6A	j
11	0B	VT	43	2B	+	75	4B	K	107	6B	k
12	0C	FF	44	2C	,	76	4C	L	108	6C	l
13	0D	CR	45	2D	−	77	4D	M	109	6D	m
14	0E	SO	46	2E	.	78	4E	N	110	6E	n
15	0F	SI	47	2F	/	79	4F	O	111	6F	o
16	10	DLE	48	30	0	80	50	P	112	70	p
17	11	DC1	49	31	1	81	51	Q	113	71	q
18	12	DC2	50	32	2	82	52	R	114	72	r
19	13	DC3	51	33	3	83	53	S	115	73	s
20	14	DC4	52	34	4	84	54	T	116	74	t
21	15	NAK	53	35	5	85	55	U	117	75	u
22	16	SYN	54	36	6	86	56	V	118	76	v
23	17	ETB	55	37	7	87	57	W	119	77	w
24	18	CAN	56	38	8	88	58	X	120	78	x
25	19	EM	57	39	9	89	59	Y	121	79	y
26	1A	SUB	58	3A	:	90	5A	Z	122	7A	z
27	1B	ESC	59	3B	;	91	5B	[	123	7B	{
28	1C	FS	60	3C	<	92	5C	\	124	7C	|
29	1D	GS	61	3D	=	93	5D	]	125	7D	}
30	1E	RS	62	3E	>	94	5E	^	126	7E	~
31	1F	US	63	3F	?	95	5F	_	127	7F	DEL

```
    ColNoToColLet = Chr$(64 + ColumnNo)
    Exit Function
End If
For ii = 1 To Int(ColumnNo / 26)
  ColNoToColLet = ColNoToColLet & "A"
Next ii
ColNoToColLet = ColNoToColLet & Chr(64 + (ColumnNo Mod 26))
End Function
```

There is one very unconventional yet highly flexible way of creating and referencing ranges, and that is through the assignment of a custom name for a contiguous range of cells. In the top left corner of the Excel Worksheet environment there is a textbox that displays the name (or location) of the currently active cell. For example, if the top left cell is selected, by default "A1" will be shown in this textbox. This textbox is referred to as the "Name Box" in the Excel environment because it displays the "name" of the currently selected cell or cells.

By default, Ranges are shown in the "Name Box" using an A1 style of addressing. However, it is possible to select a single or range of contiguous cells and type a new unique name for them in the Name Box. For example, a single cell could have its range named "SalesSubtotal" and, presumably, this cell could be referenced as such to update the Sales Subtotal for the Worksheet. Such a range could then be activated using the following code:

```
Range("SalesSubtotal").Select
```

In addition to manually setting the range, it can be set from VBA code by using code such as the following:

```
Range("A10:B12").Select
ActiveWorkbook.Names.Add Name:="AtenthruB12", RefersToR1C1:=
"=Sheet1!R10C1:R12C2"
```

In this code snippet, a contiguous range of cells (A10:B12) is assigned the name "AtenthruB12" using VBA code. The above code could easily be modified to reference a single cell.

Custom-named Ranges have one limitation — they cannot encompass multiple discontinuous areas of a worksheet. Thus, a Range such as that previously illustrated in Figure 1.6 (B8:D10,F8:H10), could not be assigned a custom name. Also notice that when a custom range encompasses multiple adjacent cells, and a single cell within the custom range is selected, the cell address is shown instead of the custom name range. *The entire custom range, not a subset of the range, must be selected for the custom Range name to be shown in the Name Box.*

One last topic on ranges will be covered, which is of paramount importance when constructing automated data analysis systems. When building functions, it is necessary that every cell referenced by the function contain data. Because functions are often built utilizing looping structures over a range of columns and rows (Chapter 4 covers functions in great detail), it is important to have a way of determining if a range of Cells has any empty Cells. An empty cell is regarded by an Excel function as 0, and if a division by zero operation is performed in Excel, this message will appear as the result: #DIV/0!

```
Function EmptyCellsinRange(Workbook, Worksheet, toprow,
leftcol, bottomrow, rightcol) As Boolean
'The following function uses CountA to return the number of non
empty Cells in a Range
Dim TotalCells As Long, NonEmptyCells As Long
ActivateWorkbook (Workbook)
If OneVisibleWorksheet(Workbook) = False Then
Sheets(Worksheet).Activate
TotalCells = ((bottomrow - toprow) + 1) * ((rightcol - leftcol) + 1)
NonEmptyCells = Application.CountA(Range(Cells(toprow,
leftcol), Cells(bottomrow, rightcol)))

If TotalCells - NonEmptyCells = 0 Then
    'All Cells in Range Contain SOMETHING
    EmptyCellsinRange = False
    Else
    'Some Cells in Range ARE EMPTY
    EmptyCellsinRange = True
End If
End Function
```

Whenever functions are to be built utilizing a range of data, the foregoing function should be used to ensure that the selected range does not contain any empty Cells. If empty Cells are encountered in the selected range, the function will return the Boolean value True. The function works by calculating the total number of Cells in the range bounded by the passed parameters: toprow, leftcol, bottomrow, rightcol. The CountA method is then utilized to determine the number of nonempty Cells within the bounded range. If the total number of Cells minus the nonempty Cells is equal to zero, then every cell in the range bounded by the passed parameters contains something.

1.5 USING EXPLICIT REFERENCING

Notice that, in many of the earlier examples, the Workbooks and Worksheets that a range was to focus on were indicated explicitly in the coding as in the following example.

```
Workbooks(ActiveWorkbook.Name).Worksheets(ActiveSheet.Name).Range
(CellRange$).Value = "Cell Contents"
```

In this statement, the portion of code
Workbooks(ActiveWorkbook.Name).Worksheets(ActiveSheet.Name)
is redundant because, if the Workbook and Worksheet to be accessed are not specified, the active Workbook and Worksheet will be acted upon. For example, the Range method can be called to act upon the active Workbook and Worksheet without specifying them as in this example:

```
Range("A1").Select
```

Here, row 1 column A will be selected in the active Workbook and Worksheet. So, what then is the sense in specifying the Workbook and Worksheet within the code? The obvious point is that it makes the code more readable to anyone who is trying to debug the application. Such a coding scheme has far wider implications.

The Workbook name and Worksheet name, much like the range method, can also be represented by a string variable. The significance of this is that they can be passed as a parameter to any subroutine or function. The implication of this is that subroutines can be written to perform generic functions on any Worksheet regardless of the Workbook it resides in. The next subroutine illustrates how this can be done.

```
Sub ExplicitReference(ByVal wkbook$, ByVal wksheet$)
'An Example of Using Explicit Referencing
Workbooks(wkbook$).Worksheets(wksheet$).Range("A1").Value = "A"
Workbooks(wkbook$).Worksheets(wksheet$).Range("A2").Value = "B"
Workbooks(wkbook$).Worksheets(wksheet$).Range("A3").Value = "C"
End Sub
```

Here the letters A,B,C will be written down the first column (A) of the Worksheet, and Workbook specified with the passed parameters `wkbook$` (Workbook) and `wksheet$` (Worksheet). But why specify the Workbook and Worksheet if one will be acting upon the active Workbook and Worksheet pair anyway? Although this does require a little bit of extra effort on the part of the programmer, it adds enormous utility to the code that is written in that it can be "recycled" to be utilized in other applications. Subroutines written in this manner can perform their operations on any Workbook/Worksheet pair and are not limited in their scope of operation. It is always best to think in terms of not just writing the code to accomplish the task at hand but how to write the code in such a manner that its usefulness is not limited. In the instance where the active Workbook/Worksheet pair are to be utilized, a call can be made to the subroutine in the following manner:

```
Call ExplicitReference(ActiveWorkbook.Name, ActiveSheet.Name)
```

Such a call makes it very easy to utilize an explicitly referenced subroutine for an active Workbook/Worksheet pair, and the programmer does not have to deal with remembering the exact Workbook and sheet name if they choose to utilize the Active object as shown in the foregoing statement.

1.6 ROWS AND COLUMNS

Rows and Columns are the navigational tools utilized when working with Excel. In an organized Worksheet, everything is laid out in rows and columns. If the Worksheet is a "flat" type format, each column will hold only one parameter, and the first row of the Worksheet will contain descriptive headers of the information each column is to contain. Each row will then contain a line of information about a particular entity. The entity could be a customer, a chemical compound, a location, or any discrete entry that would have data that corresponds to the fields the columns are utilized to hold. Rows are simple enough to manipulate. A row can only be referenced by a number, so, accordingly, the way to select any row is

```
Rows(5).Select
```

Or, using a more explicit reference,

```
Workbooks(ActiveWorkbook.Name).Worksheets(ActiveSheet.Name).Ro
ws(5).Select
```

In the above examples, row number 5 is selected, and all the Cells in row number 5 across all columns are selected. It is not possible to select multiple rows using the Rows method. The Range method discussed earlier in the chapter must be utilized to do this.

The row number can, of course, be replaced with a numeric variable as follows.

```
Sub SelectRows()
Dim ii as Integer
'Select Rows Utilizing Variables
For ii = 3 To 6
  Rows(ii).Select
Next ii
End Sub
```

Columns, on the other hand, can be referenced by either a letter or a number. This lends more flexibility to the referencing scheme, but can be more confusing and problematic as well. For example, to select the second column from the left (or the column labeled "B"), the following commands could be utilized:

```
Columns(2).Select
```

Or:

```
Columns("B").Select
```

Again, the referencing can be made more explictly with the syntax:

```
Workbooks(ActiveWorkbook.Name).Worksheets(ActiveSheet.Name).
Columns(2).Select
```

Or:

```
Workbooks(ActiveWorkbook.Name).Worksheets(ActiveSheet.Name).
Columns("B").Select
```

As in the case with the Rows method, the column number or letter can be replaced by a variable. The next subroutine shows how to do this using either a numeric or string reference.

```
Sub SelectCols()
Dim ii As Integer
For ii = 2 To 6
   Columns(ii).Select
   Columns(Chr(64 + ii)).Select
Next ii
End Sub
```

Here, the column is both directly referenced using a number (`ii`) and by a letter using the string character code representing the column label that is returned via the `Chr()` function. Either scheme will work just fine.

One of the nice things about the row and column methods is that they select all the entities within that particular row or column. For example, if the Cells form of the Range method is utilized, the starting points of the columns and rows must be specified. Utilizing the rows method, however, all the columns in the row are selected by default. The reverse is true when using the column method.

1.7   SEARCHING WORKSHEETS — USING FIND

The ability to search a Worksheet is probably the most critical skill an Excel VBA programmer must have. Data cannot be extracted or manipulated if it cannot be found. The key to finding data is consistancy. Ideally, every Worksheet should have "landmarks" that remain the same from report to report. The landmarks are always some form of text: a header, table label, or some identifying form that precedes the data which is to be utilized. By searching the Worksheet for these landmarks and determining their location, the programmer is then in a position to create routines to act upon that data appropriately.

In Excel, rows and columns can be hidden from view. The practical consequence of this is that the user may not wish to extract or search for data that resides in a row or column that is hidden from view. Thus, any routine that searches for data should have the option to omit hidden columns and rows from its field of search. The following function does just that:

```
Function FindPos(ByVal wkbook$, ByVal sheet, ByVal cellcts$,
ByVal SearchHidden As Boolean, ByVal MatchCase As Boolean) As
String
'Pass Contents that you are looking for in sheet via cellcts$
'Function will return the address of location of contents
'wkbook$ = Workbook name
'sheet = Worksheet name in string or numeric form (variant)
'cellcts$ = searching for
'SearchHidden = search in hidden Cells too? True/False
'MatchCase = differentiate between upper and lower case?
True/False
Dim Msg$
On Error GoTo NotFound
Call ActivateWorkbook(wkbook$)
Worksheets(sheet).Select
'Make Sure an Entire Row or Column is not selected, this will
```

```
'cause this function to only search rows or columns - not both
'which will result in failure
Range("A1").Select
'Find Results on Sheet
Select Case SearchHidden
  Case True
    'Search Includes Hidden Cells
    Cells.Find(What:=cellcts$, LookIn:=xlFormulas, LookAt _
    :=xlPart, SearchOrder:=xlByColumns, SearchDirection _
    :=xlNext, MatchCase:=MatchCase).Activate
  Case False
    'Search Omits Hidden Cells
    Cells.Find(What:=cellcts$, LookIn:=xlValues, LookAt _
    :=xlPart, SearchOrder:=xlByColumns, SearchDirection _
    :=xlNext, MatchCase:=MatchCase).Activate
End Select
FindPos = Selection.Address(ReferenceStyle:=xlR1C1)
Exit Function
NotFound:
Msg$ = "Could Not Find Instance of " & cellcts$
Select Case Application.DisplayAlerts
  Case True
  MsgBox "Not Found!", vbExclamation + vbOKOnly, Msg$
  Case False
  'Why is Application.DisplayAlerts always shown in Debug as
Application.DisplayAlerts = TRUE ?
  Debug.Print Msg$
End Select
End Function
```

The FindPos function returns the position in "R1C1" format, of the specified text string cellcts$, which contains the cell contents to be searched for. The Workbook and Worksheet to be searched must be specified as well as the Boolean switch SearchHidden. If SearchHidden is set to TRUE, then the function will include hidden rows and columns in the search range. If SearchHidden is set to FALSE, then the function will not include hidden rows and columns in the search range. This is accomplished by means of the LookIn:= parameter shown above. If set to xlFormulas, then every cell will be searched regardless of whether it is hidden or not. If set to xlValues, then only visible Cells will be searched. The function also has the capability to distinguish between capital and lowercase letters when searching for text via the MatchCase Boolean switch. If MatchCase is set to "True," then the case will be considered when comparing the Cells contents to the search string cellcts$. If MatchCase is set to "False," then the case will be irrelevant when performing a search.

Notice also this line of code right in the beginning of the function.

```
Range("A1").Select
```

The purpose of this statement is twofold. First, if a row, column, or range is selected, then the search will be limited to the selected range! Selecting the first cell in the Worksheet ensures that no ranges are inadvertantly preselected in the Worksheet. Also, by selecting cell "A1" as the active cell ensures that the search will begin at the top of the Worksheet as opposed to the middle of the Worksheet, which would be the case if cell "C46" happened to be the currently active cell in the Worksheet. If the programmer happens to know that the string to be searched for will occur several times, they may wish to alter where the search begins, especially if the desire is to find the second or third instance of the string within a Worksheet. It may, in fact, be desirable to modify the function to be able to pass the starting cell to be searched as a parameter.

There is one very nasty bug that the author has confirmed with Microsoft via a premier support case (Case ID: SRX040415600672). The `FindPos` function as coded above will not work with merged Cells. It will return an "Object variable or With block variable not set" error. The only known workaround to this problem is to replace the SearchOrder:=xlByColumns with Search Order:=xlByRows. It is unknown why searching by rows allows detection of strings in merged Cells, whereas searching by columns does not allow detection of strings in merged Cells.

Although the `FindPos` function gives the user the ability to find the location of any parameter they wish, the format in which the location is returned is less than desirable. The "R1C1" format does not lend itself well to usability within the Excel VBA framework. Recall that it is the Cells form of the Range method that is most flexible in manipulating data. Fortunately, it is fairly easy to extract the row and column information from the R1C1 format returned by the FindPos function. The following function returns the Row number from an R1C1 style string.

```
Function FindRow(addr$) As Integer
'Pass Current Address in "R1C1" Format to this Function and
it will Return the Row Number

Dim I As Integer
Dim strt As Integer, noc As Integer
For I = 1 To Len(addr$)
  If Mid$(addr$, I, 1) = "R" Then
     strt = I + 1
  End If
  If Mid$(addr$, I, 1) = "C" Then
     noc = I - strt
  End If
Next I
FindRow = Val(Mid$(addr$, strt, noc))
End Function
```

Here the row is found simply by taking the passed R1C1 address (`addr$`) and looking between the "R" and the "C" in the string to determine the row number. This function is extremely useful when used in combination with the FindPos function in a manner such as this:

```
FindRow(FindPos(ActiveWorkbook.Name, Activesheet.name, "B",
True, False))
```

Such an arrangement could even be embedded within the Cells form of the Range method to provide the row location for the Cells method.

The next function returns the Column number from an R1C1 style string.

```
Function FindCol(addr$) As Integer
'Pass Current Address in "R1C1" Format to this Function and
it will Return the Column Number
```

```
Dim I As Integer
Dim strt As Integer, noc As Integer
For I = 1 To Len(addr$)
  If Mid$(addr$, I, 1) = "C" Then
      strt = I + 1
      noc = Len(addr$) - I
  End If
Next I
FindCol = Val(Mid$(addr$, strt, noc))
End Function
```

This function simply finds the "C" in the R1C1 string and returns the numeric value of all the characters following the "C" (which is the column number). Again, this function is extremely useful when used in combination with the FindPos function in a manner such as this:

```
FindCol(FindPos(ActiveWorkbook.Name, Activesheet.name, "B",
True, False))
```

1.8 COPYING, CLEARING, AND DELETING DATA

The entire discussion up to this point has been preparing the reader for this topic, copying data. Before data in a Worksheet can be copied, it must be (1) located utilizing the find command and (2) selected using some variation of the Range method.

Copying can take the form of one of two methods. The first and most obvious would be to copy a cell or range of Cells from one section of a Worksheet to another section of a Worksheet. The second would be to copy values contained in Cells of a Worksheet into variables defined within the VBA program, or vice versa. Oftentimes, an array will be utilized to store a section of Worksheet cell values. The values are most often read into the array by means of a looping mechanism. Some examples will be given to show the reader how to accomplish both forms of copying. The Workbook shown in Figure 1.8 will be utilized for the examples in the text.

Looking at Figure 1.8, there are three data series in an 8-row × 12-column format. Suppose it is desired to read the information from the first data series into an array. This can be accomplished in the following manner. First, a Public variable should be declared at the top of the module to hold the data as follows:

```
Public WellPos(8, 12) As Single
```

The data in this spreadsheet is from a 96 well microplate, so the variable is named WellPos (for well position), and the first index 8 denotes rows, whereas the second index 12 denotes columns (as is the case with matrices). Therefore, WellPos(2,7) should contain the data in row 2 column 7 of the first data series, which in this case is 3.63. It is a fairly easy process to read the information from the Cells into an array using a looping structure, but before that can be done, a "landmark" must be found that denotes the start of the first data series. Looking at Figure 1.8, notice that the letter "A" is one column to the left of the starting column of each data series and also denotes the starting row of each data series. More importantly, notice that a capital "A" is not found on the Worksheet prior to cell A4. Therefore, searching for the start of the first data series, it is possible to use "A" as the search string because it is unique prior to the start of the first data series. The following subroutine will locate the start of the first data series and write the values from each cell into the Public variable WellPos.

```
Sub ReadIntoArray()
'Reads Well Values from "CopyExample.xls" Worksheet into Array
```

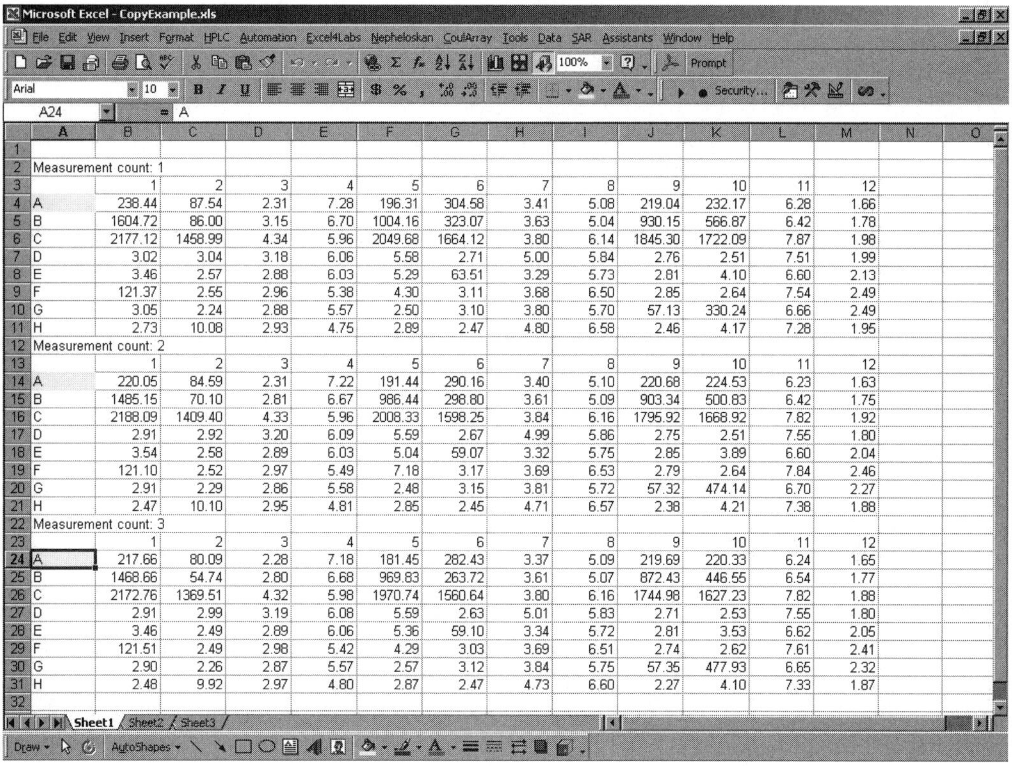

	A	B	C	D	E	F	G	H	I	J	K	L	M	N	O
1															
2	Measurement count: 1														
3		1	2	3	4	5	6	7	8	9	10	11	12		
4	A	238.44	87.54	2.31	7.28	196.31	304.58	3.41	5.08	219.04	232.17	6.28	1.66		
5	B	1604.72	86.00	3.15	6.70	1004.16	323.07	3.63	5.04	930.15	566.87	6.42	1.78		
6	C	2177.12	1458.99	4.34	5.96	2049.68	1664.12	3.80	6.14	1845.30	1722.09	7.87	1.98		
7	D	3.02	3.04	3.18	6.06	5.58	2.71	5.00	5.84	2.76	2.51	7.51	1.99		
8	E	3.46	2.57	2.88	6.03	5.29	63.51	3.29	5.73	2.81	4.10	6.60	2.13		
9	F	121.37	2.55	2.96	5.38	4.30	3.11	3.68	6.50	2.85	2.64	7.54	2.49		
10	G	3.05	2.24	2.88	5.57	2.50	3.10	3.80	5.70	57.13	330.24	6.66	2.49		
11	H	2.73	10.08	2.93	4.75	2.89	2.47	4.80	6.58	2.46	4.17	7.28	1.95		
12	Measurement count: 2														
13		1	2	3	4	5	6	7	8	9	10	11	12		
14	A	220.05	84.59	2.31	7.22	191.44	290.16	3.40	5.10	220.68	224.53	6.23	1.63		
15	B	1485.15	70.10	2.81	6.67	986.44	298.80	3.61	5.09	903.34	500.83	6.42	1.75		
16	C	2188.09	1409.40	4.33	5.96	2008.33	1598.25	3.84	6.16	1795.92	1668.92	7.82	1.92		
17	D	2.91	2.92	3.20	6.09	5.59	2.67	4.99	5.86	2.75	2.51	7.55	1.80		
18	E	3.54	2.58	2.89	6.03	5.04	59.07	3.32	5.75	2.85	3.89	6.60	2.04		
19	F	121.10	2.52	2.97	5.49	7.18	3.17	3.69	6.53	2.79	2.64	7.84	2.46		
20	G	2.91	2.29	2.86	5.58	2.48	3.15	3.81	5.72	57.32	474.14	6.70	2.27		
21	H	2.47	10.10	2.95	4.81	2.85	2.45	4.71	6.57	2.38	4.21	7.38	1.88		
22	Measurement count: 3														
23		1	2	3	4	5	6	7	8	9	10	11	12		
24	A	217.66	80.09	2.28	7.18	181.45	282.43	3.37	5.09	219.69	220.33	6.24	1.65		
25	B	1468.66	54.74	2.80	6.68	969.83	263.72	3.61	5.07	872.43	446.55	6.54	1.77		
26	C	2172.76	1369.51	4.32	5.98	1970.74	1560.64	3.80	6.16	1744.98	1627.23	7.82	1.88		
27	D	2.91	2.99	3.19	6.08	5.59	2.63	5.01	5.83	2.71	2.53	7.55	1.80		
28	E	3.46	2.49	2.89	6.06	5.36	59.10	3.34	5.72	2.81	3.53	6.62	2.05		
29	F	121.51	2.49	2.98	5.42	4.29	3.03	3.69	6.51	2.74	2.62	7.61	2.41		
30	G	2.90	2.26	2.87	5.57	2.57	3.12	3.84	5.75	57.35	477.93	6.65	2.32		
31	H	2.48	9.92	2.97	4.80	2.87	2.47	4.73	6.60	2.27	4.10	7.33	1.87		
32															

FIGURE 1.8 Example Workbook for copying data.

```
Dim startrow As Integer, startcol As Integer
Dim row As Integer, col As Integer
'First Find the starting Positions of the Matrix
startrow = FindRow(FindPos("CopyExample", 1, "A", False, True))
startcol = FindCol(FindPos("CopyExample", 1, "A", False, True))
+ 1
'Populate Public array WellPos using looping structure
For col = startcol To startcol + 11
  For row = startrow To startrow + 7
    WellPos((row - (startrow - 1)), (col - (startcol - 1))) = _
    Workbooks("CopyExample.xls").Worksheets(1).Cells(row,
    col).Value
  Next row
Next col
End Sub
```

There are several comments worth making about this subroutine. First, notice that when utilizing the `FindPos` function, the `MatchCase` parameter must be set to True. Otherwise, the `FindPos` function will return cell A2 as there is a lower case "A" in the string "Measurement count: 1" that resides in cell A2. Also notice that when using the `FindCol` function to determine the location of the starting column, a one is added to the result ("+1"). This is because the cell with "A" is always one column left of the true starting position of the data matrix. Looking at the looping structure, it is apparent that 8 row loops are nested inside 12 column loops, and that the beginnings of the loops are offset by the startcol and startrow locations. Because the WellPos variable is only dimensioned 8 × 12 (WellPos(8,12)), it is necessary to remove the offset when referencing the WellPos variable in the looping structure. The offset is added to represent the true position of the data on the Worksheet so that it may be referenced by the Cells method.

Once the data have been read into the array, it can be extracted at any time rather easily. The following subroutine will pop up a message box displaying the value of any set of indexes contained within the array.

```
Sub DisplayArrayVal(ByVal row As Integer, ByVal col As Integer)
'Display the contents in the WelPos Array for row & col passed
MsgBox "Well Value in row " & Chr(64 + row) & " and col " &
Str(col) & _
" is: " & Str(WellPos(row, col)), vbOKOnly + vbInformation,
"Retrieved Value"
End Sub
```

This subroutine should only be run *after* the ReadIntoArray subroutine has been executed. Otherwise, there will be no values written into the array to extract. The subroutine will display a message box telling the value at a particular row and column within the array (Figure 1.9). Notice that, on the Worksheet, the Cells are limited to two decimal places. However, when the value in the cell is extracted to a variable, its full precision (in this case Single precision) is maintained. That is why the message box displays values with more than two decimal places.

The second and more intuitive form of copying is copying data from one Worksheet to another. Users do this all the time manually by selecting a section of the Worksheet by hand with the mouse and copying and pasting it. This is fine for an occasional task, but when a series of Worksheets must be formatted in an identical manner, it gets tiresome very quickly to keep copying and pasting the data by hand. Although it is more work up front, it is much more efficient to construct a macro that will locate the data (using landmarks) and automatically copy the data to its proper location. This is especially true if formatting Worksheets in such a manner is part of an ongoing process or task that will be carried out for months if not years. Copying a section of a Worksheet from one Worksheet to another is not a complicated process, the following subroutine accomplishes this task quite nicely.

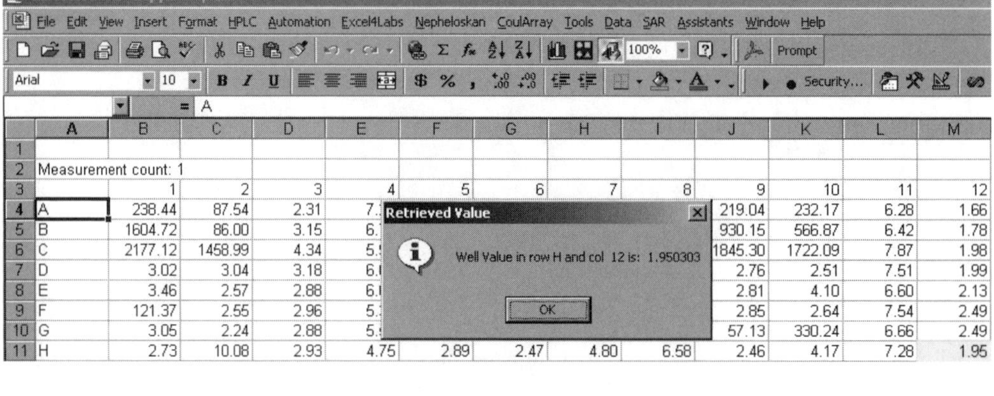

FIGURE 1.9 Displaying the value of a particular element of the Array.

```
Sub CopySection(ByVal frmwkbook, ByVal frmwksht, ByVal r1,
ByVal c1, ByVal r2, ByVal c2, ByVal towkbook, ByVal towksht,
ByVal row, ByVal col)

'Copies the "Chunk" Bounded by r1,c1,r2,c2 in frmwkbook to
towkbook starting at row,col

'Select "from" Workbook and Worksheet
ActivateWorkbook (frmwkbook)
Worksheets(frmwksht).Select
Range(Cells(r1, c1), Cells(r2, c2)).Select
Selection.Copy
'Select "to" Workbook and Worksheet
ActivateWorkbook (towkbook)
Worksheets(towksht).Select
Range(Cells(row, col), Cells(row, col)).Select
ActiveSheet.Paste
End Sub
```

One nice aspect of this subroutine is that it allows the user to specify what Workbook and Worksheet the information will be copied *from* as well as what Workbook and Worksheet the information will be copied *to*. This provides the ultimate in flexibility. Something else worth keeping in mind is that if a range is copied from one Worksheet, and the user wishes to specify a range on another Worksheet to copy that range to, the sizes of the ranges must be identical, or the process will fail. This subroutine avoids having the user check if the ranges are identical by simply specifying a starting position for the copying to begin. The selected range will be copied to the Worksheet beginning at the starting point, and extending however many rows and columns from the starting point that the original copied range encompassed.

Often, Worksheets will have a header, that is, a series of rows at the top of the Worksheet that contain information relating to all the data contained in the Worksheet. It is often advantageous to copy header information when constructing a new Worksheet. The following subroutine does just that.

```
Sub CopyHeader(ByVal frmwkbook, ByVal frmwksht, ByVal nrows,
ByVal towkbook, ByVal towksht)
'Copies a Header of n rows in frmwkbook to towkbook starting
at top of Worksheet
'Select "from" Workbook and Worksheet
Dim RRange$
ActivateWorkbook (frmwkbook)
Worksheets(frmwksht).Select
RRange$ = "1:" & Trim$(Str$(nrows))
Rows(RRange$).Select
Selection.Copy
'Select "to" Workbook and Worksheet
ActivateWorkbook (towkbook)
Worksheets(towksht).Select
Range(Cells(1, 1), Cells(1, 1)).Select
ActiveSheet.Paste
End Sub
```

As in the previous subroutine, the CopyHeader subroutine allows the user to specify the Workbook and Worksheet the header is to be copied from, and the Workbook and Worksheet the header is to be pasted to.

While copying data is the most common manipulation of Worksheet data, other manipulations are useful as well. Sometimes, it is advantageous to cut a section of data out of one Worksheet and paste it into another Worksheet. The following subroutine will accomplish this task.

```
Sub CutSection(ByVal frmwkbook, ByVal frmwksht, ByVal r1, ByVal
c1, ByVal r2, ByVal c2, ByVal towkbook, ByVal towksht, ByVal
row, ByVal col)
'Cuts the "Chunk" Bounded by r1,c1,r2,c2 in frmwkbook to
towkbook starting at row,col
'Select "from" Workbook and Worksheet
ActivateWorkbook (frmwkbook)
Worksheets(frmwksht).Select
Range(Cells(r1, c1), Cells(r2, c2)).Select
Selection.Cut
'Select "to" Workbook and Worksheet
ActivateWorkbook (towkbook)
Worksheets(towksht).Select
Range(Cells(row, col), Cells(row, col)).Select
ActiveSheet.Paste
End Sub
```

As with the previous CopySection subroutine, the user must specify what Workbook and Worksheet the information will be cut *from* as well as what Workbook and Worksheet the information will be moved *to*. Cutting data and pasting is, for all essential purposes, the same as moving data. One of the nice aspects of cutting data from a data series is that the user can visually see what pieces of data have already been extracted (or moved).

Sometimes, however, the user just wants to get rid of pieces of data, that is, remove them in their entirety from the Worksheet. There are two ways of doing this. The first is the Delete method. The only drawback of the Delete method is that, from a conceptual sense, Excel is, for all practical purposes, removing the cell itself, not just the information in it. As a consequence of this, when the cell is removed, the other Cells surrounding it must shift to fill the void left by the removed cell. The surrounding Cells will either shift to the left or shift up, depending upon how the delete method is coded. A good analogy to this is that force is applied to the Worksheet from either the right side or the bottom, and which side the force is applied determines how the Cells will shift. When the user invokes the delete method, they are able to pick whether the force is to be applied to the right side or bottom of the Worksheet. The following subroutine will delete a block of Cells specified by the user.

```
Sub DeleteSection(ByVal wkbook, ByVal wksht, ByVal r1, ByVal
c1, ByVal r2, ByVal c2, ByVal ShiftLeft As Boolean)
'Deletes the "Chunk" Bounded by r1,c1,r2,c2 in wksht in wkbook
'Select Workbook and Worksheet
ActivateWorkbook (wkbook)
Worksheets(wksht).Select
Range(Cells(r1, c1), Cells(r2, c2)).Select
Select Case ShiftLeft
   Case True
```

```
        Selection.Delete Shift:=xlToLeft
     Case False
        Selection.Delete Shift:=xlUp
   End Select
End Sub
```

The second method for getting rid of data is the Clear method. Unlike the Delete method, the Clear method just empties the Cells of their contents rather than removing them in their entirety. As a result, the Cells in the Worksheet do not shift to the left or upward. This is advantageous because tables within the Worksheet will not become skewed as can happen when Cells are deleted. The consequence of the Cells not shifting is that there are "holes" left in the Worksheet where the data have been cleared. The following subroutine will clear a section of Cells within a Worksheet.

```
Sub ClearSection(ByVal wkbook, ByVal wksht, ByVal r1, ByVal
c1, ByVal r2, ByVal c2)
'Clears the "Chunk" Bounded by r1,c1,r2,c2 in wksht in wkbook
'Select Workbook and Worksheet
ActivateWorkbook (wkbook)
Worksheets(wksht).Select
Range(Cells(r1, c1), Cells(r2, c2)).Select
Selection.Clear
End Sub
```

1.9 SORTING DATA

The ability to sort data within a Worksheet is of paramount importance. Often, the user will wish to extract the largest or smallest value from a particular subgrouping of data. Typically, data are stored down columns, and it is the rows that are sorted based on the values in one or more columns. Up to three columns may be selected for sorting purposes. The GUI utilized when manually sorting data in Excel is shown in Figure 1.10.

FIGURE 1.10 The sorting GUI in Excel.

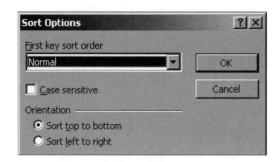

FIGURE 1.11 Additional sorting options available in Excel.

In code, the sorting columns are referred to as Keys, and the first Key has the highest precedence. Only one Key (or column) is required to sort data. However, if there is a chance that elements in the column could be of equal value, it is often essential to have a second or even third Key (or column) to sort upon when elements in the preceding Key of precedence turn up equal. In addition, each Key can be sorted in either ascending or descending order. If a header exists in the Worksheet, this information can be specified, and the header rows at the top of the Worksheet will not be sorted. The area of the Worksheet to be sorted must be specified as a range (usually of columns), which is selected prior to specifying the Keys and sorting.

The variety of parameters that can be set when sorting a Worksheet proves problematic when trying to create a generic subroutine. Which parameters should be included and which should be excluded? The following subroutine is a good compromise and lends itself to use in a variety of situations. Figure 1.11 shows additional options available when sorting data in Excel.

```
Sub SortWorksheet(ByVal wkbook, ByVal wksht, ByVal startcol
As Integer _
, ByVal endcol As Integer, ByVal Header As Integer, ByVal ColKey1, _
ByVal Key1SortAscending As Boolean, ByVal ColKey2, _
ByVal Key2SortAscending As Boolean)
'Sorts the Columns Specified in a particular Worksheet by Keys
'Use Up to 2 Keys, for only one Key set ColKey2 = "" (Null $)
'Header = # of Header Rows: 0 = No Header
Dim Key1Range$, Key2Range$, Key1Dir$, Key2Dir$, Hdr%
'Select Workbook and Worksheet
ActivateWorkbook (wkbook)
Worksheets(wksht).Select
'Select Columns to Sort
Columns(GetColRng(startcol, endcol)).Select
'Set Header Range
Select Case Header
  Case 0
  'Worksheet is devoid of Header Row
  Hdr = xlNo
  If IsNumeric(ColKey1) = False Then
      'A Letter was passed as the Col #
    Key1Range$ = Trim(ColKey1) & "1"
```

```vb
    If ColKey2 <> "" Then Key2Range$ = Trim(ColKey2) & "1"
    Else
    'A Number was passed as the Col #
    Key1Range$ = Chr(64 + ColKey1) & "1"
    If ColKey2 <> "" Then Key2Range$ = Chr(64 + ColKey2) & "1"
  End If
 Case Else
'Worksheet Contains a Header Row
 Hdr = xlGuess
 If IsNumeric(ColKey1) = False Then
    'A Letter was passed as the Col #
    Key1Range$ = Trim(ColKey1) & Trim(Str(Header))
    If ColKey2 <> "" Then Key2Range$ = Trim(ColKey2)
    & Trim(Str(Header))

    Else
    'A Number was passed as the Col #
    Key1Range$ = Chr(64 + ColKey1) & Trim(Str(Header))
    If ColKey2 <> "" Then Key2Range$ = Chr(64 + ColKey2)
    & Trim(Str(Header))
 End If
End Select
'Set Ascending/Descending Order for Keys
If Key1SortAscending = True Then
   Key1Dir$ = xlAscending
   Else
   Key1Dir$ = xlDescending
End If
If Key2SortAscending = True Then
    Key2Dir$ = xlAscending
    Else
    Key2Dir$ = xlDescending
End If
Select Case ColKey2
   Case "" 'Only a single key utilized for sorting
   Selection.Sort Key1:=Range(Key1Range$), Order1:=Key1Dir$,
   Header:=Hdr%, _

   OrderCustom:=1, MatchCase:=False,
Orientation:=xlTopToBottom

Case Else '2 Keys utilized for sorting
    Selection.Sort Key1:=Range(Key1Range$), Order1:=Key1Dir$,
Key2:=Range(Key2Range$) _

   , Order2:=Key2Dir, Header:=Hdr%, OrderCustom:=1, MatchCase:= _
```

```
            False, Orientation:=xlTopToBottom
   End Select
   End Sub
```

This subroutine allows the user to select up to two Keys to sort a group of columns on. Two Keys are sufficient for 90% of the sorting applications most users will encounter. If only one Key is to be utilized for the sorting algorithm, the parameter `ColKey2` should be set to a null (empty) string (""). Remember that the Keys in this case are, in fact, columns in the Worksheet, and thus, they can be referenced by number or by letter. It is possible to specify the keys as either a number or a letter. This is done by having the passed parameters (`ColKey1`, `ColKey2`) set to a Variant type of variable and having the subroutine check if they are a number with the `IsNumeric( )` function. If a letter was passed as a key (column reference), then the literal string passed as the key is utilized when setting up the Key Range. If a number was passed, then it is used to produce the proper alphabetical column reference using the statements `Chr(64 + ColKey1)` or `Chr(64 + ColKey2)`. Notice also the use of the `Trim( )` function when utilizing a numerical variable in building a string. This is essential because all numerical variables contain a placeholder at the beginning of the number to allow for a negative sign. When a number is converted into a string via the `Str( )` function, this placeholder is converted into a space if the number is positive and does not contain a negative sign. This leading space will cause an error if it is left in the range that is being constructed, and hence the `Trim()` function is utilized to remove it.

Each Key can be set to sort in either an ascending or descending order. This is set in the subroutine with Boolean passed parameters `Key1SortAscending` and `Key2SortAscending`, which are self explanatory. If the Worksheet contains a header, this information should be conveyed to the subroutine using the `Header` variable, which is of type integer. If `Header` is set to 0, then the Workbook is treated as if it has no header. If a header is present in the Workbook, then the number of rows at the top of the Worksheet that constitute the header should be passed to the subroutine via the `Header` parameter.

The last parameters set up which columns in what Workbook\Worksheet pair the sort function will be performed on. The Workbook and Worksheet are selected in the same manner as in any of the preceding examples. Selecting the columns to be included in the sorting algorithm is slightly more problematic. This is because the Columns method is not nearly as flexible as it should be in VBA. For example, to select multiple columns in VBA, the statement

```
Columns("A:C").Select
```

works just as expected using alphabetic column references. However, if numerical column references are utilized, the statement

```
Columns(1:3).Select or Columns("1:3").Select
```

will fail.

A single-column numerical reference, interestingly enough, will function without error as in the case of

```
Columns(3).Select
```

which works just fine.

The practical consequence of this limitation is that a function needs to be constructed to replicate a column range in terms of alphabetical references from numerical references. When coding in VBA, numerical references are almost exclusively utilized because they are easier to deal with and do not require elaborate schemes of construction as do string references. The following function, when passed a starting column and an ending column, will return an alphabetical string range that can be utilized by the Columns method to select multiple columns, as is required in the `SortWorksheet` subroutine. Notice that this function makes use of the previously defined `ColNoToColLet` function earlier in this chapter.

```
Function GetColRng(ByVal startcol As Integer, ByVal endcol As
Integer) As String

GetColRng = ColNoToColLet(startcol) & ":" & ColNoToColLet
(endcol)

End Function
```

1.10 DELETING ROWS AND COLUMNS

Invariably, when working with copious amounts of data, there will be data that are to be retained and data that are to be eliminated. The most convenient way of eliminating unwanted data is to delete the row or column it reside in. This works quite well if each row or column holds values related to a single specific parameter. Should a column or row hold values for multiple parameters, this approach is not practical. In such instances, data must be removed on a cell-by-cell basis. Often, in a Worksheet that contains a report, each row will hold values pertaining to a specific parameter. An example report to be utilized for purposes of this discussion is shown in Figure 1.12.

In the report in Figure 1.12, a number of compounds have been analyzed by a machine. Each row contains data related to a single compound, where compounds are identified by the "CMPD-" prefix. Suppose in this report there are standards that are run to check the integrity of the testing being done. In this case, the standards are chloramphenicol, nifuroxime, and bromoquinoline. At the end of the analysis it is desired to automatically remove the rows that contain the standards

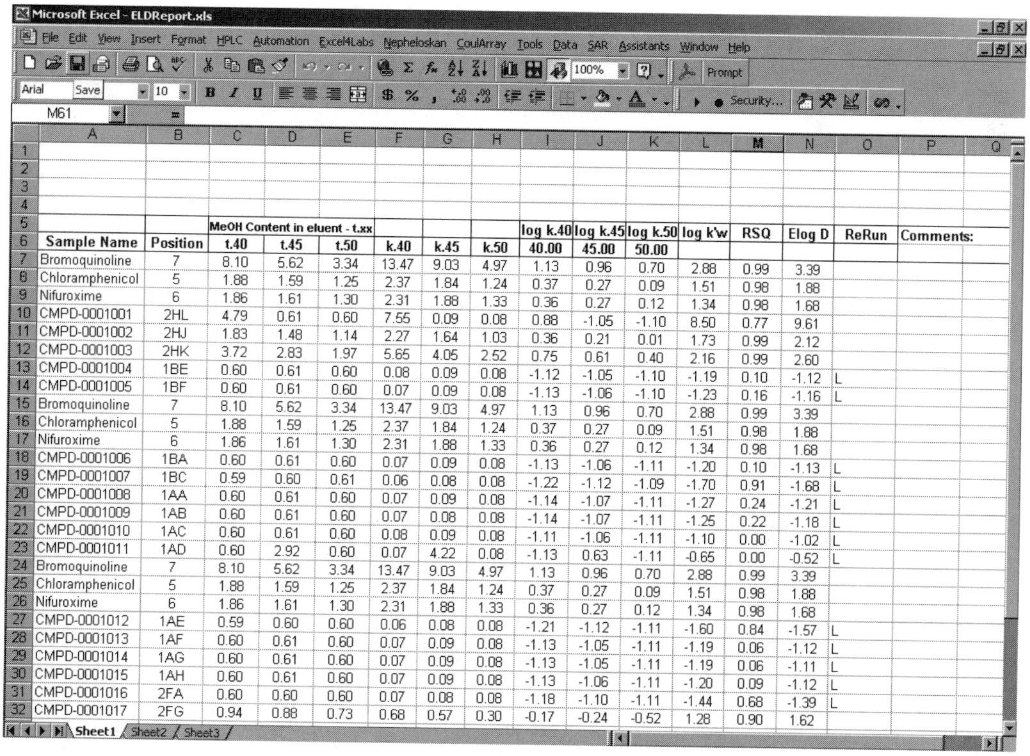

Sample Name	Position	MeOH Content in eluent - t.xx						log k.40	log k.45	log k.50	log k'w	RSQ	Elog D	ReRun	Comments:
		t.40	t.45	t.50	k.40	k.45	k.50	40.00	45.00	50.00					
Bromoquinoline	7	8.10	5.62	3.34	13.47	9.03	4.97	1.13	0.96	0.70	2.88	0.99	3.39		
Chloramphenicol	5	1.88	1.59	1.25	2.37	1.84	1.24	0.37	0.27	0.09	1.51	0.98	1.88		
Nifuroxime	6	1.86	1.61	1.30	2.31	1.88	1.33	0.36	0.27	0.12	1.34	0.98	1.68		
CMPD-0001001	2HL	4.79	0.61	0.60	7.55	0.09	0.08	0.88	-1.05	-1.10	8.50	0.77	9.61		
CMPD-0001002	2HJ	1.83	1.48	1.14	2.27	1.64	1.03	0.36	0.21	0.01	1.73	0.99	2.12		
CMPD-0001003	2HK	3.72	2.83	1.97	5.65	4.05	2.52	0.75	0.61	0.40	2.16	0.99	2.60		
CMPD-0001004	1BE	0.60	0.61	0.60	0.08	0.09	0.08	-1.12	-1.05	-1.10	-1.19	0.10	-1.12	L	
CMPD-0001005	1BF	0.60	0.61	0.60	0.07	0.09	0.08	-1.13	-1.06	-1.10	-1.23	0.16	-1.16	L	
Bromoquinoline	7	8.10	5.62	3.34	13.47	9.03	4.97	1.13	0.96	0.70	2.88	0.99	3.39		
Chloramphenicol	5	1.88	1.59	1.25	2.37	1.84	1.24	0.37	0.27	0.09	1.51	0.98	1.88		
Nifuroxime	6	1.86	1.61	1.30	2.31	1.88	1.33	0.36	0.27	0.12	1.34	0.98	1.68		
CMPD-0001006	1BA	0.60	0.61	0.60	0.07	0.09	0.08	-1.13	-1.06	-1.11	-1.20	0.10	-1.13	L	
CMPD-0001007	1BC	0.59	0.60	0.61	0.06	0.08	0.08	-1.22	-1.12	-1.09	-1.70	0.91	-1.68	L	
CMPD-0001008	1AA	0.60	0.61	0.60	0.07	0.09	0.08	-1.14	-1.07	-1.11	-1.27	0.24	-1.21	L	
CMPD-0001009	1AB	0.60	0.61	0.60	0.07	0.08	0.08	-1.14	-1.07	-1.11	-1.25	0.22	-1.18	L	
CMPD-0001010	1AC	0.60	0.61	0.60	0.08	0.09	0.08	-1.11	-1.06	-1.11	-1.10	0.00	-1.02	L	
CMPD-0001011	1AD	0.60	2.92	0.60	0.07	4.22	0.08	-1.13	0.63	-1.11	-0.65	0.00	-0.52	L	
Bromoquinoline	7	8.10	5.62	3.34	13.47	9.03	4.97	1.13	0.96	0.70	2.88	0.99	3.39		
Chloramphenicol	5	1.88	1.59	1.25	2.37	1.84	1.24	0.37	0.27	0.09	1.51	0.98	1.88		
Nifuroxime	6	1.86	1.61	1.30	2.31	1.88	1.33	0.36	0.27	0.12	1.34	0.98	1.68		
CMPD-0001012	1AE	0.59	0.60	0.60	0.06	0.08	0.08	-1.21	-1.12	-1.11	-1.60	0.84	-1.57	L	
CMPD-0001013	1AF	0.60	0.61	0.60	0.07	0.09	0.08	-1.13	-1.05	-1.11	-1.19	0.06	-1.12	L	
CMPD-0001014	1AG	0.60	0.61	0.60	0.07	0.09	0.08	-1.13	-1.05	-1.11	-1.19	0.06	-1.11	L	
CMPD-0001015	1AH	0.60	0.61	0.60	0.07	0.09	0.08	-1.13	-1.06	-1.11	-1.20	0.09	-1.12	L	
CMPD-0001016	2FA	0.60	0.60	0.60	0.07	0.08	0.08	-1.18	-1.10	-1.11	-1.44	0.68	-1.39	L	
CMPD-0001017	2FG	0.94	0.88	0.73	0.68	0.57	0.30	-0.17	-0.24	-0.52	1.28	0.90	1.62		

FIGURE 1.12 An example report.

once it has been determined that the values returned by analyzing the standards fall within an acceptable testing range (or window).

The first step required in setting up such a process is to determine the number of rows within a Worksheet that contain data. The function LastRow accomplishes this task quite nicely.

```
Function LastRow(Workbook, Worksheet) As Long
'This function will return the last ROW in the Worksheet of
Workbook
On Error GoTo Fix_it:
ActivateWorkbook (Workbook)
Sheets(Worksheet).Activate
Workbooks(Workbook).Worksheets(Worksheet).Cells.Find(What:="*
",  _
   SearchDirection:=xlPrevious, SearchOrder:=xlByRows).Activate
LastRow = FindRow(Selection.Address(ReferenceStyle:=xlR1C1))
Exit Function
Fix_it:
'Debug.Print "Error Description: "; Error; " - Error Code: ";
Err
LastRow = 0
End Function
```

Here the Workbook name and the Worksheet name (or number) is passed to the function. The methodology of this function is not readily apparent even to an experienced VBA programmer, so some explanation is in order here. The function starts searching for a nonempty cell by rows from the bottom of the Worksheet. As soon as the function encounters a nonempty cell (searching by rows from the bottom), it knows this is the maximum row value for any data point in this Worksheet. Notice that this function has an error-trapping mechanism. This is because, should a Worksheet not have any data in it, the find function will return an error (having not been able to find anything). In this particular case, the last-row value is set to 0 by the error-handling mechanism should a Worksheet be completely empty.

The final task at hand is to delete any rows in which one of the standards are present. To accomplish this, a subroutine must "visit" each row in the Worksheet, and should that row contain a standard, delete it. The problem here is that as the subroutine loops through the rows of the Worksheet, as soon as a single row is deleted, the index of the loop counter no longer references the next row of data in the Worksheet. This is because deleting a row caused all the data to shift *upward* by one row. The routine will now skip a row of data for each row that is to be deleted. Hence, some rows will not be checked for the presence of a standard if a loop counter is utilized as a row index. The solution to this problem is to use the loop only to propagate through the number of rows to be checked, and to have a separate variable to be utilized as a pointer to the row of data to be checked. If the last row checked was deleted, then the pointer is not incremented, as the next row of data will automatically pop up to the current pointer position. If the last row checked was not deleted, then the pointer is incremented.

The next subroutine DeleteRowX accomplishes this task quite nicely. The algorithm it utilizes is as follows. The Workbook and Worksheet to be acted upon must be specified in the first two passed parameters wkbook$ and wksheet, respectively. A string, cellcts$, must be specified that contains the contents which, if found in a cell, will cause the row to be deleted. In this subroutine, only a single column is searched for the specified string cellcts$, which is specified in the passed parameter cellcol%. The subroutine could just as easily be modified to search through a number of columns for the specified string, should that prove necessary. As an alternative to modification, the subroutine could be called several times if multiple columns need to be searched, perhaps utilizing a looping structure to increment the passed parameter `cellcol%`.

```
Sub DeleteRowX(wkbook$, wksheet, cellcts$, cellcol%)
'Deletes Rows in Workbook (wkbook$) Sheet (wksheet$) which
contain (cellcts$) in
'Column Number (cellcol%)
Dim Total_Rows As Integer, rowptr As Integer
Dim ii As Integer
Total_Rows = TotalRows(wkbook$, wksheet)
rowptr = 1
For ii = 1 To Total_Rows
If Workbooks(wkbook$).Worksheets(wksheet).Cells(rowptr,
cellcol%) = cellcts$ Then
    Rows(rowptr).Select
    Selection.Delete Shift:=xlShiftUp
Else
    rowptr = rowptr + 1
End If
Next ii
'ReSave the Workbook without the empty rows
Workbooks(wkbook$).Save
End Sub
```

To eliminate the standards in the preceding example, the following calls could be made within the VB environment:

```
Call DeleteRowX("ELDReport.xls",1,"Bromoquinoline",1)
Call DeleteRowX("ELDReport.xls",1,"Chloramphenicol",1)
Call DeleteRowX("ELDReport.xls",1,"Nifuroxime",1)
```

This will eliminate the standards from the report, leaving only the data for the specific compounds that have been run.

Suppose that it is also desirable to remove the "Position" and "Rerun" columns (labeled "B" and "O," respectively) from the Worksheet before it is to be uploaded to a database. In theory, no two columns in a Worksheet should be assigned the same header. The `DeleteColumnX` subroutine will delete the first column it finds with a header (`header$`) in row (`header_row%`) in Worksheet(`wksheet`) in Workbook(`wkbook$`).

```
Sub DeleteColumnX(wkbook$, wksheet, header$, header_row%)
'Deletes a column which has header "header$" in row header_row%
For ii = 1 To 26
  If Workbooks(wkbook$).Worksheets(wksheet).Cells(header_row%,
  ii) = header$ Then

    Workbooks(wkbook$).Worksheets(wksheet).Columns(ii).Delete
    Exit Sub
  End If
Next ii
End Sub
```

To delete the columns first discussed, the following two calls to the `DeleteColumnX` subroutine would have to be made.

```
Call DeleteColumnX("ELDReport.xls",1,"ReRun",5)
Call DeleteColumnX("ELDReport.xls",1,"Position",6)
```

1.11 SUMMARY

To utilize and manipulate information in Excel in any meaningful manner, a developer must know how to access the data contained within its Worksheets. As such, the programmer must have the fundamental skills required to access data contained within different Workbook/Worksheet pairs. The idiosyncrasies of accessing data using explicit referencing are discussed, as well as the methods of searching for data required to automate a particular task. The methods of copying, clearing, and deleting data from Worksheets are discussed, as well as the fundamental methods of deleting row and columnar data. With mastery of these basic skills, the reader is ready to proceed into more specialized topics.

2 Methods of Loading and Saving Data in Excel

2.1 INTRODUCTION

The purpose of this text is to empower the reader to take in information provided in some format (where that format is not customizable or alterable by the user), do something with the information that is taken in (calculations, formatting, sorting, etc.), and then create a Worksheet, report, or file that contains the information in a format desired by the end user. The middle portion of this statement is the bulk of what this text is dedicated to, performing automated operations on information contained within data files or Worksheets. Such operations can be something as simple as formatting changes, or as complex as curve fitting. The first and last statements within the purpose just stated, taking data in to have an operation performed on it, and writing the data out after the given operations have been performed on it, are covered only in this chapter. It is here that the reader will learn how to bring data into Excel using a manual file-loading dialog box, or by using an algorithm to load all the files present in a certain location (folder) that have certain characteristics. It is also within this section that the reader will learn how to write data out in a format that will be of use to their end user or customer.

2.2 USING THE STANDARD OPEN FILE DIALOG BOX TO LOAD A FILE

Many times the user of a macro will be required to only load a few files for an automated procedure to be run on them. Often, such files will not be distinguishable or predictable by means of some type of algorithm. In such a case, it will be necessary for the user to load the files by hand. This is best done by means of the standard open file dialog box built within VBA.

When constructing a means of opening data files in Excel, it is best to keep a few things in mind. Foremost is that the Workbook to be loaded into Excel must be referenced by its name, which is identical to its filename. Therefore, it makes sense to utilize a function to load files in Excel and have it return the filename of the file the user picked to load. The returned string containing the filename should be set to a Public variable that can be referenced by any subroutine in any module of the Workbook.

When bringing up the file open dialog box, there are a few things the developer should specify to optimize the utility of the file open dialog box. The most obvious would be that an appropriate title should be placed at the top of the dialog box so the user knows which file they should load. It is also greatly advantageous to specify what the default directory will be when the file open dialog box is activated. Nothing is more irritating to an end user than having to click through a complex directory structure over and over again to get to the same set of data files. Later in this section it will be shown how to save the last directory a user visited when the macro was last run to the Windows registry, where it can be recalled the next time the macro is run and utilized to open the dialog box to the path it visited the last time the macro was run. Using such an approach, the end user will only have to navigate to the path where the data files are present the first time the macro is run, as the program will "remember" its last path by means of the Windows registry for subsequent macro executions. The last parameter that should be set is the number of filters the

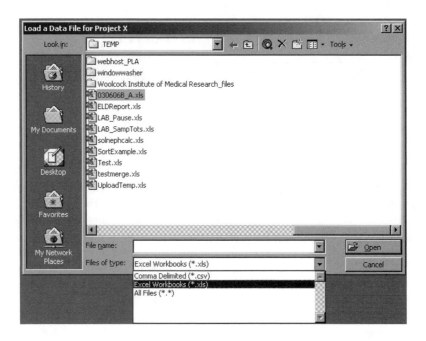

FIGURE 2.1 A standard open file dialog box.

file open dialog box will display upon opening (Figure 2.1). Common filters utilized in addition to the standard *.xls are: (*.xlt, *.csv, *.txt). The following function allows all of the above parameters to be set, which affords the developer a great deal of flexibility.

```
Function LoadWithDialog(ByVal filters$, ByVal defaultdir$,
ByVal WindowCaption$) As String
'Function to Load files of type specified with Dialog Box
'Returns the name of the Workbook Loaded
Dim vFileName As String
Dim nochoice As Integer
Dim fillen As Integer
Dim strike As Integer
Dim ErrorBoxCaption$, ErrorBoxMsg$
On Error GoTo FileOpenErr
'Set Up Captions and Messages
ErrorBoxCaption$ = "You must choose a File"
filters$ = LCase(filters$)
fillen = Len(filters$)
If fillen Mod 3 = 0 Then
    'if valid filters then process
    For ii = 1 To fillen Step 3
     Search$ = Mid(filters, ii, 3)
     'Stop
     GoSub Build_Filters
   Next ii
```

```
End If
'Add default
FileFilter = FileFilter & "All Files (*.*),*.*,"
'remove trailing ','
FileFilter = Left$(FileFilter, Len(FileFilter) - 1)
'If the passes default directory is valid change the current
directory to it
If DirectoryExists(defaultdir$) = True Then
    ChDir defaultdir$
End If
Try_Again:
vFileName = Application.GetOpenFilename(FileFilter, ,
WindowCaption$)
'Bug - Tries to Open "False.xls" on 2nd Pass Only
If vFileName = "False" Then GoTo FileOpenErr
If vFileName <> "" Then
    Workbooks.Open Filename:=vFileName
End If
'File was chosen and loaded into Worksheet - now exit
LoadWithDialog = ActiveWorkbook.Name
Exit Function
Build_Filters:
Select Case Search$
  Case "xls"
    FileFilter = FileFilter & "Excel Workbooks (*.xls), *.xls,"
  Case "xlt"
    FileFilter = FileFilter & "Excel Templates (*.xlt), *.xlt,"
  Case "csv"
    FileFilter = FileFilter & "Comma Delimited (*.csv), *.csv,"
  Case "txt"
    FileFilter = FileFilter & "Text Files (*.txt), *.txt,"
End Select
Return
FileOpenErr:
'If no file is chosen you end up here
strike = strike + 1
Select Case strike
  Case 1
   ErrorBoxMsg$ = "You will get one more chance!"
   nochoice = MsgBox(ErrorBoxMsg$, vbOKOnly + vbExclamation,
ErrorBoxCaption$)
     GoTo Try_Again
  Case 2
    ErrorBoxMsg$ = "Macro will terminate"
```

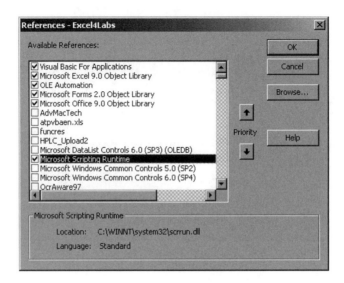

FIGURE 2.2 Setting the Microsoft scripting runtime reference.

```
    nochoice = MsgBox(ErrorBoxMsg$, vbOKOnly + vbExclamation,
    ErrorBoxCaption$)

            End
  End Select
  End Function
```

The Function `LoadWithDialog` relies upon a second function named `DirectoryExists`, which checks to make sure the default directory passed to the `LoadWithDialog` function actually exists before trying to set it as the default directory. (If the specified default directory is invalid, trying to set it as the default directory will result in an error.) The `DirectoryExists` function will not work properly unless the Library "Microsoft Scripting Runtime" (C:\WINNT\system32\scrrun.dll) is checked in References. Note that this Reference will need to be checked for EVERY Workbook in which the `DirectoryExists` function is to be utilized. To set the reference, choose from the VBA menu bar Tools->References and check the reference "Microsoft Scripting Runtime" as shown in Figure 2.2.

'To utilize Directory Exists Function the following Library MUST be installed

```
' "Microsoft Scripting Runtime" must be checked in References!
' C:\WINNT\system32\scrrun.dll
Function DirectoryExists(full_path As String) As Boolean
Dim m_FileSys As FileSystemObject
Set m_FileSys = New FileSystemObject
  DirectoryExists = m_FileSys.FolderExists(full_path)
End Function
```

The open file dialog box illustrated in Figure 2.1 can be reproduced by typing the following code within the immediate (debug) window *provided a module with the function LoadWithDialog is active.*

```
Debug.Print LoadWithDialog("csvxls", "c:\temp","Load a Data
File for Project X")
```

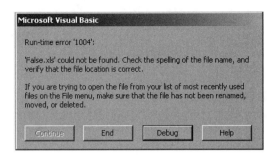

FIGURE 2.3 A bug in Excel VBA will cause this error without corrective action.

The name of the Workbook selected by the user will then be printed to the immediate (debug) window. Notice that the filters desired are specified in the first passed parameter as a string with the filter suffixes listed together one after another without spaces. The next two parameters are the default directory and dialog box title, respectively.

Notice that the user is given two opportunities to choose a valid file before the subroutine will abort. There is one bug within the Excel VBA system that is particularly troublesome in this regard. The line of code:

```
If vFileName = "False" Then GoTo FileOpenErr
```

should not be required. It was necessary to add this line of code because if the user presses the "CANCEL" button on the open file dialog box twice, the error will not be caught and the second time the Cancel button is pressed (for reasons unknown), and Excel will attempt to open a file named "False.xls" instead of triggering an error. This will result in an error occurring as shown in Figure 2.3.

2.3 USING THE STANDARD SAVE AS DIALOG BOX TO SAVE A FILE

Setting up a generic function to save a Workbook utilizing the standard save as dialog box is slightly more complicated than the previous instance using the standard open file dialog box because there are a few additional considerations to keep in mind when saving files (Figure 2.4). As in the previous instance, it is still necessary to pick which filters will be available, the default directory the dialog box opens with, and a title for dialog box. It is also desirable to have the function return a string that is the name of the Workbook the save as dialog box just saved. Two additional considerations are the following. The Workbook to be saved should be specified, in the event it is not the active Workbook. Also, it would be desirable to specify if the Workbook should be closed or left open immediately after it is saved. The next function allows all of these options to be set utilizing passed parameters.

```
Function SaveWithDialog(ByVal Workbook, ByVal filters$, ByVal
defaultdir$, ByVal WindowCaption$, ByVal CloseAfterSave As
Boolean) As String
'Function to Save a Workbook as a Particular Type
' With a User Defined FileName
'The First Filter in the string filters$ will be the default
filter
Dim vFileName As String
Dim nochoice As Integer
Dim fillen As Integer
```

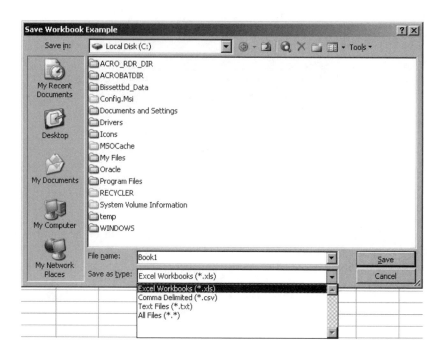

FIGURE 2.4 Standard SaveAsDialog box example.

```
Dim strike As Integer
Dim ErrorBoxCaption$, ErrorBoxMsg$
Set BookToSave = Workbooks(Workbook)
On Error GoTo FileOpenErr
'Set Up Captions and Messages
ErrorBoxCaption$ = "You Have Chosen NOT to Save this File!"
filters$ = LCase(filters$)
fillen = Len(filters$)
If fillen Mod 3 = 0 Then
    'if valid filters then process
    For ii = 1 To fillen Step 3
      Search$ = Mid(filters, ii, 3)
      'Stop
      GoSub Build_Filters
    Next ii
End If
'Add default
FileFilter = FileFilter & "All Files (*.*),*.*,"
'remove trailing ','
FileFilter = Left$(FileFilter, Len(FileFilter) - 1)
'If the passes default directory is valid change the current
directory to it
If DirectoryExists(defaultdir$) = True Then
    ChDir defaultdir$
```

```
End If
Try_Again:
vFileName = Application.GetSaveAsFilename(, FileFilter, ,
WindowCaption$)
If vFileName = "False" Then GoTo FileOpenErr
If vFileName <> "" Then
    'Valid FileName - Save Workbook
    BookToSave.SaveAs Filename:=vFileName
    'Optional Close On Save
    If CloseAfterSave = True Then
        BookToSave.Close SaveChanges:=False
    End If
    'Valid Filename Given and Workbook Saved - now exit
    SaveWithDialog = Extract_FileName(vFileName, True)
    Exit Function
End If
MsgBox "File Not Saved but Valid Filename Given?!!!", vbOKOnly
+ vbCritical, "Unexpected Condition!"
Exit Function
Build_Filters:
Select Case Search$
  Case "xls"
    FileFilter = FileFilter & "Excel Workbooks (*.xls),
    *.xls,"
  Case "xlt"
    FileFilter = FileFilter & "Excel Templates (*.xlt), *.xlt,"
  Case "csv"
    FileFilter = FileFilter & "Comma Delimited (*.csv), *.csv,"
  Case "txt"
    FileFilter = FileFilter & "Text Files (*.txt), *.txt,"
End Select
Return
FileOpenErr:
'If no file is chosen you end up here
strike = strike + 1
Select Case strike
  Case 1
    ErrorBoxMsg$ = "You will get one more opportunity to Save
    this File!"
    nochoice = MsgBox(ErrorBoxMsg$, vbOKOnly + vbExclamation,
    ErrorBoxCaption$)
        GoTo Try_Again
  Case 2
    ErrorBoxMsg$ = "File will not be saved!"
```

```
    nochoice = MsgBox(ErrorBoxMsg$, vbOKOnly + vbExclamation,
    ErrorBoxCaption$)
            End
    End Select
End Function
```

Notice that the last passed parameter `CloseAfterSave` (Boolean) determines if the Workbook that is saved will remain open (False) or will be closed after saving (True). Also note that the first passed parameter is the Workbook name, and that the Workbook is referenced from the variable `BookToSave` using the `Set` command.

In this instance, `vFileName` returns the complete path, not just the filename. Therefore, to return the filename from the function, it is necessary to strip the path from the filename. This is done by means of the `Extract_Filename` function whose code is listed below. (In fact, `vFileName` returns the full path in the `LoadWithDialog` function, but in that function it is possible to set the return value of the function to the active Workbook name; however, in this instance, this is impractical because the Workbook saved may be closed, renamed, or not in fact be the active Workbook.)

```
Function Extract_FileName(fullpath As String, Extension As
Boolean)
'Returns the filename from the end of a fullpath
Dim ii As Integer, dotmark As Integer, slashmark As Integer
For ii = Len(fullpath) To 1 Step -1
  If Mid(fullpath, ii, 1) = "." Then
      dotmark = ii
      End If
      If Mid(fullpath, ii, 1) = "\" Then
         slashmark = ii
         'End of FileName
         Exit For
      End If
Next ii
Select Case Extension
  Case True
  'Return filename with extension
  Extract_FileName = Mid(fullpath, (slashmark + 1),
  (Len(fullpath) - slashmark))

  Case False
  'Return filename without extension
  Extract_FileName = Mid(fullpath, (slashmark + 1),
  (Len(fullpath) - slashmark) - (Len(fullpath) - (dotmark- 1)))
End Select
End Function
```

This function can be tested from the immediate window (*provided the module in which the function resides is the active module*) by typing

```
Debug.Print SaveWithDialog(ActiveWorkbook.Name,
"xlscsvtxt","c:\temp","Save Workbook Example",False)
```

2.4 AUTOMATICALLY OPENING FILES AND TEMPLATES

Although using a dialog box gives the user the ultimate in flexibility in terms of choosing a file, the price to be paid for such flexibility is speed. The user must manually specify each file to be loaded or saved by the macro. In the instance where there is no object consistency in the file names and locations to be utilized, there is no alternative to such an approach. However, when file names and locations are predictable by means of an algorithm, it makes sense to have Excel open or save them without any intervention. The next function will open a file automatically, provided a valid file path is passed to the function.

```
Function AutoLoadFile(ByVal fullpath$) As String
'Subroutine to Automatically Load a File into a Excel
On Error GoTo FileOpenErr
'Check for Existence of File
If FileExists(fullpath$) = False Then
    'File Passed Does Not Exist - Log Error and Exit
    Debug.Print fullpath$; " Does Not Exist!"
    Exit Function
End If
'If the File Exists then open it
Workbooks.Open Filename:=fullpath$
'Return Workbook Name to Public Variable via Function
AutoLoadFile = ActiveWorkbook.Name
'File was chosen and loaded into Worksheet - now exit
Exit Function
FileOpenErr:
'In the event of an error execution jumps here
'Write Error to Log File
End
End Function
```

Notice that the function first checks to see if the file passed to the function really exists, and thus will be able to be opened. If the file does exist, Excel makes an attempt to open it. If Excel is successful at opening the file, the Workbook name is returned by the function. If Excel cannot open the file, the function jumps to the FileOpenErr: location. It is after this label that an appropriate error handling mechanism could be put into place, such as generating an error log file.

The function is able to check if a file exists by means of the `FileExists` function. This function makes use of the fact that the `Dir` function will return a zero length string ("") if the path specified in the function cannot be found.

```
Function FileExists(full_path As String) As Boolean
'Note: full_path must include suffix !!! (*.xls, *.txt, etc
. . .)
  If Dir(full_path) <> "" Then
     FileExists = True
     'Debug.Print "File Exists"
     Else
     FileExists = False
```

```
      'Debug.Print "File Doesn't Exist"
      End If
End Function
```

2.5 IMPORTING DATA TO A WORKSHEET

There will be occasions when it is desirable to add data to an existing Workbook rather than opening
a new Workbook or Worksheet to access the data. This is especially true when compiling data for
a report that ideally would be saved as a single Workbook. This can be accomplished manually by
making the following selections from the Excel menu: Data -> Get External Data -> Import Text
File. This manual process can be automated by means of the VBA subroutine ImportDataToWork-
book, the code for which is as follows.

```
Sub ImportDataToWorkbook(Workbook$, NewSheetName$, fullpath$, _
ByVal ColumnTypeText As Boolean, _
ByVal TabDelimiter As Boolean, ByVal SemicolonDelimiter As
Boolean, _
ByVal CommaDelimiter As Boolean, ByVal SpaceDelimiter As
Boolean)

'This Subroutine will import data to an exisiting Workbook
'Workbook$ is the name of the Workbook to receive to data
'A new sheet named NewSheetName$ will be created to store the
data

'fullpath$ is the full path (path + filename) to the data to
be imported

'Columns Can be set to Type Text (True) or General (False)
for Cols A-Z

'Up to four delimiters can be specified
Dim Name$, ConnPath$, CDT As Integer
Call ActivateWorkbook(Workbook$)
'Set Up Properties and Methods
ConnPath$ = "TEXT;" & fullpath$
Name$ = Extract_FileName(fullpath$, False)
Select Case ColumnTypeText
  Case True
    CDT = 2
  Case False
    CDT = 1
End Select
ActiveWorkbook.Worksheets.Add
  With ActiveSheet.QueryTables.Add(Connection:=ConnPath$,
Destination:=Range("A1"))

    .Name = Name$
    .FieldNames = True
    .RowNumbers = False
```

```
    .FillAdjacentFormulas = False
    .PreserveFormatting = True
    .RefreshOnFileOpen = False
    .RefreshStyle = xlInsertDeleteCells
    .SavePassword = False
    .SaveData = True
    .AdjustColumnWidth = True
    .RefreshPeriod = 0
    .TextFilePromptOnRefresh = False
    .TextFilePlatform = xlWindows
    .TextFileStartRow = 1
    .TextFileParseType = xlDelimited
    .TextFileTextQualifier = xlTextQualifierDoubleQuote
    .TextFileConsecutiveDelimiter = False
    .TextFileTabDelimiter = TabDelimiter
    .TextFileSemicolonDelimiter = SemicolonDelimiter
    .TextFileCommaDelimiter = CommaDelimiter
    .TextFileSpaceDelimiter = SpaceDelimiter
    .TextFileColumnDataTypes = Array(CDT, CDT, CDT, CDT, CDT,
    CDT, CDT, CDT, _
    CDT, CDT, CDT, CDT, CDT, CDT, CDT, CDT, CDT, CDT, CDT,
    CDT, CDT, CDT, _
    CDT, CDT, CDT, CDT)
    .Refresh BackgroundQuery:=False
End With
ActiveSheet.Name = NewSheetName$
End Sub
```

As with any subroutine, this one is a compromise between flexibility and length. The Workbook to which the data is to be imported to must be specified, along with a sheet name to be given to the new sheet that will be created to hold the imported data. The full path to the data file to be imported must also be specified, along with the data type of the columns to be imported. In this instance, the first 26 columns (A–Z) have their type specified as either (1) general or (2) text. This is set within the subroutine by means of the CDT variable (Column Data Type). In addition, up to four delimiters can be specified for the separation of data. They are tab, semicolon, comma, and space.

If the path to the data for importation is not readily known, it would be advantageous to make use of a dialog box to obtain the path to the data to be imported. The next function returns the path a user selects from a dialog box as a string.

```
Function GetPathWithDialog(ByVal filters$, ByVal defaultdir$,
ByVal WindowCaption$) As String
'Function to Load files of type specified with Dialog Box
'Returns the name of the Workbook Loaded
Dim vFileName As String
Dim nochoice As Integer
```

```vb
Dim fillen As Integer
Dim strike As Integer
Dim ErrorBoxCaption$, ErrorBoxMsg$
On Error GoTo FileOpenErr
'Set Up Captions and Messages
ErrorBoxCaption$ = "You must choose a File"
filters$ = LCase(filters$)
fillen = Len(filters$)
If fillen Mod 3 = 0 Then
    'if valid filters then process
    For ii = 1 To fillen Step 3
    Search$ = Mid(filters, ii, 3)
    'Stop
    GoSub Build_Filters
  Next ii
End If
'Add default
FileFilter = FileFilter & "All Files (*.*),*.*,"
'remove trailing ','
FileFilter = Left$(FileFilter, Len(FileFilter) - 1)
'If the passes default directory is valid change the current
directory to it
If DirectoryExists(defaultdir$) = True Then
    ChDir defaultdir$
End If
Try_Again:
vFileName = Application.GetOpenFilename(FileFilter, ,
WindowCaption$)
'Bug - Tries to Open "False.xls" on 2nd Pass Only
If vFileName = "False" Then GoTo FileOpenErr
If vFileName <> "" Then
    'File Name Captured with DialogBox
    GetPathWithDialog = vFileName
    Exit Function
End If
Build_Filters:
Select Case Search$
  Case "xls"
   FileFilter = FileFilter & "Excel Workbooks (*.xls), *.xls,"
  Case "xlt"
   FileFilter = FileFilter & "Excel Templates (*.xlt), *.xlt,"
  Case "csv"
   FileFilter = FileFilter & "Comma Delimited (*.csv), *.csv,"
  Case "txt"
```

```
   FileFilter = FileFilter & "Text Files (*.txt), *.txt,"
End Select
Return
FileOpenErr:
'If no file is chosen you end up here
strike = strike + 1
Select Case strike
  Case 1
    ErrorBoxMsg$ = "You will get one more chance!"
    nochoice = MsgBox(ErrorBoxMsg$, vbOKOnly + vbExclamation,
ErrorBoxCaption$)
      GoTo Try_Again
  Case 2
    ErrorBoxMsg$ = "Macro will terminate"
    nochoice = MsgBox(ErrorBoxMsg$, vbOKOnly + vbExclamation,
ErrorBoxCaption$)
      End
End Select
End Function
```

It is now possible to embed the function `GetPathWithDialog` within a call to the subroutine `ImportDataToWorkbook` to obtain the path to the data for importation via a dialog box. This utility can be demonstrated within the VBA environment by typing the following in the immediate (or debug) window. (Note: the entire call must be typed on one line and then hit return.)

```
Call ImportDataToWorkbook(ActiveWorkbook.Name,
 "ImportedData",GetPathWithDialog("csvtxtxls","c:\temp","Window
Caption"),True,True,False,True,False)
```

2.6 AUTOMATICALLY SAVING FILES AND TEMPLATES

As data are gathered and analyzed, and reports are created, all this information must be stored somewhere. Often, it is desirable to store such reports in some central repository, usually on a rules-based system. When an automated data analysis routine creates only a single report, it is not a big deal for the user to save such a file manually. When tens or hundreds of such reports are created, it is extraordinarily inconvenient to have a user manually save each file to its desired location. No matter what the situation is, it is always preferable to automate every possible aspect of any given routine. It provides consistency, not to mention convenience. The next function will automatically save an entire Excel Workbook or an active Worksheet to a location the user specifies, and of a type that the user indicates (Workbook, comma delimited, text, etc.). The function will return the saved filename name as a string.

```
Function SaveWorksheet(ByRef wkbook$, wksht, ByVal Path$, ByVal
fname$, ByVal ftype As String) As String
Dim xlFormat&, fullpath$

'If no filename passed assume user wants date/time stamp
filename!

If fname$ = "" Then
    fname$ = Format(Date$, "yyyymmdd") & "_" & Format(Time$,
    "hhmmss")
```

```vba
    End If
    'Add Appropriate Extension
    Select Case ftype
      Case 1, "xls"
         'Excel Workbook
         fname$ = fname$ & ".xls"
         xlFormat& = xlWorkbookNormal
      Case 2, "csv"
         'Comma Delimited File
         fname$ = fname$ & ".csv"
         xlFormat& = xlCSV
      Case 3, "txt"
         'Text File
         fname$ = fname$ & ".txt"
         xlFormat& = xlTextWindows
      Case 4, "xlt"
         'Excel Template
         fname$ = fname$ & ".xlt"
         xlFormat& = xlTemplate
      Case 5, "xla"
         'Excel AddIn
         fname$ = fname$ & ".xla"
         xlFormat& = xlAddIn
      Case 6, "html"
         'Excel HTML
         fname$ = fname$ & ".html"
         xlFormat& = xlHtml
    End Select
    'Check for ending "\"
    If Right$(Path$, 1) <> "\" Then Path$ = Path$ & "\"
    'The full path consists of the path AND the filename
    fullpath$ = Path$ & fname$

    'If the Path doesn't exist -> make the directories!
    Call CreateDataDir ((Path$))

    'Select Workbook and Worksheet
    ActivateWorkbook (wkbook$)
    Worksheets(wksht).Select

    Select Case FileExists(fullpath$)
      Case True
       'The File already Exists - Error
       'Debug.Print "Error: File Already Exists"
       'In this instance overwrite existing files!
```

```
ActiveWorkbook.SaveAs Filename:=fullpath$,
FileFormat:=xlFormat& _
,Password:="",WriteResPassword:="",
ReadOnlyRecommended:=False, _
CreateBackup:=False
Case False
'The File Does not Exist - Write to specified Location
ActiveWorkbook.SaveAs Filename:=fullpath$,
FileFormat:=xlFormat& _
,Password:="", WriteResPassword:="",
ReadOnlyRecommended:=False, _
CreateBackup:=False
End Select
'Rename Global Workbook Variable to Match New Workbook Name
wkbook$ = fname$
'Return New Workbook Name via Function
SaveWorksheet = fname$
End Function
```

This function, like that of its earlier predecessors, returns the *new* name of the Workbook that is being saved. Whenever a Workbook is saved with a different filename, the Workbook's name changes (see Chapter 1). If a public variable is utilized to reference the Workbook, its contents are also changed to reflect the new Workbook name. This is done by passing the Workbook name parameter wkbook$ by reference as opposed to by value (this process will be explained in Chapter 3). The Worksheet is specified (wksht) only to denote which sheet will be active when the save method is initiated. This is of critical importance because if the Workbook is saved as any type other than Workbook or template, only the active sheet will be saved. For example, if a Workbook is saved as type comma delimited (*.csv) or as type text (*.txt), only data on the active sheet will be saved in that format. Also notice that if the filename is not specified (fname$ passed as an empty string ""), then a date/time stamp is assigned as the filename with the form: YearMonthDay_ HoursMinutes Seconds.extension (Example: 20031110_111159.xls -> file saved November 10 2003 at 11:11:59 AM). The date/time scheme could be modified to reflect any different number of formatting schemes via the format() function. Notice also that the path the user plans on writing the file to must exist, or an error will occur because Excel will not be able to write a file to a path that does not exist. To ensure that the given path does indeed exist, the CreateDataDir subroutine is called utilizing the path passed to the SaveWorksheet function.

```
Public Sub CreateDataDir(Path$)
Dim Result As Boolean
Dim pathstep(25) As String, ii As Integer, msp As Integer
'Debug.Print "Full Path: "; Path$
'Determine all Subpaths
For ii = 1 To Len(Path$)
 If Mid(Path$, ii, 1) = "\" Then
    pathstep(idx) = Mid(Path$, 1, (ii - 1))
    'Debug.Print idx; " : "; pathstep(idx)
    idx = idx + 1
```

```
 End If
 msp = idx - 1
 'If  ii = Len(Path$) Then
     'Debug.Print "Last Char: "; Mid(Path$, ii, 1)
 'End If
Next ii
'Change To Proper Drive
ChDrive pathstep(0)
'Change to Proper Directory
ChDir pathstep(0)
'Check Each Step in the full path, if a directory doesn't exist
'create it along the way to the final path
For ii = 1 To msp
  Result = DirectoryExists(pathstep(ii))
  Select Case Result
    Case True
    'Debug.Print "Path Found !!! Do Not Recreate!"
    'MsgBox "Path Found !!", vbExclamation, "System Error"
    'Log Error Here
    'Debug.Print ii; "of "; msp; " "; pathstep(ii); " Exists!!"
    Case False
    'Debug.Print "Path Not Found !!! Creating Dir!"
    MkDir pathstep(ii)
    ChDir pathstep(ii)
    'Debug.Print ii; "of "; msp; " "; pathstep(ii); " Created!!"
  End Select
Next ii
End Sub
```

This subroutine has a great deal of flexibility in that it will avoid errors that can invariably occur when a user tries to write a data file to an invalid path. The subroutine works by separating the passed path into a series of successive subpaths that propagate from the start of the drive mount point (usually "c:\") until the final destination path is reached. Each successive subpath is stored in the array pathstep(), and up to 25 subpaths may be utilized before reaching the final destination path. Once the subpaths have been extracted, the subroutine loops through each successive subpath on the way to the final destination directory and checks to see if it exists. If each successive subpath exists, then nothing is done. If any successive subpaths do not exist, then they are created until the final destination directory has been reached. This subroutine is especially handy if a VBA macro is to write data files to a certain location every time, and the macro is to be distributed to a variety of users. By utilizing this subroutine, it is not necessary to manually create any data file locations prior to running the macro; they will all be created on the fly as necessary.

The SaveWorksheet function must be called each time a Workbook is to be saved as a different file type. If the user wishes to save the Workbook in both an Excel and comma delimited format, two calls would have to be made to the SaveWorksheet function. This is a common occurrence as oftentimes a report is saved as an Excel-type file to preserve specific formatting features inherent in the Workbook, and a comma delimited file is also created of the same Workbook to be utilized for data uploading purposes.

There is one additional instance when automatically saving a Worksheet has a great deal of utility. This occurs when templates are utilized within macros. When a template is first loaded into Excel, it is always desirable to resave the template as a Workbook with a specific filename so as not to corrupt the original template file.

The above-discussed routines can be tested by typing the following in the immediate (debug) window (*provided there is at least one Workbook open in Excel, and the module in which the function resides is the active module*):

```
Debug.Print SaveWorksheet(activeWorkbook.Name,
activesheet.name,"c:\temp","ldat","xls")
ldat.xls
Debug.Print SaveWorksheet(activeWorkbook.Name,
activesheet.name,"c:\temp","","xls")
20031110_111159.xls
```

2.7 ALLOWING THE USER TO BROWSE FOR A DIRECTORY

Having the ability to let a user search for a certain directory adds a great deal of flexibility to any application. Most applications need to write files to certain locations for a variety of purposes. Some of the files will no doubt be reports of the results generated by the algorithms executed within the macro. Others may be temporary or transitional files written for purposes such as to rename a Workbook, to back up data that is being acquired during an ongoing process, or to convert an Excel template into a Workbook. No matter what the purpose is of the files that must be saved during an ongoing process, the point is that they must be saved somewhere. In some applications, the developer may wish to specify exactly where each and every file will be written without giving the user any choice about the matter. More likely than not, the end user will be given a choice as to where files generated by an application should be written. Later in this chapter it will be discussed how to store such choices or locations in the Windows registry so that they can be remembered for subsequent macro executions.

File locations can make or break a macro in terms of ease of use. If the user is constantly loading files from the same directory, it is advantageous to store this location to the Windows registry and have the file-loading dialog box open to this default location with each subsequent macro execution. This saves the user the annoyance of having to browse the same location over and over again. It is also extremely practical in that if a new user tries to utilize the data analysis macros, and is ignorant of where the data files are stored, the macro will automatically open the file-loading dialog box to the proper location. The same holds true when it comes time for files to be saved by a macro as well.

As in the case of any process, there are superior and inferior methods of having users choose file locations. Probably the least desirable method, which is employed far too often, is to simply have the user type the desired file location into a text box within a GUI. The problem with this approach is that it is very easy for the user to make a mistake when typing out the full path for a file location. One small spelling mistake or one left-out intermediate directory toward the destination directory is all it takes for disaster to occur. A superior method of choosing file paths in an application is to have a text box with a button next to it. When the button is pressed, a "Browse for Folder" dialog box appears. The user then selects a folder from a Windows Explorer type interface, and the complete path to this folder is then written into the textbox on the form. This approach has several appealing factors. Most importantly, the path returned to the textbox is always valid. In addition, if the user wishes to modify the returned path in any way, shape, or form, they may do so by simply altering the path within the textbox. For example, if a user were storing forms to a path such as "~\2003\Nov\" and the current month is now December, he or she might wish to alter the path to "~\2003\Dec\." When the GUI is closed or the macro is started, it is good practice

FIGURE 2.5 The directory choice example GUI.

to check the validity of any user-specified paths, and should they not exist, have the macro create them. A sample application has been created to demonstrate this process. It can be run in Excel by selecting from the menu <u>A</u>DA -> Chapter <u>2</u> -> <u>B</u>rowse. When the sample macro is executed from the menu, the GUI shown in Figure 2.5 will appear.

When the "Choose Directory" button is pushed on the GUI shown in Figure 2.5, a standard Windows "Browse for Folder" dialog box appears as shown in Figure 2.6.
The code snippet below is what activates the GUI shown in Figure 2.5.

```
Sub BrowseExample()
  'Show the Directory Choosing Example GUI
  frmBrowse.Show
End Sub
```

When the "Choose Directory" button is pushed the following code is executed from within the form.

```
Private Sub Browse_Button_Click()
  'Utilize the Browse Dialog Box to Return Directory into
  TextBox
  frmBrowse.Path_TextBox.Text = BrowseForDirectory
End Sub
```

Notice that it is the `BrowseForDirectory` function that activates a standard Windows "Browse for Folder" dialog box and returns the chosen path to the text box on the form. In order

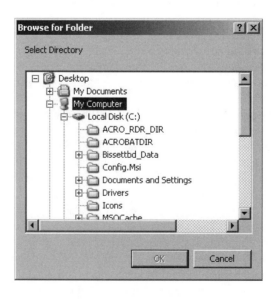

FIGURE 2.6 Standard Windows Browse for Folder dialog box.

for the `BrowseForDirectory` function to work, the following code must be placed at the top of the module the `BrowseForDirectory` function resides in.

```
'****************************************************************
```

'The following declaration must be at the top of the module to use the `BrowseForDirectory` function

```
'****************************************************************
```

```
Private Type BROWSEINFO
hWndOwner As Long
pidlRoot As Long
sDisplayName As String
sTitle As String
ulFlags As Long
lpfn As Long
lParam As Long
iImage As Long
End Type
Private Declare Function SHBrowseForFolder Lib "shell32.dll"
(bBrowse As BROWSEINFO) As Long
Private Declare Function SHGetPathFromIDList Lib "shell32.dll"
(ByVal lItem As Long, ByVal sDir As String) As Long
```

Once these functions and handles have been declared within a module, the `BrowseForDirectory` function can be executed within that module. The code for the `BrowseForDirectory` function is as follows.

```
Function BrowseForDirectory() As String
Dim browse_info As BROWSEINFO
Dim item As Long
Dim dir_name As String
 browse_info.hWndOwner = hwnd
 browse_info.pidlRoot = 0
 browse_info.sDisplayName = Space$(260)
 browse_info.sTitle = "Select Directory"
 browse_info.ulFlags = 1 ' Return directory name.
 browse_info.lpfn = 0
 browse_info.lParam = 0
 browse_info.iImage = 0
 item = SHBrowseForFolder(browse_info)
 If item Then
    dir_name = Space$(260)
    If SHGetPathFromIDList(item, dir_name) Then
       BrowseForDirectory = Left(dir_name, InStr(dir_name,
       Chr$(0)) - 1)
    Else
```

```
        BrowseForDirectory = ""
      End If
    End If
End Function
```

2.8 SETTING THE STARTING DIRECTORY FOR A USER TO BROWSE FROM

The previous "browse for directory" example works well enough for most purposes, but it is missing one key feature: the ability to set the starting directory from which the user is to start browsing from. In some applications, this may not be that critical. It becomes a huge convenience factor when the directory of choice is several layers deep, buried within numerous subdirectories. The end user will invariably get tired of clicking through all the top-level directories, especially when all the directories that would most likely be chosen are buried deep within a complex path structure.

It would seem that setting such a starting point for the "Browse for Directory" dialog box would be a trivial matter, but in fact it is not. This is because Microsoft Office VBA does not support the "AddressOf" operator that is required to tell Windows where the "call back" functions are. The following workaround utilizes methods first published by Ken Getz and Michael Kaplan in the May 1998 issue of *Microsoft Office & Visual Basic for Applications Developer*.

Even if the command ChDir "c:\temp" is placed at the beginning of the BrowseForDirectory function, the "Browse for Directory" dialog box will still start at the point "My Computer." The code for the workaround to fix this deficiency within Microsoft Excel is far too long to publish within this chapter (in fact, it encompasses an entire module "BrowseSpecDir"), but is listed within Appendix A.

In addition to the ability to pick the starting directory, numerous other options can be selected when utilizing the "Browse for Folder" dialog box. They include allowing direct entry of file paths (not a good idea), validating entries (good idea), showing the status bar and associated text, showing files contained within the directories, utilizing the newer "Browse for Folder" dialog box style, and whether to center the "Browse for Folder" dialog box on the screen. The next function, SteroidBrowse, makes use of the code in module BrowseSpecDir to utilize the additional options available when utilizing the "Browse for Folder" dialog box just discussed.

```
Function SteroidBrowse(ByVal initialpath$, ByVal
AllowDirectEntry As Boolean, ByVal ValidateEntry As Boolean, _
ByVal ShowStatusText As Boolean, ByVal ShowFilesToo As Boolean,
ByVal UseNewDialogStyle As Boolean, _
ByVal CenterOnScreen As Boolean) As String
'Set up Browse Options via Boolean Flags
  Dim RetStr As String, Flags As Long, DoCenter As Boolean
  Flags = BIF_RETURNONLYFSDIRS
  If AllowDirectEntry = True Then
      Flags = Flags + BIF_EDITBOX
  End If
  If ValidateEntry = True Then
      Flags = Flags + BIF_VALIDATE
  End If
  If ShowStatusText = True Then
      Flags = Flags + BIF_STATUSTEXT
```

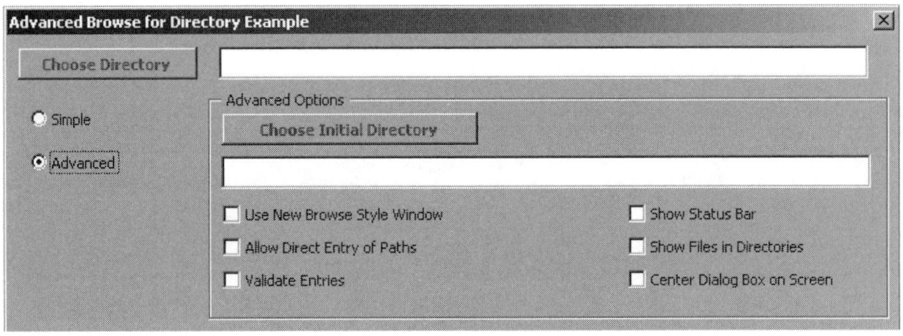

FIGURE 2.7 The Advanced Browse sample application.

```
   End If
   If ShowFilesToo = True Then
      Flags = Flags + BIF_BROWSEINCLUDEFILES
   End If
   If UseNewDialogStyle = True Then
      Flags = Flags + BIF_NEWDIALOGSTYLE
   End If
   If DirectoryExists(initialpath$) = False Then
      'An Invalid initial path was specified - Change to Root
      initialpath$ = "c:\"
   End If
  SteroidBrowse = GetDirectory(initialpath$, Flags,
  CenterOnScreen, "Please select a location to store data files")
 End Function
```

An example application complete with GUI has been created to illustrate the use of the function just presented. It can be run by selecting from the menu ADA -> Chapter 2 -> Advanced Browse. Doing so will bring up the GUI illustrated in Figure 2.7. The GUI can be run in one of two modes, simple and advanced, which are chosen by means of the radio buttons on the left-hand side of the GUI. Simple is the default mode, and when run in this mode, the example functions exactly as the Browse for Directory example given in the previous section of this chapter. The simple mode is illustrated in Figure 2.8.

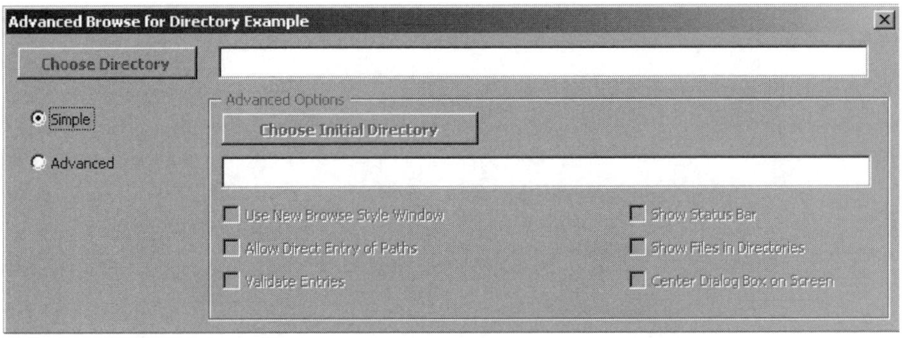

FIGURE 2.8 The Advanced Browse GUI in simple mode.

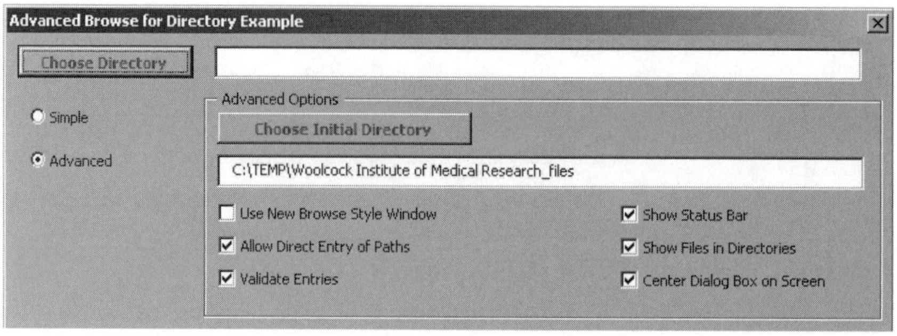

FIGURE 2.9 The Advanced Browse GUI in advanced mode.

When the advanced mode is chosen, however, things begin to get interesting. When the advanced radio button is clicked, all the options available within the advanced options frame on the GUI become available as shown in Figure 2.9. Notice that, in the advanced mode, a starting directory must be specified using the Browse for Folder dialog box in the simple mode by pressing the "Choose Initial Directory" button within the advanced options frame. The initial directory chosen will be written to the text box immediately below the "Choose Initial Directory" button. After this has been done, when the user presses the "Choose Directory" button, the "Browse for Folder" dialog box will open to the initial directory specified. Notice also that there are two possible styles for the "Browse for Folder" dialog box, the original style of which is shown in Figure 2.10.

A newer style dialog box is also available, which is illustrated in Figure 2.11. The big difference between the two styles is that the newer style dialog box has a new folder button for the creation of new folders, which is extremely convenient. One drawback of the newer style dialog box is that

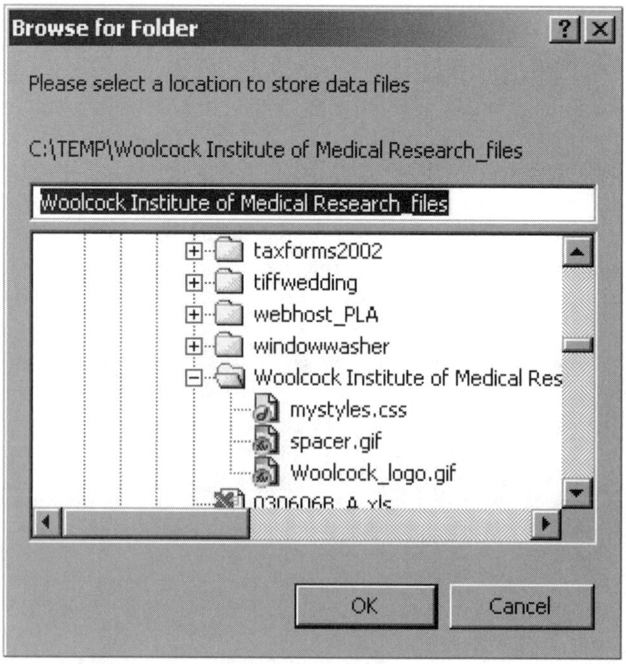

FIGURE 2.10 The "Old Style" Browse for Folder dialog box.

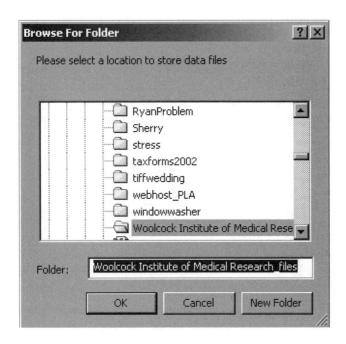

FIGURE 2.11 The "New Style" Browse for Folder dialog box.

it is unable to display a status bar, so this option is irrelevant if the "use new browse style" option is checked.

As with all applications, there is a lot of behind-the-scenes code that enable this application to function in the manner that it does. When the GUI is first shown, the simple option button is chosen as default by means of the following code snippet.

```
Private Sub UserForm_Activate()
  frmAdvBrowse.SimpleButton.Value = True
  End Sub
```

When the simple button is clicked, manually or in code as is the case with the UserForm_Activate subroutine just shown, the following subroutine is executed to disable the advanced options on the GUI.

```
Private Sub SimpleButton_Click()
'Disable Frame, Button, & Textbox
frmAdvBrowse.AdvOptFrame.Enabled = False
frmAdvBrowse.Initial_Button.Enabled = False
frmAdvBrowse.StartPath_Text.Enabled = False
'Disable all Checkboxes
frmAdvBrowse.AllowDirEntryCheck.Enabled = False
frmAdvBrowse.CenterDBCheck.Enabled = False
frmAdvBrowse.NewStyleCheck.Enabled = False
frmAdvBrowse.ShowFilesCheck.Enabled = False
frmAdvBrowse.ShowStatusCheck.Enabled = False
frmAdvBrowse.ValidateEntryCheck.Enabled = False
End Sub
```

When the advanced button is pressed, the advanced options are enabled on the GUI by means of the next subroutine.

```
Private Sub AdvancedButton_Click()
'Enable Frame, Button, & Textbox
frmAdvBrowse.AdvOptFrame.Enabled = True
frmAdvBrowse.Initial_Button.Enabled = True
frmAdvBrowse.StartPath_Text.Enabled = True
'Enable all Checkboxes
frmAdvBrowse.AllowDirEntryCheck.Enabled = True
frmAdvBrowse.CenterDBCheck.Enabled = True
frmAdvBrowse.NewStyleCheck.Enabled = True
frmAdvBrowse.ShowFilesCheck.Enabled = True
frmAdvBrowse.ShowStatusCheck.Enabled = True
frmAdvBrowse.ValidateEntryCheck.Enabled = True
End Sub
```

Setting the initial path for the advanced browse mode is accomplished in the same manner as the original directory choosing the example, with the following subroutine.

```
Private Sub Initial_Button_Click()
  'Utilize the Browse Dialog Box to Return Directory into TextBox
  frmAdvBrowse.StartPath_Text.Text = BrowseForDirectory
End Sub
```

Using the example for the advanced method of "Browse for Folder" is accomplished with the following subroutine, which is activated when the "Choose Directory" button is pressed. The subroutine utilizes information obtained from the GUI to make advanced calls to the BrowseSpecDir module.

```
Private Sub Browse_Button_Click()
'Check for Valid Initial Directory
If DirectoryExists(frmAdvBrowse.StartPath_Text.Text) = False Then
  MsgBox "Initial Directory is not valid!", vbExclamation +
  vbOKOnly, "Invalid Directory"
    Exit Sub
End If
  'Utilize the Advanced Options with the Browse Dialog Box to
  Return Directory into TextBox
  frmAdvBrowse.Path_TextBox.Text =
  SteroidBrowse(frmAdvBrowse.StartPath_Text.Text, _
  frmAdvBrowse.AllowDirEntryCheck.Value, _
  frmAdvBrowse.ValidateEntryCheck.Value, _
  frmAdvBrowse.ShowStatusCheck.Value, _
  frmAdvBrowse.ShowFilesCheck.Value, _
  frmAdvBrowse.NewStyleCheck.Value, _
  frmAdvBrowse.CenterDBCheck.Value)
End Sub
```

2.9 USING THE WINDOWS REGISTRY TO SAVE SETTINGS

In the advanced "browse for directory" example shown previously, there are numerous options that can be set in the graphical user interface. The example is designed to start in the "simple" mode with no initial directory set, and no options selected. Suppose, however, that it was desired to have an application "remember" what its past settings were and how to automatically select those settings upon startup. There are two means of accomplishing this task. The first is to store the settings prior to exiting the program in a file and then to have the file read immediately upon executing the program to determine what the initial settings should be set to. Such initialization (*.ini) files were utilized for years under the Windows operating systems. This method has fallen into disfavor with the advent of the Windows Registry. The Windows Registry is essentially a large database that stores information on the programs that reside in the operating system. This section details how to write information to, and read information from, the Windows Registry within the Excel VBA framework.

The Windows Registry is not an easy application to access. There is no icon within Windows to run the program, nor is there a link from the start menu. This was done on purpose to keep people from tampering with the Windows Registry. Elements stored in the Windows Registry are called keys, and if a key is deleted or changed to an invalid value, this can cause a program to become inoperable, or, in the very worst case, disable the Windows operating system itself. Therefore, great care must be taken when modifying elements in the Windows Registry. To access the Windows Registry, type "regedit.exe" in the Run box under the Start menu as illustrated in Figure 2.12.

Elements in the Windows Registry are viewed in the registry editor much like files are viewed in the Windows Explorer, except in this case the "files" are keys and they have a value associated with them. The value associated with a key can be thought of as a variant because keys can contain nearly any value such as string, integer, single, double, or Boolean. In VB and VBA, keys are stored at this location within the Windows Registry:

HKEY_CURRENT_USER\Software\VB and VBA Program Settings\
Under the VB and VBA Program Settings folder, subfolders will be present. The subfolders are denoted as Application Names, as these folders are where all the keys for a single application should be stored. Folders beneath the Application Name folders are known as Sections, and different Section subfolders can be utilized to file keys that are common to a certain aspect of an application. For example, one section subfolder might hold the keys to one of many GUIs utilized in the application. Applications and Sections follow the VB and VBA Program Settings folder as shown:

\VB and VBA Program Settings\Application Name\Section
The take-home message here is that, under the VB and VBA Program Settings location, there can be numerous applications, each of which can have an Application Name subfolder. Each application subfolder can have numerous section subfolders to hold groupings of related keys utilized in the application.

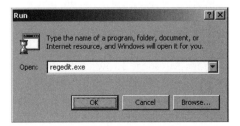

FIGURE 2.12 Running the Windows Registry editor.

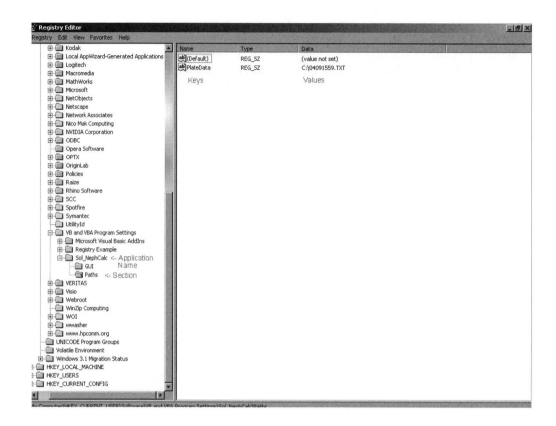

FIGURE 2.13 How sections, keys, and values are viewed in the Registry editor for different applications.

Looking at Figure 2.13, it can be seen that, for the application named Sol_NephCalc, there are two sections that have been created in the Windows Registry, named GUI and Paths (each of which has a subfolder under the application name). Notice that on the right-hand side of the screen is a listing of the key names (under the name header), with the corresponding value to each listed key shown under the data header. To demonstrate how to create sections in the Windows Registry and how to store keys with values within the various sections pertaining to a particular application, the advanced browse example in the previous section of this chapter will be modified. It will be shown how to write code to store the last settings the user chose when the application was last run, as well as how to retrieve and set those settings for subsequent executions of the program.

Notice that the GUI for the advanced browse application shown in Figure 2.14 is populated with three components: checkboxes, radio buttons, and textboxes. For purposes of this example, three sections will be created in the Windows Registry for this application, and they will be called Check, Radio, and Path to represent the components contained within the GUI. The application name will be termed "Advanced Browse." Therefore, within the Windows Registry, it is desired to see the following three structures created:

\VB and VBA Program Settings\Advanced Browse\Check
\VB and VBA Program Settings\Advanced Browse\Path
\VB and VBA Program Settings\Advanced Browse\Radio

The question now becomes: at what point is it advantageous to save the settings contained within the GUI? Sometimes, the decision is quite obvious, as in the case where a user makes a few selections from the GUI and then presses a button that triggers the program to begin execution

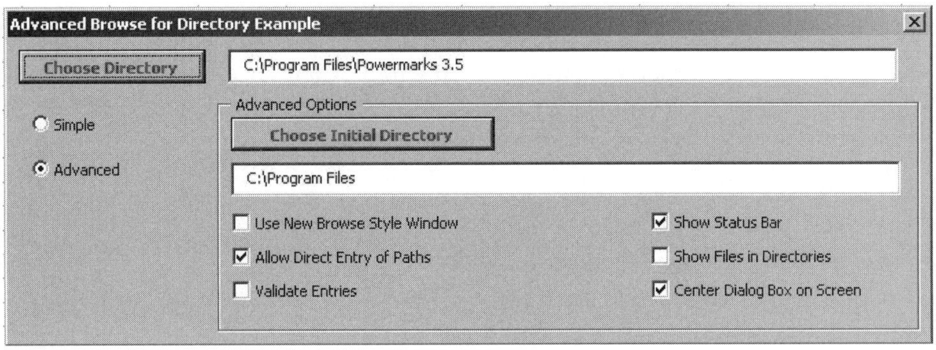

FIGURE 2.14 GUI for the Advanced Browse application.

based on the selections made within the GUI. In such an occurrence, it would be most logical to store the settings to the system registry when the button is pressed to execute the program, or, if the form is hidden *and* unloaded after such a button is pressed, the constructs to save the program's settings to the Windows Registry can be executed from the Terminate method of the user form.

This discussion prompts an explanation of a topic that is often poorly understood in the utilization of user forms, which is the subtle but important differences between show, hide, load, and unload methods. When the (.Show) method is invoked with respect to a form, the form pops up while the application is executing. What is neither obvious nor intuitive, however, is that, for the form to be shown, it must first be loaded into memory. This is accomplished by means of the load method. However, if the show method is executed without the form first being loaded into memory, the compiler will *automatically* load the form into memory. Thus:

```
frmRegExample.Show
```

is equivalent to:

```
Load frmRegExample
frmRegExample.Show
```

The problem that occurs is with the utilization of the (.Hide) method. When the hide method is invoked, the form disappears during program execution, and this is clearly a case of "out of sight, out of mind." When a form is hidden, however, the form still resides in memory and can be utilized again (with the previous settings unchanged) by invoking the (.Show) method. If this form is never to be used again after it is hidden, it should be unloaded from memory with the unload method. If too many forms are merely hidden in an application and not unloaded, the program will have a "memory leak." In other words, it will be allocating system memory and not returning it. For small applications, this may not be a problem. However, as projects become more complex in scope, it is important, and also good practice, to deallocate memory whenever it is no longer needed. What will happen is that should "enough" memory leak from an application, the pool of available memory left for the operating system and other applications to function on becomes smaller and smaller. At some point, the amount of memory leaked reaches a critical mass where applications or even the operating system are no longer able to function correctly, and the operating system must be rebooted to recapture the "leaked" memory. Thus:

```
frmRegExample.Hide
```

is *not* equivalent to:

```
frmRegExample.Hide
Unload frmRegExample
```

For the Advanced Browse application, it would make the most sense to store the settings on the GUI when one of three conditions occurs: (1) the "Choose Directory" button is pressed, (2) the "Choose Initial Directory" button is pressed, (3) the form is unloaded (terminated) by pressing the X in the upper right corner of the form. Knowing this, a subroutine can be called that stores the system settings whenever one of the three events occurs.

```
Sub SaveRegSettings()
'Save all GUI Options to Windows Registry
'Save to Checkbox Section
SaveSetting "Advanced Browse", "Check", "NewStyleCheck",
frmRegExample.NewStyleCheck.Value
SaveSetting "Advanced Browse", "Check", "AllowDirEntryCheck",
frmRegExample.AllowDirEntryCheck.Value
SaveSetting "Advanced Browse", "Check", "ValidateEntryCheck",
frmRegExample.ValidateEntryCheck.Value
SaveSetting "Advanced Browse", "Check", "ShowStatusCheck",
frmRegExample.ShowStatusCheck.Value
SaveSetting "Advanced Browse", "Check", "ShowFilesCheck",
frmRegExample.ShowFilesCheck.Value
SaveSetting "Advanced Browse", "Check", "CenterDBCheck",
frmRegExample.CenterDBCheck.Value
'Save to Radio Section
SaveSetting "Advanced Browse", "Radio", "SimpleButton",
frmRegExample.SimpleButton.Value
SaveSetting "Advanced Browse", "Radio", "AdvancedButton",
frmRegExample.AdvancedButton.Value
'Save to Path Section
SaveSetting "Advanced Browse", "Path", "ChosenPath",
frmRegExample.Path_TextBox.Text
SaveSetting "Advanced Browse", "Path", "InitialPath",
frmRegExample.StartPath_Text.Text
End Sub
```

The `SaveSetting` command is utilized in the `SaveRegSettings` subroutine to save the last known condition of each element on the GUI (Figure 2.15, Figure 2.16, Figure 2.17). The SaveSetting command has the following elements:

SaveSetting *application name, section, key, value*

In the example, the `SaveRegSettings` subroutine saves all the elements contained within the GUI to the "Advanced Browse" application space within the Windows Registry under the three sections named Path, Radio, and Checkbox. To save the settings when either the "Choose Directory" or "Choose Initial Directory" button are pressed, add the code `Call SaveRegSettings` to the click event. To save the current GUI settings when the form is unloaded, add the same call to the terminate event of the form.

```
Private Sub UserForm_Terminate()
'First Save GUI settings for subsequent executions
Call SaveRegSettings
End Sub
```

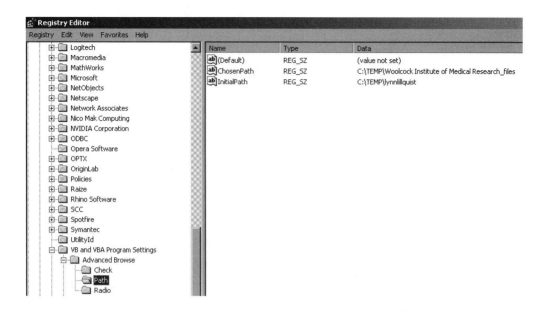

FIGURE 2.15 The SaveRegSettings subroutine creates these registry entries (shown with Path key values).

When the GUI is loaded into memory, the settings stored in the Windows Registry using the above routines must be read and assigned back to the objects that reside in the GUI. However, there is one important factor to think about here, and it is literally a "which came first, the chicken or the egg" scenario. Consider this: if the application is installed on a machine and has never been run, then there are no settings in the Windows Registry for the program to read. Trying to read keys that do not exist from the Windows Registry will result in empty strings being returned. Setting the property of an object to an empty string will result in an error.

What makes the most sense is to create a function in which the property to be set is passed by Reference, and if the key exists for that object, set the property for that object within the function. Because the property is passed by reference, it *should* propagate back to the original object on the form. *Unfortunately, it will not.* Thus, a function such as the one shown in the following text will not work correctly.

```
Function SetPropFromRegKey(ByRef Property, ByVal appname$,
ByVal section$, ByVal key$) As Variant
'This Function will not work. A Bug in VBA prevents Properties of
'Objects from being set when being passed by reference
```

Name	Type	Data
(Default)	REG_SZ	(value not set)
AllowDirEntryCheck	REG_SZ	True
CenterDBCheck	REG_SZ	True
NewStyleCheck	REG_SZ	False
ShowFilesCheck	REG_SZ	False
ShowStatusCheck	REG_SZ	True
ValidateEntryCheck	REG_SZ	False

FIGURE 2.16 The Checkbox key values stored in the Windows Registry.

Name	Type	Data
(Default)	REG_SZ	(value not set)
AdvancedButton	REG_SZ	True
SimpleButton	REG_SZ	False

FIGURE 2.17 The Radio key values stored in the Windows Registry.

```
If GetSetting(appname$, section$, key$, "") <> "" Then
    Property = GetSetting(appname$, section$, key$, "")
    Else
    SetPropFromRegKey = "No Registry Key for " & key$ & " key!"
End If
End Function
```

In VB 6.0 and VBA, there is an exception to the pass by reference rule — if the property of a control is passed into a function by reference, meaning that its value *should* be able to be permanently changed, it nevertheless will not be. It will, instead, behave as if the property were passed into the function by value. This bug has been corrected under the newest flavor of Visual Basic termed VB.NET.

The consequence of this bug is that each and every key read from the Windows Registry must be checked with an If–Then statement to determine if the key does indeed exist. If the key does exist, then its value can be set to the corresponding property of the object it represents. If the key does not exist, then the property of the object is left with its default value. A very common error when deploying programs it to have an error generated when the code is first executed because the program tries to set the property of an object to the value of a key in the Windows Registry that does not yet exist (because the program has never been run on the machine before, and hence the setting has never been saved).

Keys can be read from the Windows Registry by utilizing the GetFromRegistry function that follows. If the key exists, its value is returned. If the key does not exist, an empty string is returned. The function could be easily modified to take a different action or return some sort of statement should a key not exist.

```
Function GetFromRegistry(ByVal appname$, ByVal section$, ByVal
key$) As Variant
'Read the specified Key from the Windows Registry - Return as
Variant
If GetSetting(appname$, section$, key$, "") <> "" Then
   GetFromRegistry = GetSetting(appname$, section$, key$, "")
     Else
     GetFromRegistry = ""
End If
End Function
```

With the ability to extract the information from the registry using the GetFromRegistry function, it is now possible to read the stored state for each object in the GUI when the form is activated. If an objects state is not an empty (or null) string, the property is then assigned to the object via an if–then statement. The If–Then statements prevent the errors discussed earlier of trying to set the property of an object to a null (or empty) string that will occur should the key not exist. The subroutine UserForm_Activate shows how this is accomplished for the Advanced Browse sample application.

```vba
Private Sub UserForm_Activate()
'Load Previously Utilized Settings from the Registry - if they
exist!
'Load Checkbox Values
If GetFromRegistry("Advanced Browse", "Check", "NewStyleCheck")
<> "" Then
   frmRegExample.NewStyleCheck.Value =
   GetFromRegistry("Advanced Browse", "Check", "NewStyleCheck")
End If
If GetFromRegistry("Advanced Browse", "Check",
"AllowDirEntryCheck") <> "" Then
   frmRegExample.AllowDirEntryCheck.Value =
   GetFromRegistry("Advanced Browse", "Check",
   "AllowDirEntryCheck")
End If
If GetFromRegistry("Advanced Browse", "Check",
"ValidateEntryCheck") <> "" Then
   frmRegExample.ValidateEntryCheck.Value =
   GetFromRegistry("Advanced Browse", "Check",
   "ValidateEntryCheck")
End If
If GetFromRegistry("Advanced Browse", "Check",
"ShowStatusCheck") <> "" Then
   frmRegExample.ShowStatusCheck.Value =
   GetFromRegistry("Advanced Browse", "Check",
   "ShowStatusCheck")
End If
If GetFromRegistry("Advanced Browse", "Check",
"ShowFilesCheck") <> "" Then
   frmRegExample.ShowFilesCheck.Value =
   GetFromRegistry("Advanced Browse", "Check",
   "ShowFilesCheck")
End If
If GetFromRegistry("Advanced Browse", "Check", "CenterDBCheck")
<> "" Then
   frmRegExample.CenterDBCheck.Value =
   GetFromRegistry("Advanced Browse", "Check", "CenterDBCheck")
End If
'Load Radio Section
If GetFromRegistry("Advanced Browse", "Radio", "SimpleButton")
<> "" Then
   frmRegExample.SimpleButton.Value = GetFromRegistry("Advanced
   Browse", "Radio", "SimpleButton")
End If
If GetFromRegistry("Advanced Browse", "Radio",
"AdvancedButton") <> "" Then
```

```
    frmRegExample.AdvancedButton.Value =
    GetFromRegistry("Advanced Browse", "Radio",
    "AdvancedButton")
End If
'Load Path Section
If GetFromRegistry("Advanced Browse", "Path", "ChosenPath") <>
"" Then
    frmRegExample.Path_TextBox.Text = GetFromRegistry("Advanced
Browse", "Path", "ChosenPath")
End If
If GetFromRegistry("Advanced Browse", "Path", "InitialPath")
<> "" Then
    frmRegExample.StartPath_Text.Text =
GetFromRegistry("Advanced Browse", "Path", "InitialPath")
End If
End Sub
```

2.10 DETERMINING SUBFOLDERS OF A CHOSEN FOLDER

There was a time not so very long ago that it was a problem for engineers and scientists to capture enough valid data to make a critical measurement or determination. In today's world, a different quandary exists, and that is, with modern machines it is possible to generate enormous quantities of data. The problem now is not getting enough points just to make a determination, but with all the data that has been captured, *which* points should be utilized in making a determination. Subsequent chapters in this book discuss methods that can be employed to make such determinations, but the focus for the moment is simply how to get to the data.

As is often the case with machines, as they have grown more complex, the methods utilized for storing the information they gather have grown more complex as well. Many instruments will give the user the option of storing data in a wide variety of formats such as: comma, tab, or space delimited, ASCII text, or a proprietary format such as an Excel spreadsheets. It is also common for instruments to have the ability to organize measurements of a particular item or nature to a specified subfolder. Such a subfolder will often reside below a folder selected for storage of the test data. For example, if compound XYZ is run on an HPLC with four methods, each of which has a different percentage of methanol in solution, say 40, 50, 60, and 70%, the following directory structure might be created (Figure 2.18).

Top folder: XYZ
Sub folders under XYZ: method40, method50, method60, method70

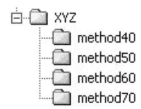

FIGURE 2.18 Example of possible subfolder nomenclature.

Under such a scheme, all files generated by the instrument under the 40% methanol run (to include data files, log files, error files, etc.) would be stored in the method40 folder under the corresponding compound name folder (in this instance XYZ). Although this makes complete sense from an organizational standpoint, it presents unique problems to the developers should they wish to pull out all the data files related to compound XYZ and do some sort of analysis on them.

To accomplish such an analysis, it becomes necessary to be able to accomplish three tasks: (1) the ability to determine the names of subfolders that reside beneath a specified "parent" folder, (2) the ability to extract the names of all files that reside within a specified folder, and (3) the ability to open those files relevant to the task that is to be performed from the list.

The following subroutine will determine all the subfolders under a specified path and return the names of all the subfolders in an array named SubDirList().

```vba
Sub ListDirectories(Path$, SubDirList() As String)
'Pass path$ to desired directory, Sub will return array
SubDirList
'which will have all the subdirectories contained within Path$
'Make Sure path$ ends in "\"!
If Right$(Path$, 1) <> "\" Then
    Path$ = Path$ & "\"
End If
Dim DirRetStr As String, I As Integer
'get the first subdirectory in the passed directory
DirRetStr = Dir(Path$, vbDirectory)
'loop until the Dir functions returns ""
Do While DirRetStr <> ""
 'Ignore the current directory and the encompassing directory
 If DirRetStr <> "." And DirRetStr <> ".." Then
    'Make sure DirRetStr is a Directory and Not a File.
    If (GetAttr(Path$ & DirRetStr) And vbDirectory) =
    vbDirectory Then

        I = I + 1
        'ReSize the Array (+1 to reflect most recent directory added)
        ReDim Preserve SubDirList(1 To I)
        'add the Directory to the array of subdirectories
        (SubDirList())

        SubDirList(I) = Path$ & DirRetStr
        Debug.Print I, SubDirList(I)
   End If
 End If
'Call Dir function again.  Could return a File or Directory
'Above code qualifies its property of returned value on each
loop pass

DirRetStr = Dir()
Loop
End Sub
```

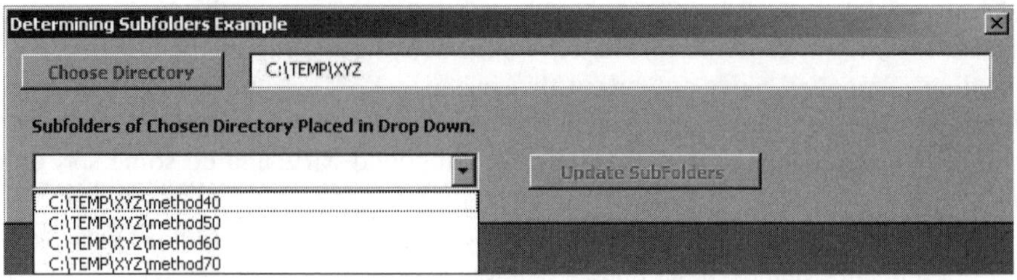

FIGURE 2.19 The Find SubDirectories Example application.

A sample application has been created, which shows how to utilize this application. It can be run by selecting from the menu ADA->Chapter 2->Find SubDir Example. When run, the GUI shown in Figure 2.19 will appear.

There are only two buttons on the GUI. The first (with red text) allows the user to choose the directory from which the subfolders will be extracted. It uses the `BrowseForDirectory` function previously discussed in this chapter.

```
Private Sub Browse_Button_Click()
   'Utilize the Browse Dialog Box to Return Directory into TextBox
   frmListSubDir.Path_TextBox.Text = BrowseForDirectory
End Sub
```

The second button calls the `ListDirectories` subroutine and utilizes the array `SubFolderList` to populate a drop-down box (also known as Combo Box) with all the subdirectories of the chosen directory. This is accomplished by means of a For Each In loop, the mechanism of which will be explained in Chapter 3.

```
Private Sub GetSubfoldersButton_Click()
Dim SubFolderList() As String
'Erase any PreExisting Entries in Combobox Control
SubDirBox.Clear
'Create a List of all Subdirectories in Specified Path
Call ListDirectories(Path_TextBox.Text, SubFolderList())
'Add Each Subfolder to the Combobox Control
For Each ii In SubFolderList
   SubDirBox.AddItem ii
Next ii
End Sub
```

2.11 DETERMINING FILES WITHIN A CHOSEN FOLDER

The next task that needs to be solved for automatic data extraction is the ability to determine the names of all the files that reside within a selected folder. Once the names of the files residing within a directory or group of directories is known, algorithms can be put to work to extract the necessary parameters from the required files to complete a particular task. A sample application has been constructed to read the filenames of files residing in a particular directory into an array. The application can easily be run from the menu by selecting ADA->Chapter 2->List Files Example. Doing so will make the GUI shown in Figure 2.20 appear.

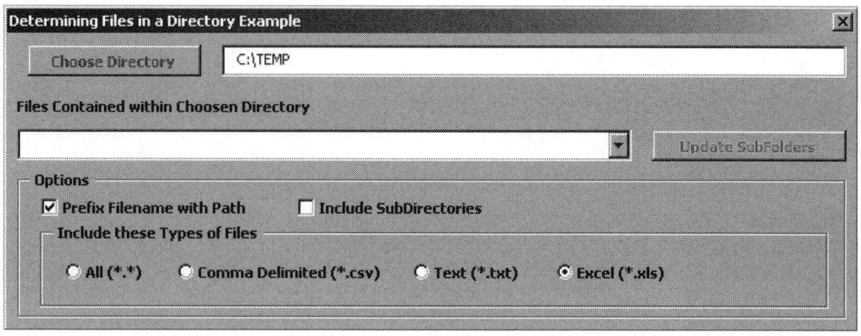

FIGURE 2.20 The List Files in Directory example.

The function `Create_File_List` not only reads the files residing in a particular folder (indicated by `DirPath$`) into an array, but also has the ability to limit files written to the array to those with a certain suffix (*.csv, *.txt, etc.). If all files are to be chosen, then an asterisk (*) should be passed as `sFileType`. The function can also return filenames from subfolders within the chosen folder (specified by DirPath$) by setting the passed parameter `bDoSubs` equal to true. An additional option is the ability to prefix the filename returned with the path indicating its location. This can be accomplished by setting the passed parameter `bPrefixFileNameWithPath` to True. This is an especially attractive option should the user choose to return the filenames contained within subfolders of the specified folder, as it will be readily apparent whether the file returned lies in the "top most" folder, or is contained within a subfolder.

Two points are worth mentioning about this function that make it unique compared with the discussions that have been entertained so far. First, and most important, is that this function utilizes recursive calls. This means, in layman's terms, that, under the right conditions, this function will call *itself* while the code is executing *within itself*. In this instance, recursion is utilized to process any subdirectories that may reside under the chosen directory specified by `DirPath$`. The second point is that this function utilizes what is called a static variable. Normally, when a function or subroutine is called, all the variables within the function or subroutine are cleared and reset to their default states. Because this function utilizes recursion to process any subdirectories that may fall beneath the chosen directory, the number of files counted (stored in the variable `fileCount`) must be preserved after each recursive call. This is done by declaring the variable `fileCount` utilizing the `Static` statement. When the function is called, the Boolean passed parameter RecursiveCall is set to True if the function is to be called recursively, that is, called from within itself. If the function is called outside of itself (or normally), the Boolean passed parameter `RecursiveCall` is set to False. The Boolean passed parameter `RecursiveCall` simply indicates to the function whether or not it should reset the static variable `fileCount` to 0, which must be done when a new file list is to be created.

```
Function Create_File_List(DirPath$, sFileType As String,
FileList() As String, _
bDoSubs As Boolean, bPrefixFileNameWithPath As Boolean, _
Optional RecursiveCall As Boolean = False) As Integer
'Subroutine Extracts Selected Files from a given path "dirpath$"
'SFileType to specify certain kinds of files only (csv, xls,
etc.) use * for all
'FileList is the Array of Files Returned
'Search and Extract from SubDirectories set BDoSubs = True
'Include path in filename set bPrefixFileNameWithPath = True
```

```
Dim subDirectories() As String, fileAndPathList() As String
Dim subDirCount As Integer
Static fileCount As Integer
Dim SearchString As String
Dim fname
Dim i As Long

'If this is NOT a recursive call then reset the fileCount Variable
If RecursiveCall = False Then fileCount = 0

'make certain the directory path ends in a \
If Right(DirPath$, 1) <> "\" Then
    DirPath$ = DirPath$ & "\"
End If
SearchString = DirPath$ & Dir(DirPath$ & "*.*", vbDirectory)
Do While SearchString <> DirPath$
 If Not (Right(SearchString, 2) = "\." Or Right(SearchString, 3)
 = "\..") Then
  If GetAttr(SearchString) = vbDirectory Then
     'do this if a subdirectory
     If bDoSubs Then
        'add to array of directories
        subDirCount = subDirCount + 1
        ReDim Preserve subDirectories(1 To subDirCount)
        subDirectories(subDirCount) = SearchString
     End If
Else
  'do this if a file of the correct filetype
  If UCase(Right(fname, Len(sFileType))) = UCase(sFileType) Or
  sFileType = "*" Then
  fileCount = fileCount + 1
  ReDim Preserve FileList(1 To fileCount)
  ReDim Preserve fileAndPathList(1 To 2, 1 To fileCount)
  fileAndPathList(1, fileCount) = DirPath$
  fileAndPathList(2, fileCount) = fname
  'include path in fileList if option to do so set
  If bPrefixFileNameWithPath Then
     FileList(fileCount) = SearchString
     'Debug.Print fileCount, FileList(fileCount)
     Else
     FileList(fileCount) = fname
     Debug.Print fileCount, FileList(fileCount)
  End If
 End If
End If
```

```
End If
fname = Dir()
SearchString = DirPath$ & fname
Loop
'call recursively to process subdirectories
If subDirCount > 0 And bDoSubs Then
    For i = 1 To subDirCount
        Create_File_List subDirectories(i), sFileType, FileList(),
        bDoSubs, bPrefixFileNameWithPath, True
          Next i
End If
Create_File_List = fileCount
End Function
```

In the example given, the user can choose to return all files or files of type text, Excel, or comma delimited. If the user chooses to return files in subfolders of the chosen folder, then the files returned are automatically prefixed with the path indicating location. If the user does not wish to include filenames contained in subfolders, then prefixing filenames with the path is optional. The user must simply choose a directory to search for filenames with the "Choose Directory" button. When the "Update Subfolders" button is pushed, the drop-down (or combo box) is repopulated utilizing the parameters specified in the GUI.

2.12 PRACTICAL STRATEGIES FOR DEALING WITH LARGE AMOUNTS OF DATA

When automated processes are put into place to handle large volumes of data, one problem manifests itself time and time again. The problem is how to store large amounts of data in a manner such that the information that has been processed will not be lost forever. This problem can be further subdivided into two categories, which are: (1) how such information can be stored in such a manner locally (on an individual's computer) and (2) how such information can be transferred to a database where it can be shared with a multitude of groups and users within the organization.

Storing information locally on one individual machine can be problematic for a variety of reasons, the most obvious of which is, should the machine experience a hardware malfunction, precious data could be lost forever. It is also more difficult to share data that is stored on a local machine for a variety of reasons. It is far more practical to upload data analyzed at the local level to a corporate database where it will be backed up (or at least should be) to protect against the possibility of loss due to equipment malfunctions, viruses, etc. An additional benefit of storing data to a database is that there are all kinds of software tools on the market that allow people to view the information contained within a database without giving them the ability to change or corrupt it.

Chances are the data storage solution employed at any given site will involve both local storage of analyzed data, and remote storage of data within a database. Usually, the data will be stored locally for some period of time, after which the data will be uploaded to a database. Once the data has been uploaded to a database, it really does not matter if it continues to be stored locally anymore. One very effective way of keeping track of data stored locally is to give the files a date-and-time stamp within the filename. Utilizing such a method gives anyone the ability to immediately see when a file was created. For example, a file might have a format such as

YYYYMMDD_hhmmss.xls <-> 20040310_164500.xls

In this example, a date_time stamp is utilized as a filename. In the example, the file was created on March 10, 2004, at 4:45 PM (the time is in military 24-h format, and no seconds have elapsed: "00").

A format such as this can be made even more effective by utilizing prefixing and suffixing of other pertinent information. For example, it might be helpful to prefix the filename with the type of machine the data were acquired on, and suffix the filename with the type of measurement taken. For example:

HPLC_20040310_164500_LogD.xls
ElecChemArray_20040310_164500_Stability.xls
UVplatereader_20040310_164500_pKa.xls

It is also very practical to include the serial number of the machine the data were acquired on within a filename. In the event that a particular machine begins generating erroneous data after a certain point in time, it is possible with such a file-naming convention to delete all of the data files created by such a machine after a certain point in time.

Table 2.1 shows all of the User-Defined Date/Time Formats that can be utilized with the Format Function. This table can be referenced in Visual Basic help by typing "User-Defined Date/Time Formats (Format Function)" in the index tab of Microsoft Excel's online help.

A sample application has been constructed to show how various date–time stamps can be constructed utilizing the Format function in Excel VBA. This application writes out the selected format to the "preview" textbox located at the bottom of the GUI shown in Figure 2.21. The preview GUI is updated to reflect the changes made to the GUI whenever the "Generate" button in the lower right corner of the GUI is pressed. The Preview pane will automatically update if some of the option buttons or checkboxes are selected. It is important to realize while this sample application may seem to be overkill, it can be utilized "behind the scenes" to generate any type of date-time stamp that the user wishes to employ in any given project. The form shown in the sample application does not need to be displayed in order to generate a date-time stamp. Also consider that the resulting date-time stamp can be written into a variable, or utilized as a filename, which has infinitely more practical applications.

The sample application shown in Figure 2.21 can be run by selecting from the menu ADA-> Chapter 2->Date Time Example. In this example, all the code is generated at the form level with no module level code being called or otherwise utilized. The code that drives this example is listed below:

```
Dim mc As String  'Month Component
Dim dc As String  'Day Component
Dim yc As String  'Year Component
Dim tc As String  'Time Component
Private Sub CheckBox_MonthILZ_Click()
  If frmDateTime.OptionButton_MonthNumb = True Then Call
  OptionButton_MonthNumb_Click

End Sub
Private Sub CheckBox_NoSpaces_Click()
   Call Preview_Button_Click
End Sub
Private Sub CheckBox_Prefix_Click()
   'If frmDateTime.CheckBox_Prefix = True Then Call
   Preview_Button_Click

End Sub
Private Sub CheckBox_Suffix_Click()
```

TABLE 2.1
User-Defined Date–Time Formats (Format Function)

(:)	Time separator. In some locales, other characters may be used to represent the time separator. The time separator separates hours, minutes, and seconds when time values are formatted. The actual character used as the time separator in formatted output is determined by your system settings.
(/)	*Date separator.* In some locales, other characters may be used to represent the date separator. The date separator separates the day, month, and year when date values are formatted. The actual character used as the date separator in formatted output is determined by your system settings.
c	Display the date as ddddd and display the time as ttttt, in that order. Display only date information if there is no fractional part to the date serial number; display only time information if there is no integer portion.
d	Display the day as a number without a leading zero (1–31).
dd	Display the day as a number with a leading zero (01–31).
ddd	Display the day as an abbreviation (Sun–Sat).
dddd	Display the day as a full name (Sunday–Saturday).
ddddd	Display the date as a complete date (including day, month, and year), formatted according to your system's short date format setting. The default short date format is m/d/yy.
dddddd	Display a date serial number as a complete date (including day, month, and year) formatted according to the long date setting recognized by your system. The default long date format is mmmm dd, yyyy.
aaaa	The same as dddd, only it is the localized version of the string.
w	Display the day of the week as a number (1 for Sunday through 7 for Saturday).
ww	Display the week of the year as a number (1–54).
m	Display the month as a number without a leading zero (1–12). If m immediately follows h or hh, the minute rather than the month is displayed.
mm	Display the month as a number with a leading zero (01–12). If m immediately follows h or hh, the minute rather than the month is displayed.
mmm	Display the month as an abbreviation (Jan–Dec).
mmmm	Display the month as a full month name (January–December).
oooo	The same as mmmm, only it is the localized version of the string.
q	Display the quarter of the year as a number (1–4).
y	Display the day of the year as a number (1–366).
yy	Display the year as a 2-digit number (00–99).
yyyy	Display the year as a 4-digit number (100–9999).
h	Display the hour as a number without leading zeros (0–23).
Hh	Display the hour as a number with leading zeros (00–23).
N	Display the minute as a number without leading zeros (0–59).
Nn	Display the minute as a number with leading zeros (00–59).
S	Display the second as a number without leading zeros (0–59).
Ss	Display the second as a number with leading zeros (00–59).
ttttt	Display a time as a complete time (including hour, minute, and second), formatted using the time separator defined by the time format recognized by your system. A leading zero is displayed if the leading zero option is selected and the time is before 10:00 A.M. or P.M. The default time format is h:mm:ss.
AM/PM	Use the 12-h clock and display an uppercase AM with any hour before noon; display an uppercase PM with any hour between noon and 11:59 P.M.
am/pm	Use the 12-h clock and display a lowercase am with any hour before noon; display a lowercase pm with any hour between noon and 11:59 P.M.
A/P	Use the 12-h clock and display an uppercase A with any hour before noon; display an uppercase P with any hour between noon and 11:59 P.M.
a/p	Use the 12-h clock and display a lowercase a with any hour before noon; display a lowercase p with any hour between noon and 11:59 P.M.
AMPM	Use the 12-h clock and display the AM *string literal* as defined by your system with any hour before noon; display the PM string literal as defined by your system with any hour between noon and 11:59 P.M. AMPM can be either uppercase or lowercase, but the case of the string displayed matches the string as defined by your system settings. The default format is AM/PM.

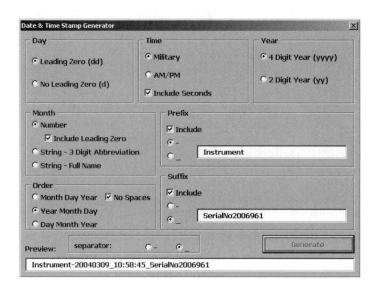

FIGURE 2.21 The date–time stamp sample generator application.

```
    'If frmDateTime.CheckBox_Suffix = True Then Call
    Preview_Button_Click
End Sub
Private Sub CheckBox_TimeSec_Click()
    If frmDateTime.OptionButton_TimeAMPM = True Then Call
    OptionButton_TimeAMPM_Click

    If frmDateTime.OptionButton_TimeMil = True Then Call
    OptionButton_TimeMil_Click

End Sub
Private Sub OptionButton_DayLZ_Click()
'Day With Leading Zeros
dc = Format(Date, "dd")
End Sub

Private Sub OptionButton_DayNLZ_Click()
'Day Without Leading Zeros
dc = Format(Date, "d")
End Sub
Private Sub OptionButton_mmm_Click()
mc = Format(Date, "mmm")
End Sub
Private Sub OptionButton_mmmm_Click()
mc = Format(Date, "mmmm")
End Sub
Private Sub OptionButton_MonthNumb_Click()
If frmDateTime.CheckBox_MonthILZ = True Then
    mc = Format(Date, "mm")
```

```
      Else
        mc = Format(Date, "m")
    End If
    End Sub
    Private Sub OptionButton_SepDash_Click()
        Call Preview_Button_Click
    End Sub
    Private Sub OptionButton_SepUndscr_Click()
      Call Preview_Button_Click
    End Sub
    Private Sub OptionButton_TimeAMPM_Click()
    If frmDateTime.CheckBox_TimeSec = True Then
        tc = "hh:mm:ss AM/PM"
        Else
        tc = "hh:mm AM/PM"
    End If
    End Sub
    Private Sub OptionButton_TimeMil_Click()
    If frmDateTime.CheckBox_TimeSec = True Then
     tc = "hh:mm:ss"
     Else
     tc = "hh:mm"
    End If
    End Sub
    Private Sub OptionButton_YY_Click()
    yc = Format(Date, "yy")
    End Sub
    Private Sub OptionButton_YYYY_Click()
    yc = Format(Date, "yyyy")
    End Sub
    Private Sub Preview_Button_Click()
    Dim DateStr As String, TimeStr As String, DateFormat$
    Dim Preview$
    TimeStr = Format(Time, tc)
    ' Returns current system date in the system-defined long date
    format.
    If frmDateTime.CheckBox_NoSpaces = True Then
        If frmDateTime.OptionButton_OrderDMY = True Then
            DateStr = dc & mc & yc
        End If
        If frmDateTime.OptionButton_OrderMDY = True Then
            DateStr = mc & dc & yc
        End If
        If frmDateTime.OptionButton_OrderYMD = True Then
```

```
         DateStr = yc & mc & dc
     End If
     Else
     If frmDateTime.OptionButton_OrderDMY = True Then
         DateStr = dc & " " & mc & " " & yc
     End If
     If frmDateTime.OptionButton_OrderMDY = True Then
         DateStr = mc & " " & dc & " " & yc
     End If
     If frmDateTime.OptionButton_OrderYMD = True Then
         DateStr = yc & " " & mc & " " & dc
     End If
 End If
 'DateStr = Format(Date, DateStr) This will add unwanted "/"'s
 to date!

 If frmDateTime.OptionButton_SepUndscr = True Then
     Preview$ = DateStr & "_" & TimeStr
     Else
     Preview$ = DateStr & "-" & TimeStr
 End If
 'Add Prefix if desired
 If frmDateTime.CheckBox_Prefix = True Then
     If frmDateTime.OptionButton_PrefixUndscr = True Then
         Preview$ = frmDateTime.TextBox_Prefix & "_" & Preview$
     Else
         Preview$ = frmDateTime.TextBox_Prefix & "-" & Preview$
     End If
 End If
 'Add Suffix if desired
 If frmDateTime.CheckBox_Suffix = True Then
     If frmDateTime.OptionButton_SuffixUndscr = True Then
         Preview$ = Preview$ & "_" & frmDateTime.TextBox_Suffix
     Else
         Preview$ = Preview$ & "-" & frmDateTime.TextBox_Suffix
   End If
 End If
 frmDateTime.TextBox_Preview = Preview$
 End Sub
 Private Sub UserForm_Initialize()
 'Initialize Form Elements, must click option buttons to set
 Format Components
 Call OptionButton_DayLZ_Click
 frmDateTime.OptionButton_DayLZ = True
 Call OptionButton_TimeMil_Click
```

```
frmDateTime.OptionButton_TimeMil = True
Call OptionButton_YYYY_Click
frmDateTime.OptionButton_YYYY = True
Call OptionButton_MonthNumb_Click
frmDateTime.OptionButton_MonthNumb = True
OptionButton_OrderYMD.Value = True
frmDateTime.CheckBox_MonthILZ.Value = True
frmDateTime.CheckBox_NoSpaces.Value = True
Call OptionButton_SepUndscr_Click
End Sub
```

Although this code seems simple and straightforward enough, there are some things here worth calling attention to. Of particular noteworthiness is how the string that comprises the `Format` statement for the date is constructed. When this application was first constructed, instead of utilizing the `Format` function to return each date component individually like this:

```
yc = Format(Date, "yyyy")
```

it was thought that the string utilized to return the date in the format desired could be constructed like this:

```
yc = "yyyy"
```

Then each of the three string components could be concatenated together within the `Format` function to produce the date required in a manner like this:

```
FormatString$ = yc & mc & dc
Dummy$ = Format(Date, FormatString$)
```

Unfortunately, such a scheme does not work when dumping the result of the `Format` function into a textbox within a GUI. What ends up happening is that Excel will insert "\" marks within the date thinking it is "helping" you. Ordinarily, this would not be a big deal, but if the date–time stamp is to be utilized as a filename, this behavior is catastrophic. If the result of the format statement were not to be written to a textbox but a string variable within the macro, such a scheme would work just fine (no additional "\"s would be added). It is unforeseen pitfalls like these that eat up extraordinary amounts of a programmer's time.

When a check box is selected on the form that changes settings defined by option buttons, it is necessary to call the code associated with selected option button's click event in order to process the change associated with check box. For example, if the "Include Leading Zero" checkbox in the month frame is changed within the GUI, it is necessary to call the code associated with the option button labeled "Number" for a leading zero to be included or omitted when the month is to be specified in a numerical format. This is done in the `Private Sub CheckBox_MonthILZ_Click()` subroutine.

Similarly, it is worth mentioning that when a form is loaded, even if some option buttons and check boxes have their values set to True (or selected as the defaults at design time), the component variables for the date and time (yc, mc, dc, tc) will not be set unless the resultant code is executed within the click event that corresponds to the option buttons and check boxes associated with the date and time variable components (yc,mc,dc,tc). It is also important to note that if the click event for a check box or option button is run from code, it does not result in the option button or check box being selected or deselected as it would have if the user would have clicked the control on the form. (This is really counterintuitive.) The only way to set an objects state from code is to change its `Value` property to True. The default settings for the form in this example are set in the `UserForm_Initialize` subroutine.

One final point concerns the initialization of objects on a form. This can be accomplished in one of two manners utilizing either the `Initialize` Event or the `Activate` Event. It is

	A	B	C	D	E	F
1	Product	Size	Sleeve	Inseam	Waist	Color
2	Shirt	17	37			White
3	Shirt	18	36			Blue
4	Pants			32	36	Blue
5	Shorts	XL				Brown
6	Shirt	16	33			Green
7	Shirt	15 1/2	32			Yellow

FIGURE 2.22 A typical file suitable for database uploading.

important to be aware of the differences between the two events. The `Initialize` Event Occurs after an object is loaded but before it is shown. The `Activate` event occurs when an object becomes the active window. Conversely, the `Deactivate` event occurs when an object is no longer the active window. When should the initialization parameters for a form be put into the `Initialize` Event subroutine, and when should they be placed in the `Activate` Event subroutine?

In most instances, it is best to initialize parameters within a form with the `Initialize` Event. This is because the `Initialize` event is only called when the form is first loaded into memory. The `Activate` Event can also be utilized to initialize the parameters within a form, and seemingly will give the same effect. The difference is that every time the form loses focus and is then given focus again (usually by clicking on the form or an element within it), the code in the `Activate` Event subroutine will be run again. This will cause the form to be reset to its default settings if it should lose and regain focus within the application. In some circumstances, this may be desirable behavior, but in most, it will not.

2.13 CREATING DATABASE-"FRIENDLY" FILES

Now that it has been shown how to store files in a manner that can be dealt with on the local level, it is time to discuss how to format files such that they can be readily exported to a database. Files that are "database friendly" tend to be in a flat format. What this means is that each row of data in the Worksheet consists of a separate database entry, and each column in the Worksheet represents a single parameter stored within the chosen database.

Looking at Figure 2.22, the Worksheet contains a single row at the very top that has the "header" information, or names of the individual parameters that collectively comprise a single database entry. In this example, the header names are Product, Size, Sleeve, Inseam, Waist, and Color. Notice that, for each product, not all of the parameters are populated. For example, shirts do not have inseam or waist values, and pants do not have sleeve lengths. It is perfectly legitimate to leave a parameter null or empty in a database. It is also perfectly legitimate to have a parameter that is the same for every database entry. For example, if a particular shirt only came in the color blue, every color entry for that make of shirt would be populated with blue. Why even include the parameter if it is always the same? Because, if someone makes a query (or looks at the information contained within the database), they will want to know the color of the shirt. Flat Files such as those shown in Figure 2.22 can be readily imported into a database. To do so, such files are usually saved in a delimited format, with comma delimited (*.csv) being the most common. Such files can be written into a network folder that is polled by the database at regular intervals. When a file with a legitimate (or database readable) format is detected within the network folder, the file can be automatically loaded into the database.

2.14 OBTAINING DRIVE, DIRECTORY, AND FILE INFORMATION

This entire chapter has been dedicated to the processes involved in reading and writing data. The data that is to be read and written must, of course, come from somewhere. It is only fitting to conclude with some routines that allow the programmer to correct errors that may occur when trying to access files, drives, or directories on any given system. The routines discussed make use of the `FileSystemObject` object to extract various properties about drives, files, and directories.

The most common error that can occur when trying to read or write data is trying to access data from a drive that is not ready. A drive may not be ready for a variety of reasons. If it is a network drive, the network could be down. If the drive requires media, such as a CD-ROM or a disk, the media may not be present in the drive. In any event, it is useful to be able to check the status of a drive that is to be utilized prior to accessing the drive. The `DriveIsReady` function provides this ability.

```
Function DriveIsReady(ByVal Drive$) As Boolean
Dim FSO As FileSystemObject
Dim SpecDrive As Drive

On Error GoTo Drive_Unavailable

Set FSO = New FileSystemObject
Set SpecDrive = FSO.GetDrive(Drive$)
DriveIsReady = SpecDrive.IsReady
'We are done here
Exit Function

'Error Handling
Drive_Unavailable:
DriveIsReady = False
End Function
```

The `DriveIsReady` function will return True if the drive can be accessed, or False if the drive cannot be accessed. The drive letter of interest must be passed to the function via the `Drive$` variable. Before issuing a command such as `ChDir`, it is always prudent to check and see if the drive is in fact ready. The `DriveIsReady` function allows any macro to accomplish this with ease. It may only be necessary to call the `DriveIsReady` function if the drive to be accessed is of a certain type such as a network drive, CD-ROM, or floppy disk. The function `TypeOfDrive` allows the user to determine what kind of drive a given drive letter represents.

```
Function TypeOfDrive(ByVal Drive$, ReturnName As Boolean) As
String
Dim FSO As FileSystemObject
Dim SpecDrive As Drive

On Error GoTo Drive_Unavailable
Set FSO = New FileSystemObject
Set SpecDrive = FSO.GetDrive(Drive$)
Select Case ReturnName
  Case True
    Select Case SpecDrive.DriveType
      Case 1
        TypeOfDrive = "Floppy Disk"
      Case 2
        TypeOfDrive = "Hard Disk"
      Case 3
```

```
        TypeOfDrive = "Network Drive"
      Case 4
        TypeOfDrive = "CD Rom"
      Case Else
        TypeOfDrive = "Unknown Drive Type = " &
        SpecDrive.DriveType
  End Select
Case False
        TypeOfDrive = SpecDrive.DriveType
End Select
'We are done here
Exit Function

'Error Handling
Drive_Unavailable:
TypeOfDrive = "Drive Not Available!"
End Function
```

Two parameters are passed to the `TypeOfDrive` function. The `DriveType` property of the Drive Object returns an integer that represents the type of drive being queried. The `TypeOfDrive` function is capable of returning either the numerical representation of the drive (as a string) or a string description of the drive that is queried utilizing the passed parameter `Drive$`. The passed parameter `ReturnName` determines if the numerical representation or the string description of the queried drive is returned. A sample application utilizing the two preceding functions has been created, which can be run by selecting from the menu ADA -> Chapter 2 -> Drive Status Example.

Figure 2.23 shows how these functions can be utilized to determine when a CD-ROM drive is unavailable (usually from not having a disk in the tray). Figure 2.24 shows how the status of a network drive can also be determined utilizing the preceding two functions.

Although the preceding two functions provide the most key and critical attributes required for successful I/O operations, there are some other useful properties of drives, directories, and files that can also be obtained. When a drive is available for use, it is always nice to be able to check how much free space is available and what the total capacity of the drive is. The next two functions provide this capability.

```
Function DriveFreeSpace(ByVal Drive$) As Single
Dim FSO As FileSystemObject
Dim SpecDrive As Drive

On Error GoTo Drive_Unavailable
```

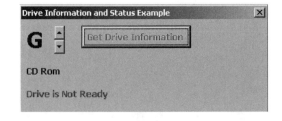

FIGURE 2.23 The drive status example on a CD-ROM drive without a disk.

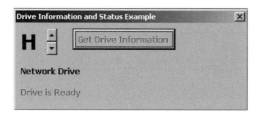

FIGURE 2.24 The status of a network drive.

```
Set FSO = New FileSystemObject
Set SpecDrive = FSO.GetDrive(Drive$)
DriveFreeSpace = SpecDrive.FreeSpace
'We are done here
Exit Function

'Error Handling
Drive_Unavailable:
DriveFreeSpace = 0
End Function

Function DriveTotalSize(ByVal Drive$) As Single
Dim FSO As FileSystemObject
Dim SpecDrive As Drive

On Error GoTo Drive_Unavailable
Set FSO = New FileSystemObject
Set SpecDrive = FSO.GetDrive(Drive$)
DriveTotalSize = SpecDrive.TotalSize
'We are done here
Exit Function

'Error Handling
Drive_Unavailable:
DriveTotalSize = 0
End Function
```

Notice that, if the drive selected has become unavailable, the values returned for free space or total size are set to 0 by the error-handling mechanisms of the functions. When dealing with files, some properties that a user might wish to read are: (1) attributes, (2) when last modified, (3) type, and (4) size. If a routine were required to append information to files, a function to ensure that a file's attributes were not set to read only would be useful. If a file were to be replaced by a routine, perhaps the routine should check when that file was last modified. Functions to perform these functions are listed below.

```
Function FileSize(ByVal FileWithPath$) As Single
Dim FSO As FileSystemObject
Dim FileSpec As File
On Error GoTo File_Unavailable
Set FSO = New FileSystemObject
```

```
Set FileSpec = FSO.GetFile(FileWithPath$)
FileSize = FileSpec.Size
'We are done here
Exit Function
'Error Handling
File_Unavailable:
FileSize = 0
End Function

Function FileType(ByVal FileWithPath$) As String
Dim FSO As FileSystemObject
Dim FileSpec As File
On Error GoTo File_Unavailable
Set FSO = New FileSystemObject
Set FileSpec = FSO.GetFile(FileWithPath$)
FileType = FileSpec.Type
'We are done here
Exit Function

'Error Handling
File_Unavailable:
FileType = "Unknown"
End Function

Function FileLastModified(ByVal FileWithPath$) As String
'Returns the Date-Time File Last Modified like this "02/22/1999
8:29:42 PM"

Dim FSO As FileSystemObject
Dim FileSpec As File

On Error GoTo File_Unavailable
Set FSO = New FileSystemObject
Set FileSpec = FSO.GetFile(FileWithPath$)
FileLastModified = FileSpec.DateLastModified
'We are done here
Exit Function

'Error Handling
File_Unavailable:
FileLastModified = "Unknown"
End Function
```

The three functions just presented are derivatives of the earlier functions and fairly self explanatory in nature. Notice that unlike the functions that operated on a *drive*, these functions that operate on a *file* must include the full path to the file to be examined. They also utilize the file object as opposed to the drive object. A function that returns the attributes of a particular file is a bit more complicated.

```
Function FileAttrib(ByVal FileWithPath$) As String
```

```vba
Dim FSO As FileSystemObject
Dim FileSpec As File, Returned_Attrib As Integer
Dim ii As Integer
On Error GoTo File_Unavailable
Set FSO = New FileSystemObject
Set FileSpec = FSO.GetFile(FileWithPath$)
Returned_Attrib = FileSpec.Attributes

'Loop through Binary Attributes and Append as Required
For ii = 7 To 0 Step -1
  If 2 ^ ii <= Returned_Attrib Then
  Returned_Attrib = Returned_Attrib - 2 ^ ii
  'Set File Attribute Based on Binary Component
  Select Case 2 ^ ii
    Case 0
      'Normal 0 Normal file. No attributes are set.
      FileAttrib = FileAttrib & "Normal"
    Case 1
      'ReadOnly 1 Read-only file. Attribute is read/write.
      FileAttrib = FileAttrib & " ReadOnly"
    Case 2
      'Hidden 2 Hidden file. Attribute is read/write.
      FileAttrib = FileAttrib & " Hidden"
    Case 4
      'System 4 System file. Attribute is read/write.
      FileAttrib = FileAttrib & " System"
    Case 8
      'Volume 8 Disk drive volume label. Attribute is read-only.
      FileAttrib = FileAttrib & " Volume"
    Case 16
      'Directory 16 Folder or directory. Attribute is read-only.
      FileAttrib = FileAttrib & " Folder or directory"
    Case 32
      'Archive 32 File has changed since last backup. Attribute
      is read/write.
      FileAttrib = FileAttrib & " Archive"
    Case 64
      'Alias 64 Link or shortcut. Attribute is read-only.
      FileAttrib = FileAttrib & " Alias"
    Case 128
      'Compressed 128 Compressed file. Attribute is read-only.
      FileAttrib = FileAttrib & " Compressed"
    End Select
  End If
Next ii
```

```
'We are done here
Exit Function

'Error Handling
File_Unavailable:
FileAttrib = "Unknown"
End Function
```

The function `FileAttrib` is a bit more problematic to code because a file can have many attributes. A drive, for example, can only be of one type; it cannot be both a hard disk and a CD-ROM. A file, however, can be both a hidden file and a system file. Because a file can have many attributes, it becomes more problematic to develop a unique number to represent all the possible attributes a file may have. The attributes property utilizes binary numbers to overcome this difficulty. This function decodes the binary number returned utilizing the attributes property, and constructs a string for function returns that contains all the file's attributes separated by spaces.

2.15 SUMMARY

The purpose of this chapter is to teach the reader how to move data in and out of Microsoft Excel utilizing the many tools available within the Visual Basic for the Applications development environment. Loading and saving data both automatically and with dialog boxes is covered, as well as the creation of files suitable for export to a database. More advanced facets of file manipulation are also covered, including how to determine the subfolders at a given location, create an array containing the names of files and subfolders at a given location, and how to store last-known file locations to the Windows Registry. It is also shown how to utilize the Windows Registry to store information pertaining to file input–output (I/O) operations in a manner such that the information can be retrieved and utilized whenever a load or save dialog box is opened. This section also covers how to obtain drive, directory, and file properties utilizing the `FileSystemObject`, `Drive`, and `File` objects.

3 Control and Manipulation of Worksheet Data

3.1 INTRODUCTION

The founding principle upon which this text is based is that it give the reader the tools to construct useful applications. In order to accomplish that task, the user must have the ability to control and manipulate data that reside in any given Worksheet within any given Workbook. For this, the programmer must have the ability to make determinations about what any given Workbook or Worksheet contains, and further, must have the ability to determine where specific pieces of data actually reside within a specific Worksheet. The programmer must also have the ability to copy information to variables where it can be stored, examined, modified, and manipulated. Finally, the developer must also have the ability to create new Workbooks and Worksheets in which to store information that has been analyzed and processed by a macro, most often for reporting purposes. This chapter covers these kinds and types of processes.

3.2 SCOPE AND USE OF VARIABLES IN EXCEL VBA

Variables are probably the single most important facet of programming in any language. Without variables there would be no programs; it is just that simple. How a programmer uses variables determines how "clean" the code they write will be. Variables, like any other computing resource, require space or take-up memory. Different kinds of variables take up different amounts of memory.

Looking at Table 3.1, it is clear that Variables that are of type Variant are the worst offenders in terms of utilizing memory. If a programmer uses a variable in Excel VBA and does not declare the variable to be of a certain type, then the variable is by default set to type Variant. This is because such a variable could conceivably be utilized for any purpose and, therefore, must be allocated enough memory to serve any purpose. The single biggest mistake any programmer can make with respect to the use of variables in Excel VBA is not to declare their use. The next biggest mistake a programmer can make with respect to variables is to "over"-declare a particular variable. For example, if a variable is to store 3 significant digits, and the programmer has declared the variable to be of type Double, they have just wasted 4 bytes of memory. Examine Table 3.1 to see why. The variable could have been declared as type Single and accomplished the same function.

One way to avoid the use of variables with undeclared data types is to utilize the `Option Explicit` statement. When the `Option Explicit` appears in a module, all variables must be explicitly declared utilizing the `Dim`, `Private`, `Public`, `ReDim`, or `Static` statements. If a macro is run and attempts to use an undeclared variable name, an error will occur.

When declaring a variable, some thought should be given to what scope a variable should have. Scope refers to the range of which a variable will be recognized. Basically, VBA supports two kinds of variables: Public and Private. A Public Variable can be used in any procedure anywhere within the project it is declared in (hence the name). If a Public Variable is declared in a standard module or a class module, it can also be used in any projects that reference the project where the Public Variable is declared. Examples:

```
Public ApproachCode as String
Public BondCoefficient as Double
```

TABLE 3.1
Variable Types in VBA and Their Storage Requirements

Data Type	Storage Size	Range
Byte	1 byte	0 to 255
Boolean	2 bytes	**True** or **False**
Integer	2 bytes	32,768 to 32,767
Long (long integer)	4 bytes	2,147,483,648 to 2,147,483,647
Single (single-precision floating-point)	4 bytes	3.402823E38 to 1.401298E-45 for negative values; 1.401298E-45 to 3.402823E38 for positive values
Double (double-precision floating-point)	8 bytes	1.79769313486231E308 to 4.94065645841247E-324 for negative values; 4.94065645841247E-324 to 1.79769313486232E308 for positive values
Currency (scaled integer)	8 bytes	922,337,203,685,477.5808 to 922,337,203,685,477.5807
Decimal	14 bytes	+/79,228,162,514,264,337,593,543,950,335 with no decimal point; +/7.9228162514264337593543950335 with 28 places to the right of the decimal; smallest nonzero number is +/0.0000000000000000000000000001
Date	8 bytes	January 1, 100 to December 31, 9999
Object	4 bytes	Any **Object** reference
String (variable-length)	10 bytes + string length	0 to approximately 2 billion
String (fixed-length)	Length of string	1 to approximately 65,400
Variant (with numbers)	16 bytes	Any numeric value up to the range of a **Double**
Variant (with characters)	22 bytes + string length	Same range as for variable-length **String**
User-defined (using **Type**)	Number required by elements	The range of each element is the same as the range of its data type

Private Variables can only be utilized by procedures that reside in the same module as the Private Variable was declared in. The `Private` statement is used to declare private module-level variables. Examples:

```
Private CompoundNumber as String
Private AreaUnderCurve as Single
```

The `Dim` statement is also utilized to declare variables. When used at the module level, the `Dim` statement is equivalent to the `Private` statement. It is best to utilize the `Private` Statement to make macro code more easy to read and interpret. When a variable is declared with the `Private` or `Dim` statement within a function or subroutine, it can only be utilized within the function or subroutine in which it was declared. Examples:

```
Dim LastName as String
Dim Age as Integer
```

Although this all seems simple enough, there is one clincher to watch out for, and it is especially good at getting those used to programming in a lower level language such as C. It is possible to make multiple variable declarations on a single line such as in the following example:

```
Dim Integer1, Integer2, Integer3 As Integer
```

In the preceding example, Integer1 and Integer2 *are declared as type Variant*, and only Integer3 is declared as type Integer. Why? Because Excel VBA requires the programmer to explicitly declare the type of each and every variable even if it is declared on the same line (unlike in C or many other languages). To declare all three variables as type Integer would require the statement to be written like this:

```
Dim Integer1 As Integer, Integer2 As Integer, Integer3 As
Integer
```

Each time a subroutine or function is called, the variables within that subroutine or function are reset to their default values (typically, a null string for string variables or variants, and zero for numerical variables). Sometimes it is necessary to preserve the value of a variable declared within a subroutine between calls. This can be done by means of the Static statement. When the Static statement is utilized instead of the Dim statement, the declared variable will retain its value between subsequent calls to the subroutine or function. Thus a statement like this:

```
Static NumberofTimesRun as Integer
```

within a subroutine (or function) will not be reset to 0 each time the subroutine is called by the macro. When the macro terminates, however, any Static variables will be reset to their default values.

Constants can also be declared within the VBA environment for values that will be utilized over and over again within an algorithm. Constants are particularly useful if algorithms are being constructed for scientific types of applications. Like any other variable, constants can be of scope Private or Public, but it really is hard to picture the utility of a Private constant. Thus, statements like this can be made:

```
Public Const SqrRootof2 as Single = 1.4142135
Private Const NoOfScans As Integer = 20
Public Const Comment_1 As String = "Unreasonable Value for
Parameter"
```

Arrays are the final topic of discussion for this section. An array is simply a matrix of variables. An array can have many dimensions, much like the known universe. Most sane people will not dimension arrays with more than three dimensions for obvious reasons — it gets very confusing very quickly. Here are some examples of differently dimensioned arrays:

```
Option Base 1 'This statement will be explained later
Public SerialNumbers(2000) as Single 'One Dimension - 2000
elements
Public XYValues(10,10) as Single 'Two Dimensions - 10 × 10 =
100 elements
Public PolarCoord(10,10,10) as Single '3 Dimensions - 1000
elements
```

The number of dimensions an array has is the number of numbers separated by commas contained within the parentheses following the variable declaration. The number of elements an array has is the total number of elements variables declared within the matrix. In the case of

```
Public XYValues(10,10) as Single
```

there are 100 variables named XYValues, and each separate XYValues variable has two dimensions (or indexes) that range in value from 1 to 10. Thus, these are some of the variables that compromise the XYValues variable matrix:

```
XYValues(1,1)  'First Matrix Element
XYValues(3,7)  'In the middle range of Elements
XYValues(10,10) 'Last Matrix Element
```

The statement `Option Base 1` serves one purpose. By default, the lower range of each element in an array statement is 0. Thus, if the above declarations were not preceded by the statement `Option Base 1`, the number of elements for each dimension would increase by 1. For example:

```
Option Base 0 'This statement is never necessary, it is the default
Public XYValues(10,10) as Single
```

creates a two-dimensional array with 121 elements (0 to 10 = 11 elements per dimension; 11 elements × 11 elements = 121 elements). The `Option Base` statement is placed at the top of a module to change the default lower index of all array elements to either 0 or 1.

3.3 OPERATING IN EXCEL'S ENVIRONMENT FROM VBA

Chapter 1 was dedicated to the discussion of how to access data in Excel. That discussion was focused primarily at the cellular level (i.e., access to individual cells within a Worksheet). In this chapter, the discussion will focus on controlling and manipulating data at the Workbook and Worksheet level within the Excel environment. Before a meaningful application can be constructed, it is of paramount importance that the programmer understands how to gain access to Workbooks and Worksheets within the Excel environment. At a minimum, the programmer must be able to determine what Workbooks exist in the Excel environment at any given moment and the vital parameters of each Workbook. Parameters such as (1) Is the Workbook visible or hidden? (2) How many Worksheets does the Workbook contain and what are their names? (3) How many sheets in a given Workbook are hidden and how many are visible? (4) Which sheets in a given Workbook contain data and which are empty? (5) What is the extent to which a Worksheet is populated? Or, in other words, how many rows and columns of data exist within a Worksheet?

Workbooks in Excel can easily be hidden from view by selecting <u>W</u>indow->H<u>i</u>de from the menu. Once this is done, the Workbook will still reside in Excel's memory, and all of the elements of the Workbook (including any code or macros that reside in it) can be utilized from Excel VBA. To unhide a Workbook, simply choose <u>W</u>indow->Unhide from the menu. A small dialog box will pop up like that depicted in Figure 3.1, from which the Workbook to unhide can be selected.

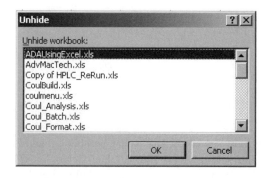

FIGURE 3.1 Unhiding a Workbook in Excel.

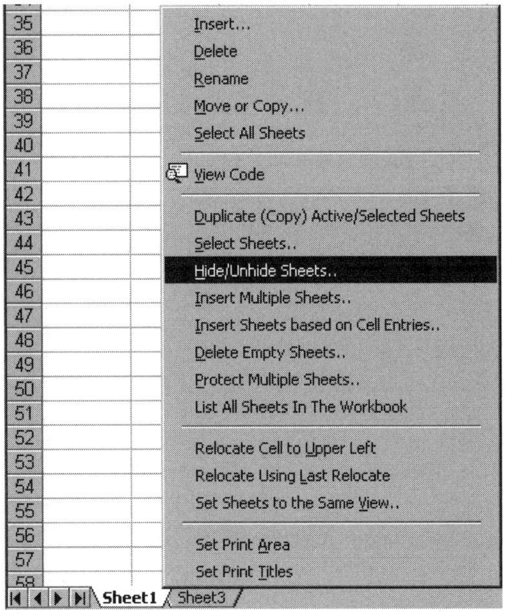

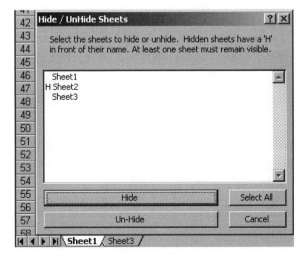

FIGURE 3.2 Hiding and unhiding sheets in Excel.

Worksheets, like Workbooks, can also be hidden or shown. Regardless of their visibility, all elements of a Worksheet can be accessed via Excel VBA code. To hide or unhide a Worksheet, simply right click on any Worksheet tab and select Hide/Unhide Sheets as shown in Figure 3.2.

Selecting this option will bring up the dialog box shown in Figure 3.3.

To show the reader how to ascertain the workspace that exists within Excel at any given moment, a sample application has been constructed that takes a "snapshot" of the Excel environment at any given moment. This application can be run by selecting from the menu ADA->Chapter 3->System Snapshot. Doing so will bring up the GUI shown in Figure 3.4.

This application does the following. When the "Take System Snapshot" button is pressed for the first time, the Excel environment is examined, and every visible and hidden Workbook is populated

FIGURE 3.3 Dialog box to hide/unhide sheets in Excel.

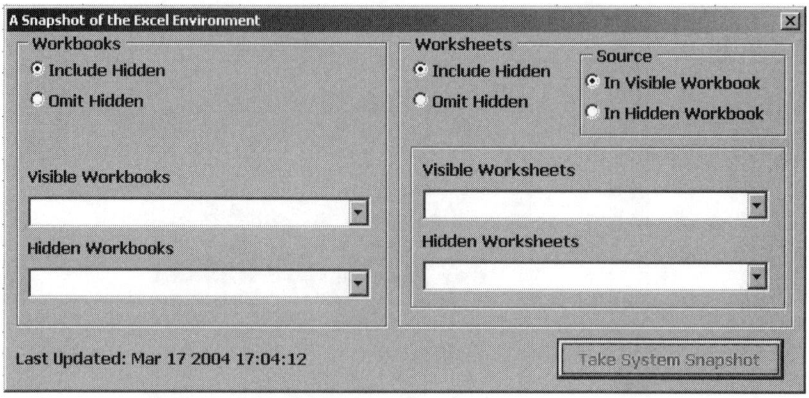

FIGURE 3.4 The System Snapshot sample application.

in the two combo boxes on the left-hand side of the GUI (unless the user chooses to omit the hidden Workbooks by means of the option button). Every time the "Take System Snapshot" button is pressed, the date and time the system was last examined is printed to the label in the left corner of the GUI. Once the Workbooks residing within the system have been determined, the user can specify a Workbook for a closer examination. This is done by selecting the Workbook in one of the combo boxes on the left-hand side of the GUI. The user must then select the source (or which) Workbook is to be analyzed (choose either a hidden or visible Workbook). Once the source Workbook has been specified via the combo box and option buttons, the user can press the "Take System Snapshot" button, and the two combo boxes on the right-hand side of the GUI will be populated with all the visible and hidden sheets that reside in the selected Workbook. This is shown in Figure 3.5.

This is a nice sample application for a variety of reasons. First, it shows the user how to capture all the information pertaining to the Excel environment at any given moment. Second, it shows how to store this information into variables (in this case arrays), where it can be readily manipulated later. Third, it shows how to transfer the stored information into the GUI that was constructed for the project.

One point worth considering when evaluating how to construct any project that will utilize forms is "how will the form be utilized" by the user within the project. In many instances, the user will simply select some options and press a button to execute the macro. In such a case, after the options have been set, the form can simply be made to disappear. (If this is done correctly, the

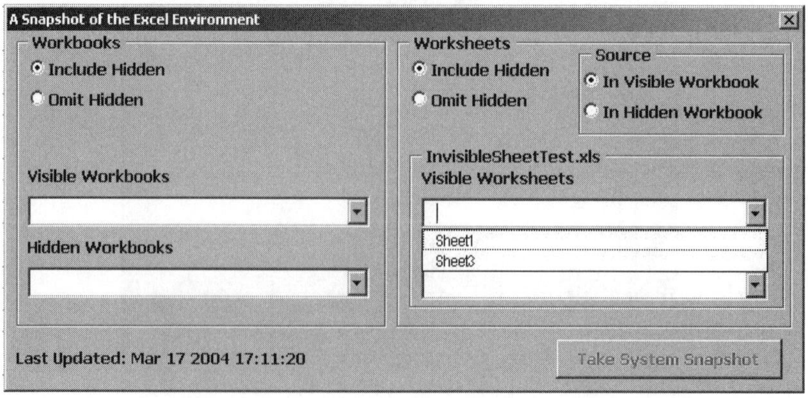

FIGURE 3.5 A System Snapshot with captured visible sheet names.

form will be unloaded and not simply hidden.) In such a case, the form created in the project can be made as a modal form. A modal form will not allow the user to do anything else in the Excel environment (open and close Workbooks, change cell values, etc.) until the form has been unloaded or hidden. A modal form is the proper form to use if the user MUST choose something and have the results returned prior to the program continuing.

In this application, however, it is desirable to have the ability to have the form open and be able to make changes within the Excel environment. Hence, the forms ShowModal property must be set to false. This can be done at design time in the Properties Window as shown in Figure 3.6.

Properties - frm3Snapshot	
frm3Snapshot UserForm	

Alphabetic | Categorized

(Name)	frm3Snapshot
BackColor	&H8000000F&
BorderColor	&H80000012&
BorderStyle	0 - fmBorderStyleNone
Caption	A Snapshot of the Excel Environm
Cycle	0 - fmCycleAllForms
DrawBuffer	32000
Enabled	True
Font	Tahoma
ForeColor	&H80000012&
Height	228.75
HelpContextID	0
KeepScrollBarsVisible	3 - fmScrollBarsBoth
Left	0
MouseIcon	(None)
MousePointer	0 - fmMousePointerDefault
Picture	(None)
PictureAlignment	2 - fmPictureAlignmentCenter
PictureSizeMode	0 - fmPictureSizeModeClip
PictureTiling	False
RightToLeft	False
ScrollBars	0 - fmScrollBarsNone
ScrollHeight	0
ScrollLeft	0
ScrollTop	0
ScrollWidth	0
ShowModal	False
SpecialEffect	0 - fmSpecialEffectFlat
StartUpPosition	1 - CenterOwner
Tag	
Top	0
WhatsThisButton	False
WhatsThisHelp	False
Width	489.75
Zoom	100

FIGURE 3.6 Setting the ShowModal form property at design time.

This application can continue running in the background and, as the user makes changes in the environment, the user can click on the form and press the "Take System Snapshot" button at any time to update the information contained within the GUI (or form).

The snapshot application updates when the user presses the "Take System Snapshot" button. To give the reader an understanding of how the application can extract all of this information from the Excel environment, the fundamental tasks that comprise the application will be discussed in order of operations.

3.4 UTILIZING ARRAYS TO STORE DATA

The first thing that happens each time the "Take System Snapshot" button is pressed is that a series of variables are declared that will store the information extracted from the Excel environment at that particular moment. The code to do this is:

```
Dim HiddenWkbooks() As String, VisibleWkbooks() As String
Dim HiddenWorksheets() As String, VisibleWorksheets() As String
Dim TotalHidden As Integer, TotalVisible As Integer, ii As
Integer
Dim ChosenWorkbook As String
```

Notice that many arrays are declared without dimensions! It is possible to do this when the number of dimensions that will be utilized is unknown. However, before the array can be utilized, its formal size must be declared. This is done by means of the ReDim statement. One caveat of the ReDim statement is that, when an array is resized utilizing the ReDim statement, all of the elements are reset to their default values *unless* the Preserve statement is used.

This statement resizes the array and erases all the elements:

```
ReDim visibleWorkbooks(1 To TotalVisibleWorkbooks)
```
This statement resizes an array and preserves all the elements:

```
ReDim Preserve visibleWorkbooks(1 To TotalVisibleWorkbooks)
```
Notice also that, in this instance, the lower bound and upper bound of each array element is specified, in this case from 1 to a variable named TotalVisibleWorkbooks. This is an exceptionally nice feature when one considers that it is often nice to begin a variable at a certain index other than 1 or 0, which can be prespecified with the Option Explicit statement.

In the snapshot example, four arrays have been defined, each of which will hold the contents to be dumped into the combo boxes (or drop-down boxes utilized on the form). A variable is also specified to keep track of the Chosen Workbook to be examined and the total number of visible and hidden elements.

While on the subject of arrays, it is appropriate to discuss the functions that actually discern the names and states of Workbooks and Worksheets that reside within the Excel environment at any given time. These functions utilize arrays to return the names of Workbooks and Worksheets that meet a certain criteria. Functions will be discussed at length in Chapter 4, but it is necessary to utilize them in this example to develop a robust application. First, the Workbook functions:

```
Function TotalVisibleWorkbooks(visibleWorkbooks() As String)
As Integer
Dim bookCounter As Workbook
For Each bookCounter In Application.Workbooks
   If Windows(bookCounter.Name).Visible = True Then
      TotalVisibleWorkbooks = TotalVisibleWorkbooks + 1
```

```
      ReDim Preserve visibleWorkbooks(1 To
      TotalVisibleWorkbooks)
      visibleWorkbooks(TotalVisibleWorkbooks) =
      bookCounter.Name

      'Debug.Print TotalVisibleWorkbooks,
      visibleWorkbooks(TotalVisibleWorkbooks)

  End If
Next bookCounter
End Function

Function TotalHiddenWorkbooks(hiddenWorkbooks() As String) As
Integer
Dim bookCounter As Workbook
For Each bookCounter In Application.Workbooks
  If Windows(bookCounter.Name).Visible <> True Then
      TotalHiddenWorkbooks = TotalHiddenWorkbooks + 1
      ReDim Preserve hiddenWorkbooks(1 To TotalHiddenWorkbooks)
      hiddenWorkbooks(TotalHiddenWorkbooks) = bookCounter.Name
      'Debug.Print TotalHiddenWorkbooks,
      hiddenWorkbooks(TotalHiddenWorkbooks)
      End If
Next bookCounter
End Function
```

The most striking thing about these two functions is probably the statement declaring the variable `bookCounter` as type `Workbook`. What does this mean? This is an object variable declaration statement. Excel is made up essentially of a collection of objects such as Workbooks, Worksheets, forms, objects that reside in forms, etc. Utilizing objects variables allows access to all of an object's properties and methods, but there is one barrier to utilizing this functionality. The use of object variables is very poorly documented, and it is not intuitive on how to accomplish many tasks utilizing them. However, utilizing object variables can allow the developer to access the entire range of properties and methods related to an object and take action on them with ease. Such is the case with these functions.

In these two functions, a loop is constructed utilizing the `bookCounter` object variable that loops through every Workbook present (or loaded) in the Excel environment (even if it is a hidden Workbook). Every Workbook is then checked as to its visible status. If the Workbook is visible and the function is to report visible Workbooks, then the total number of visible Workbooks is incremented, and the array that stores the names of the visible Workbooks is redimensioned to a size reflecting the total number of visible Workbooks. The name of the Workbook is then stored in the array. When all of the Workbooks have been looped through, the function ends, returning the total number of visible Workbooks as an integer. *It also returns the array that contains all of the Workbook names.* This is accomplished by passing an array to the function *By Reference*, which will be discussed in detail in the next section of this chapter. The functions to discern the names and states of the visible and hidden Worksheets that are present within the Excel environment work in a similar fashion.

```
Function TotalVisibleWorksheets(VisibleWorksheets() As String,
wkbook As String) As Integer
Dim wks As Worksheet
```

```
Workbooks(wkbook).Activate
For Each wks In Worksheets
  'Debug.Print wks.Name, wks.Visible
  If wks.Visible = -1 Then
     TotalVisibleWorksheets = TotalVisibleWorksheets + 1
     ReDim Preserve VisibleWorksheets(1 To
     TotalVisibleWorksheets)
     VisibleWorksheets(TotalVisibleWorksheets) = wks.Name
     'Debug.Print TotalVisibleWorksheets,
     VisibleWorksheets(TotalVisibleWorksheets)
  End If
Next wks
'Debug.Print TotalVisibleWorksheets
End Function

Function TotalHiddenWorksheets(HiddenWorksheets() As String,
wkbook As String) As Integer

Dim wks As Worksheet
Workbooks(wkbook).Activate
For Each wks In Worksheets
  If wks.Visible = 0 Then
     TotalHiddenWorksheets = TotalHiddenWorksheets + 1
     ReDim Preserve HiddenWorksheets(1 To
     TotalHiddenWorksheets)

     HiddenWorksheets(TotalHiddenWorksheets) = wks.Name
     Debug.Print TotalHiddenWorksheets,
     HiddenWorksheets(TotalHiddenWorksheets)
  End If
Next wks
End Function
```

Here, a Worksheet object variable is used to access the Worksheet properties and loop through all the Worksheets in a particular Workbook. The Workbook to be examined is passed into the function as the string parameter wkbook. Each Workbook to be examined is then made into the active Workbook in the Excel environment. Similar to the previous two functions, an array is utilized to store the names of Worksheets that are either hidden or visible, depending on the function.

3.5   PASSING PARAMETERS BY VALUE OR BY REFERENCE?

In the previous section, the four functions utilized arrays to carry information extracted in the functions back to the main body of the program, or alternatively, it could be said that arrays were utilized to store information extracted in the functions. In the previous four functions, these arrays were passed by reference. This section will articulate the meaning of passing a parameter by reference and passing a parameter by value. To illustrate the difference between the two parameter-passing paradigms, a simple example will be given. In this example, index cards will represent a passed parameter (variable), and people will represent the subroutines.

EXAMPLE 1: PASSING BY VALUE

Person 1 hands an index card with a value on it to Person 2. Person 2 *reads* the value on the index card, *transfers* it to a piece of scratch paper, and adds 2 to the value *read* from the index card. *Person 2 then hands Person 1 the original index card back with the original value unchanged.* This is an example of passing by value. *When a parameter is passed by value, the parameter passed cannot be altered by the function or subroutine it is passed to in any way.*

EXAMPLE 2: PASSING BY REFERENCE

Person 1 hands an index card with a value on it to Person 2. Person 2 *reads* the value on the index card, *erases* the original value on the index card, and *replaces* it with the value read from the index card plus 2. Person 2 then hands Person 1 the original index card back with a *new value* written on it. This is an example of passing by reference. *When a parameter is passed by reference, the parameter passed can, and often is, altered by the function or subroutine it is passed to.*

Parameters passed to a function or subroutine are by default passed by reference. This is somewhat dangerous because, if the programmer is unaware of the difference, the values passed to the function or subroutine could be altered by the function or subroutine that is called! To specify whether a parameter is passed by value or by reference, the `ByVal` or `ByRef` keywords are utilized within the function itself. Alternatively, when calling a function or subroutine, if parentheses are placed around a parameter to be passed, it will force the parameter to be treated as if being passed by value. The following snippet of code should further clarify the foregoing discussion:

```
Sub ByValByRefExample()
Dim PassValue As Integer
PassValue = 2
Call PassByVal(PassValue)
Debug.Print "Value Unchanged When Passed By Value: "; PassValue
Call PassByRef(PassValue)
Debug.Print "Value Is Changed When Passed By Reference: ";
PassValue
'Remember Default is Pass By Ref
Call UnspecifiedPass(PassValue)
Debug.Print "Value Is Changed When Passed By Reference: ";
PassValue
'Unless () force pass by Value
Call UnspecifiedPass((PassValue))
Debug.Print "Value Unchanged When Passed By Value: "; PassValue
End Sub

Sub PassByRef(ByRef Numb As Integer)
Numb = Numb + 2
End Sub

Sub PassByVal(ByVal Numb As Integer)
Numb = Numb + 2
End Sub

Sub UnspecifiedPass(Numb As Integer)
Numb = Numb + 2
```

```
End Sub
```

When the ByValByRefExample subroutine is run, it will write the following to the immediate or debug window:

```
Value Unchanged When Passed By Value:2
Value Is Changed When Passed By Reference:4
Value Is Changed When Passed By Reference:6
Value Unchanged When Passed By Value:6
```

3.6 ARRAY LOOPING STRUCTURES

When the "Take System Snapshot" button is pressed for the first time, the information containing the names of the visible and hidden Workbooks is placed into the arrays VisibleWkbooks and HiddenWkbooks, respectively. A method is required to extract this information from these arrays and place the information into the combo boxes within the forms. The following section of code accomplishes this task:

```
'Get Workbook Information
'Get Visible Workbook Information
TotalVisible = TotalVisibleWorkbooks(VisibleWkbooks())
If TotalVisible > 0 Then
    For ii = LBound(VisibleWkbooks()) To TotalVisible
      frm3Snapshot.ComboBox_Workbooks.AddItem
      VisibleWkbooks(ii)
    Next ii
End If
'Get Hidden Workbook Information if desired
TotalHidden = TotalHiddenWorkbooks(HiddenWkbooks())
If TotalHidden > 0 And frm3Snapshot.OptionButton_WkbookIH.Value
= True Then
    For ii = LBound(HiddenWkbooks()) To UBound (HiddenWkbooks())
      frm3Snapshot.ComboBox_HWorkbooks.AddItem HiddenWkbooks(ii)
Next ii
End If
```

In both cases, first the functions are called that return the number of visible or hidden Workbooks. These functions also return an array that contains the names of the visible or hidden Workbooks. Because the array is passed by reference, the functions are able to change the values within the array to represent the Workbook names that pertain to the particular case being sought (visible or hidden Workbook names). Once the total number of Workbooks is returned, it is necessary to read the items in the array and copy them into the combo box on the form. There are two easy ways to do this.

The first utilizes the LBound function, which returns the lower index of the array utilized in the for-loop to loop through each element of the returned array. The upper index of the array is determined by utilizing the value returned by the function TotalVisibleWorkbooks. In the second example, the upper index of the array is determined by utilizing the UBound function. Either method works fine, and the method to be utilized is really a matter of personal preference. Notice that, before either loop begins, an if-statement checks to make sure that the number of Workbooks for either case is at least 1 (> 0). This is because, if no hidden or visible Workbooks existed when the snapshot was taken, the arrays will not have been redimensioned (using ReDim). If the arrays have not been properly dimensioned, utilizing the Lbound or UBound on them will result in an error. (The lower bound of an array index or the upper bound of an array index cannot be returned if the array indexes have not been defined!) The looping structure is self explanatory in that it only adds the elements (names) from the arrays to the combo boxes in the form.

Once the names of the Workbooks have been populated into the combo boxes on the form, the combo boxes that hold the Worksheet names must be populated.

```
'Get Worksheet Information if applicable
'If No Workbook is Specified in the Chosen Combobox then Exit!
If ChosenWorkbook = "" Then Exit Sub
'Populate Visible Worksheet Combobox
TotalVisible = TotalVisibleWorksheets(VisibleWorksheets(),
ChosenWorkbook)
If TotalVisible > 0 Then
    For ii = LBound(VisibleWorksheets()) To TotalVisible
        frm3Snapshot.ComboBox_Worksheets.AddItem
        VisibleWorksheets(ii)
    Next ii
End If
'Populate Hidden Worksheet Combobox
TotalHidden = TotalHiddenWorksheets(HiddenWorksheets(),
ChosenWorkbook)
If TotalHidden > 0 Then
    For ii = LBound(HiddenWorksheets()) To
    UBound(HiddenWorksheets())
        frm3Snapshot.ComboBox_HWorksheets.AddItem
        HiddenWorksheets(ii)
Next ii
End If
```

The combo boxes that contain the Worksheet names are populated in exactly the same manner as the combo boxes that contained the Workbook names. The only difference here is that, before the Worksheet names can be obtained, the Workbook of focus must be specified. If a Workbook has been chosen, its value is stored in the variable ChosenWorkbook. Before running the routines to extract the Worksheet names, the macro first checks to see if a Workbook has been chosen. If a Workbook has not been chosen (or the variable ChosenWorkbook equals null or empty string), then the macro terminates. The first time the "Take System Snapshot" button is pushed, the macro will always exit the subroutine at this point because it takes one pass of the subroutine to populate the combo boxes with the Workbook names, and a Workbook name could not be chosen from the combo boxes unless they were populated.

3.7 USING OBJECT VARIABLES

Each time the "Take System Snapshot" button is pushed, the combo boxes for both the Workbook and Worksheet names have the potential to be populated. If the combo boxes are not cleared prior to executing the code for the "Take System Snapshot" button, they will retain the old Workbook and Worksheet names as well as the newly obtained Workbook and Worksheet names. A variety of methods could be utilized to clear out the contents of the combo boxes, but the cleanest and most efficient way is to utilize an object variable. As previously mentioned, the use of object variables is not very well documented, so the routine utilized to clear all the combo boxes will be explained at length.

```
Sub ClearComboBoxes()
'This will clear every ComboBox in the active GUI
Dim c As Control
For Each c In Controls
  If TypeOf c Is ComboBox Then
    'Debug.Print c.Name
    'The Clear Method is not Present in Autocomplete!
    c.Clear
  End If
Next c
End Sub
```

Notice that object variable c is dimensioned as type Control. What this means is that c will be representative of any controls found on the active form or which are active in the Excel environment. A For loop is then set up to loop through all the controls. This For loop will loop through every control that exists in the form. However, only the combo boxes are of interest as they are the only elements to have action taken on them. So, within the loop body, the TypeOf statement is utilized to check and see if the type of control is a combo box. If the current control in the loop is a combo box, then its contents are cleared utilizing the Clear method. What is important to keep in mind here is that this methodology could be utilized for any type of control or any type of object within Excel. Such routines can be utilized to set or check any properties or methods of any control, group of controls, or objects. Notice that the clear method is not supported by the autocomplete functionality, yet it does in fact exist. (Figure 3.7).

A logical person might think: Why should the routine loop through all the controls when the only control of interest is the combo box? It is tempting to utilize a routine such as the following to clear the combo boxes, but for some reason it does not seem to work. This may or may not be a bug in the Excel VBA paradigm; however, no information the author has found explains why such a routine should not work.

```
Sub ClearComboBoxes_NoWork()
'One would think this would loop through all comboboxes
'and clear them - It will not.
Dim cmbx As ComboBox
For Each cmbx In Controls
  cmbx.Clear
Next cmbx
End Sub
```

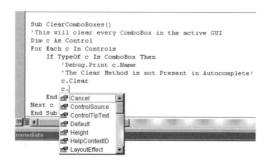

FIGURE 3.7 Not all object methods are supported by Autocomplete.

Once the combo boxes have been cleared, no selected information can be extracted from them. Therefore, if a Workbook has been chosen in which the user wishes to examine its Worksheets, it is important that this information be captured and stored before the combo boxes are cleared utilizing the just-presented routine. The following section of code accomplishes this by storing any selected Workbooks to the string variable `ChosenWorkbook`. The Workbook name stored is dependant upon whether the user wishes to examine a visible or invisible Workbook, which is determined via an option button setting (`OptionButton_inVisible.Value`).

```
'Capture Specified Workbook Information Prior to Clearing Comboboxes
Select Case frm3Snapshot.OptionButton_inVisible.Value
  Case True
  'Use Visible Workbook
  ChosenWorkbook = frm3Snapshot.ComboBox_Workbooks.Text
  Case False
  'Use Hidden Workbook
  ChosenWorkbook = frm3Snapshot.ComboBox_HWorkbooks.Text
End Select
```

The last item of interest is writing the last time the system snapshot was updated to the GUI. This is accomplished by setting the value of the label in the GUI equal to following simple function.

```
frm3Snapshot.Label_LastUpdated = LastUpdated
```

Where:

```
Function LastUpdated() As String
'Returns String with Last Updated Information
LastUpdated = "Last Updated: " & Format(Date, "mmm d yyyy ")
& Format(Time, "hh:mm:ss")
End Function
```

Also of interest is that if a Workbook has its Worksheet names listed in the combo boxes on the right-hand side of the GUI, the Workbook name those Worksheet names are associated with is written into the frame enclosing the combo boxes that contain the Worksheet names. This is accomplished with one simple line of code.

```
frm3Snapshot.Frame_Workbook.Caption = ChosenWorkbook
```

The sample application described here can be reverse engineered by the reader to gain an understanding of how to obtain information about the Excel environment necessary to manipulate information within Workbooks and Worksheets. The methods utilized within the sample application can then be customized to a particular user's requirements.

3.8 AN IN-DEPTH LOOK AT WORKSHEETS

It is within the Worksheet that all the information contained within a Workbook or project is stored. In order to program in Excel efficiently, it is a basic requirement to understand the mechanisms and peculiarities involved in extracting information from Workbooks. At a minimum, an effective Excel VBA programmer must know how to perform the following tasks: (1) how to determine if a Worksheet contains data, (2) how to determine which Worksheets in a Workbook contain data, (3) how to determine how many rows of data are present in a particular Worksheet, (4) how to determine how many columns of data are present in a particular Worksheet, (5) how to determine how many rows of data are present in a particular column of a particular Worksheet, and (6) how

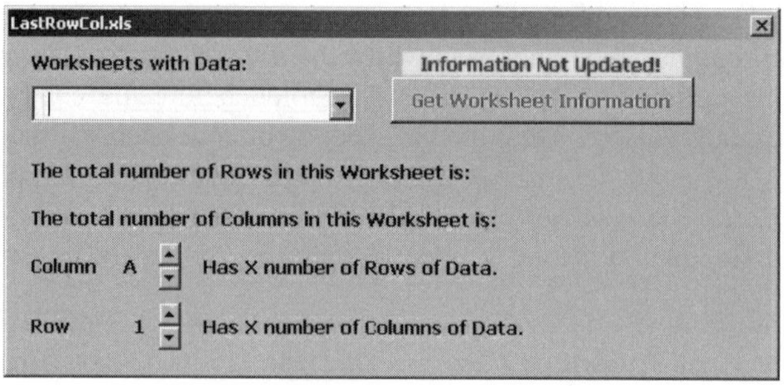

FIGURE 3.8 The Worksheet info sample application.

to determine how many columns of data are present in a particular row of a particular Worksheet. A sample project has been constructed to show the reader how to accomplish all of these tasks with relative ease. This sample application can be run by selecting ADA->Chapter 3->Worksheet Info, which will launch the GUI shown in Figure 3.8.

Notice that this application contains one combo box that stores the names of all Worksheets that contain data within the active Workbook. When the "Get Worksheet Information" button is pressed, a number of vital statistics about the Worksheet chosen in the combo box are written to the GUI. (If a Worksheet name is not chosen, the warning message "Information Not Updated!" in Red with a yellow background is shown in the GUI after pressing the "Get Worksheet Information" button.) The first two parameters written are the maximum number of rows and the maximum number of columns in the Worksheet. Immediately below this, the user may choose a particular row and column by means of spin button controls. The GUI will then also return the maximum number of rows of data for a particular column, as well as the maximum number of columns of data for a particular row.

The "LastRowCol.xls" Workbook depicted in Figure 3.9 is utilized to demonstrate the functionality of the Worksheet info sample application. In this Workbook there are three Worksheets. One contains no data and is appropriately named "empty." The other two contain identical information and are titled "Test" and "Test2," with the only difference being that "Test" has some special formatting. Looking at Figure 3.9, notice that cells B9 and F8 are colored with a red background. This is because these two cells define the "limits" of where data can possibly reside in this Worksheet. For example, the maximum row value that may contain a nonempty cell (or a cell that

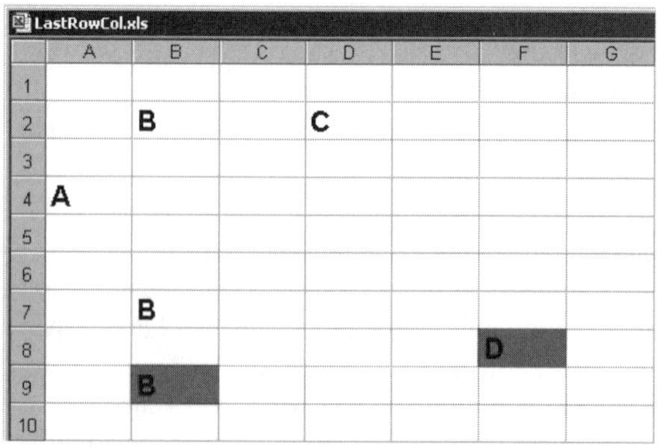

FIGURE 3.9 Sample data Worksheet for Worksheet info sample application.

has data) is 9. The maximum column value that may contain a nonempty cell (or a cell that has data) is 6 (where F represents column number 6). Looking at particular instances of rows and columns, row 2 has a maximum column of 4 (column D is the last column with data for row 2), and column 2 has a maximum row of 9 (row 9 is the last possible row that could contain data for column number 2 — which is also the maximum row value for this Worksheet).

The first task required to build this application is to have the ability to tell if a particular Worksheet is devoid of any data (empty). This can be accomplished quite easily using the `CountA` Worksheet function, which will return the number of cells that are not empty.

```
Function EmptySheet(Workbook, Worksheet) As Boolean
'Returns True if Worksheet in Workbook contains nothing (is empty)
If Application.CountA(Workbooks(Workbook).Worksheets(Worksheet).Cells)
= 0 Then
   EmptySheet = True
   Else
   EmptySheet = False
End If
End Function
```

Two functions are now required to determine the last row in a Worksheet that contains data, and the last column within a Worksheet that contains data. The following two functions accomplish this task.

```
Function LastCol(Workbook, Worksheet) As Long
'This function will return the last COLUMN in the Worksheet
of Workbook
On Error GoTo Fix_it:
ActivateWorkbook (Workbook)
Sheets(Worksheet).Activate
Workbooks(Workbook).Worksheets(Worksheet).Cells.Find(What:="*", _
  SearchDirection:=xlPrevious,
  SearchOrder:=xlByColumns).Activate
LastCol = FindCol(Selection.Address(ReferenceStyle:=xlR1C1))
Exit Function
Fix_it:
'Debug.Print "Error Description: "; Error; " - Error Code: "; Err
LastCol = 0
End Function

Function LastRow(Workbook, Worksheet) As Long
'This function will return the last ROW in the Worksheet of
Workbook
On Error GoTo Fix_it:
ActivateWorkbook (Workbook)
Sheets(Worksheet).Activate
Workbooks(Workbook).Worksheets(Worksheet).Cells.Find(What:="*", _
  SearchDirection:=xlPrevious, SearchOrder:=xlByRows).Activate
LastRow = FindRow(Selection.Address(ReferenceStyle:=xlR1C1))
Exit Function
```

```
Fix_it:
'Debug.Print "Error Description: "; Error; " - Error Code: "; Err
LastRow = 0
End Function
```

The methodology these functions employ to accomplish their tasks is not readily apparent to even an experienced VBA programmer, so some explanation is in order here. What these functions do is as follows. For finding the last row, the function starts searching for a nonempty cell by rows *from the bottom of the Worksheet*. As soon as the function encounters a nonempty cell (searching by rows starting from the bottom), it knows this is the maximum row value for any data located in this Worksheet. For finding the last column, the function starts searching for a nonempty cell by columns *from the right side of the Worksheet*. As soon as the function encounters a nonempty cell (searching by columns starting from the right), it knows this is the maximum column value for any data located in this Worksheet.

Notice also that both of these functions have error-trapping mechanisms. This is because, should a Worksheet not have any data in it, the find function will return an error (having not been able to find anything). In this particular case, the last row or column value is set to 0 by the error-handling mechanism, should they be tripped.

At the start of these functions, the Workbook and Worksheets to be acted upon are made the active Workbook and the active Worksheet. The reason for this is that if the Worksheet and Workbook are not active, the functions will not work. This is most likely a bug in the VBA environment. Usually, if the Workbook and Worksheet to be acted upon are specified within the command, it is not necessary that they be active. Case in point: the previous EmptySheet function will work fine even if the sheet to be examined is not the active sheet.

Observe also that, once the position of the last row or column is found, it is selected and then utilized in an R1C1 type of format. The previously defined functions (from Chapter 1) of FindRow and FindCol are able to then return the appropriate row or column number.

Finally, notice that these functions are declared as type Long. This is because Excel can have over 64,000 rows in a Worksheet, and the integer data type has a maximum value of 32,767. It is being mindful of little details such as this that make the difference between robust routines and routines that are subject to failure.

It is just as likely to have the need to determine the last possible row that will contain data in a particular column, or the last possible column that will contain data in a particular row. This can easily be done by applying a *range* to the search for a nonempty cell as is shown in the next two functions. Here, it is also necessary to add a third passed parameter to indicate which column or row the determination will be made in.

```
Function LastRowinColX(Workbook, Worksheet, columnX As Integer)
As Long
'This function will return the last ROW in the Worksheet of Workbook
On Error GoTo Fix_it:
ActivateWorkbook (Workbook)
Sheets(Worksheet).Activate
Workbooks(Workbook).Worksheets(Worksheet) _
    .Range(Cells(1, columnX), Cells(Rows.Count, columnX))
    . Find(What:="*", _
    SearchDirection:=xlPrevious, SearchOrder:=xlByRows).Activate
LastRowinColX=FindRow(Selection.Address(ReferenceStyle:=xlR1C1))
Exit Function
Fix_it:
```

```
'Debug.Print "Error Description: "; Error; " - Error Code: "; Err

LastRowinColX = 0
End Function

Function LastColinRowX(Workbook, Worksheet, RowX As Long) As Long

'This function will return the last COLUMN in the Worksheet
of Workbook

On Error GoTo Fix_it:
ActivateWorkbook (Workbook)
Sheets(Worksheet).Activate
Workbooks(Workbook).Worksheets(Worksheet) _
  .Range(Cells(RowX, 1), Cells(RowX, Columns.Count))
  .Find(What:="*", _
  SearchDirection:=xlPrevious,
  SearchOrder:=xlByColumns).Activate

LastColinRowX =
FindCol(Selection.Address(ReferenceStyle:=xlR1C1))

Exit Function
Fix_it:
'Debug.Print "Error Description: "; Error; " - Error Code: "; Err

LastColinRowX = 0
End Function
```

Figure 3.10 shows the Worksheet info sample application in action, where it returns the maximum row value and the maximum column value for the entire Worksheet, the maximum row value for any specific column in the active worksheet and the maximum column value for any specific row in the active Worksheet. The reader can utilize the functions contained in this example to aid in the construction of their own custom programs that require information pertaining to specific Worksheet parameters.

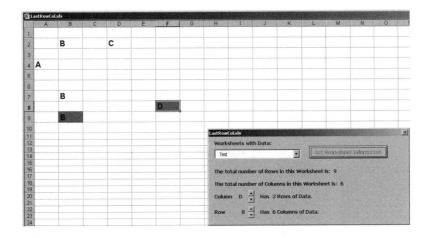

FIGURE 3.10 The Worksheet info sample application in action.

3.9 EXTRACTION OF DATA USING LANDMARKS AND LOOPING STRUCTURES

The reason so much emphasis has been placed on determining the vital parameters of a Worksheet is because without being able to determine certain Worksheet characteristics, it is impossible to extract information from a particular Worksheet. In order to do any meaningful analysis, it is necessary to read information from a Worksheet and take action upon it, often using functions that will be discussed at length in Chapter 4. Here, an example will be given regarding how to utilize some of the previously discussed information in a practical example.

Figure 3.11 depicts a fictitious data file that might be generated by some sort of machine. This data file contains labels that are denoted by being surrounded by brackets []. This file contains four principal sections; they are: [Screen Title], [Samples], [RawData], and [Results]. The [Screen Title] area contains a single item in the row following it, which is the name of the instrument the data was acquired on. The [Samples] area contains information about each sample run. This includes the position (Row and Column) on the plate the sample was taken from, as well as the molecular weight (MW) of the sample. The [RawData] section contains the raw data acquired by this instrument for this run. In this file, 10 samples were run. Notice that, in the [RawData] section, up to 14 measurements and a baseline may be acquired for each sample.

	A	B	C	D	E	F	G	H	I	J	K	L	M	N	O	P
1	[ScreenTitle]															
2	Electron Scanner 01															
3																
4	[Samples]	[Row]	[Col]	[MW]												
5	Sample-01	A	1	250.729												
6	Sample-02	A	3	310.781												
7	Sample-03	A	4	340.81												
8	Sample-04	A	7	363.849												
9	Sample-05	A	8	389.887												
10	Sample-06	A	9	363.849												
11	Sample-07	A	11	324.812												
12	Sample-08	B	1	377.794												
13	Sample-09	B	2	374.872												
14	Sample-10	B	3	353.854												
15																
16	[RawData]	B	1	2	3	4	5	6	7	8	9	10	11	12	13	14
17	Sample-01	0.092	0.102	0.112	0.122											
18	Sample-02	0.092	0.102	0.112	0.122	0.132	0.142	0.152	0.162	0.172	0.182	0.192				
19	Sample-03	0.091	0.101	0.111	0.121	0.131	0.141	0.151								
20	Sample-04	0.091	0.101	0.111	0.121	0.131	0.141	0.151	0.161	0.171	0.181	0.191	0.201	0.211		
21	Sample-05	0.090	0.100	0.110	0.120	0.130	0.140	0.150	0.160	0.170	0.180	0.190	0.200	0.210	0.220	0.230
22	Sample-06	0.090	0.100	0.110	0.120	0.130	0.140	0.150	0.160	0.170						
23	Sample-07	0.091	0.101	0.111	0.121	0.131	0.141	0.151	0.161	0.171	0.181	0.191	0.201			
24	Sample-08	0.090	0.100	0.110	0.120	0.130	0.140	0.150	0.160	0.170	0.180	0.190	0.200	0.210	0.220	0.230
25	Sample-09	0.090	0.100	0.110	0.120	0.130										
26	Sample-10	0.096	0.106	0.116	0.126	0.136	0.146	0.156	0.166							
27																
28	[Results]	[Max]														
29	Sample-01	0.122														
30	Sample-02	0.192														
31	Sample-03	0.151														
32	Sample-04	0.211														
33	Sample-05	0.230														
34	Sample-06	0.170														
35	Sample-07	0.201														
36	Sample-08	0.230														
37	Sample-09	0.130														
38	Sample-10	0.166														

FIGURE 3.11 Sample Worksheet for extraction example.

The [Results] area contains a list of the samples and the maximum value of the measurements taken for that sample (which will always be the last measurement acquired for a sample). In this particular report, there are only 10 samples, but suppose a report could contain any number of samples up to a certain maximum. Reports such as this are routinely generated in industry, and knowing how to navigate through such a Worksheet to obtain the information required for an analysis is critical for the construction of useful macros.

As an example, a macro to accomplish the following will be constructed. Suppose the application requires the [Results] section of the report to be enhanced. Perhaps it would be desirable not only to have the maximum measurement listed for each sample but also the number of measurements, the molecular weight, and the plate position (Row and Column) for each sample to be listed in the [Results] section of the report. The extraction example will demonstrate how to accomplish all of these.

At first, it might seem that constructing a routine to accomplish these tasks would be a major undertaking, but, in reality, nearly all the code to accomplish this task has already been written in both this section and the previous chapters. The correct way to attack data extraction and placement problems such as this is to first determine the "landmarks" that will be present in the data files (Figure 3.12). Landmarks are text or numerical references that will always be present within a

	A	B	C	D	E	F	G	H	I	J	K	L	M	N	O	P
1	[ScreenTitle]															
2	Electron Scanner 01															
3																
4	[Samples]	[Row]	[Col]	[MW]												
5	Sample-01	A	1	250.729												
6	Sample-02	A	3	310.781												
7	Sample-03	A	4	340.81												
8	Sample-04	A	7	363.849												
9	Sample-05	A	8	389.887												
10	Sample-06	A	9	363.849												
11	Sample-07	A	11	324.812												
12	Sample-08	B	1	377.794												
13	Sample-09	B	2	374.872												
14	Sample-10	B	3	353.854												
15																
16	[RawData]	B	1	2	3	4	5	6	7	8	9	10	11	12	13	14
17	Sample-01	0.092	0.102	0.112	0.122											
18	Sample-02	0.092	0.102	0.112	0.122	0.132	0.142	0.152	0.162	0.172	0.182	0.192				
19	Sample-03	0.091	0.101	0.111	0.121	0.131	0.141	0.151								
20	Sample-04	0.091	0.101	0.111	0.121	0.131	0.141	0.151	0.161	0.171	0.181	0.191	0.201	0.211		
21	Sample-05	0.090	0.100	0.110	0.120	0.130	0.140	0.150	0.160	0.170	0.180	0.190	0.200	0.210	0.220	0.230
22	Sample-06	0.090	0.100	0.110	0.120	0.130	0.140	0.150	0.160	0.170						
23	Sample-07	0.091	0.101	0.111	0.121	0.131	0.141	0.151	0.161	0.171	0.181	0.191	0.201			
24	Sample-08	0.090	0.100	0.110	0.120	0.130	0.140	0.150	0.160	0.170	0.180	0.190	0.200	0.210	0.220	0.230
25	Sample-09	0.090	0.100	0.110	0.120	0.130										
26	Sample-10	0.096	0.106	0.116	0.126	0.136	0.146	0.156	0.166							
27																
28	[Results]	[Max]														
29	Sample-01	0.122														
30	Sample-02	0.192														
31	Sample-03	0.151														
32	Sample-04	0.211														
33	Sample-05	0.230														
34	Sample-06	0.170														
35	Sample-07	0.201														
36	Sample-08	0.230														
37	Sample-09	0.130														
38	Sample-10	0.166														

FIGURE 3.12 Landmarks in the sample data file.

specific type of data file. The landmarks may not always be in the same place (row- and columnwise) for every data file, but the important point is that they will always be present at a specific juncture in every data file of the type to be examined. To accomplish this task, the following landmarks will be utilized: [Samples], [RawData], and [Results]. Although [Screen Title] is also a landmark in this data file (and there are some other landmarks as well), it is not necessary to determine its location for this task.

The first step is to declare some variables to hold the vital parameters that will be needed to manipulate information around the Worksheet.

```
Dim ExtractBook As String
Dim ResultsRow As Integer, ResultsCol As Integer
Dim RawDataRow As Integer, RawDataCol As Integer
Dim SamplesRow As Integer, SamplesCol As Integer
Dim ptrRow As Integer, LoopCounter As Integer
```

`ExtractBook` is a string variable utilized to hold the name of the Workbook being accessed. Utilizing global string variables suffixed with "Book" is an excellent way to manage the names of Workbooks that will be utilized in a project. The next six variables are set up to hold the positions (Row and Column) of each landmark. Finally, a variable called `ptrRow` is declared. This nomenclature denotes a "pointer" to a "row." The function of this variable is to "point" (or store) the row number of interest (or row number to be acted upon). Such pointers will be utilized a great deal in this text, and most errors in automated data analysis result when pointers are not pointing to where the designer intended them to point. `LoopCounter` is a variable that simply keeps track of how many times a `While` or `Do` loop has cycled.

The first task is obviously to load a file to be acted upon. This is easily accomplished utilizing the code discussed in Chapter 1.

```
'Load Data File
ExtractBook = LoadWithDialog("xls", "c:\", "Load the
Extract.xls file for this demo!")
```

Once the file is loaded, it is necessary to determine the location of the landmarks within the file. Location here meaning the starting row and column positions of the landmarks. Again, this is easily accomplished utilizing the code in Chapter 1.

```
'Determine Landmark Positions
ResultsRow = FindRow(FindPos(ExtractBook, ActiveSheet.Name,
"[Results]", False, False))
ResultsCol = FindCol(FindPos(ExtractBook, ActiveSheet.Name,
"[Results]", False, False))
RawDataRow = FindRow(FindPos(ExtractBook, ActiveSheet.Name,
"[RawData]", False, False))
RawDataCol = FindCol(FindPos(ExtractBook, ActiveSheet.Name,
"[RawData]", False, False))
SamplesRow = FindRow(FindPos(ExtractBook, ActiveSheet.Name,
"[Samples]", False, False))
SamplesCol = FindCol(FindPos(ExtractBook, ActiveSheet.Name,
"[Samples]", False, False))
```

With the locations of the landmarks established, the modifications to the [Results] area of the Worksheet can now be performed. First, the number of measurements is determined for each sample. A header "[No.]" is written into the column above where the number of measurements is to be stored.

```
'Determine and Write In Number of Measurements
Workbooks(ExtractBook).Worksheets(ActiveSheet.Name).Cells(Results
Row, 3) = "[No.]"
'Loop Structure to Determine number of measurements
ptrRow = RawDataRow + 1: LoopCounter = 1

While
Workbooks(ExtractBook).Worksheets(ActiveSheet.Name).Cells(ptr
Row, 1).Value <> ""
'Integrity Check - Only Add Data if Samples are IDENTICAL

    If
    Workbooks(ExtractBook).Worksheets(ActiveSheet.Name).Cells
    (ResultsRow + LoopCounter, 1) = _

    Workbooks(ExtractBook).Worksheets(ActiveSheet.Name).Cells
    (RawDataRow + LoopCounter, 1) Then

    Workbooks(ExtractBook).Worksheets(ActiveSheet.Name).Cells
    (ResultsRow + LoopCounter, 3) = _

    LastColinRowX((ExtractBook), (ActiveSheet.Name), (ptrRow)) - 2

    Else

    MsgBox
Workbooks(ExtractBook).Worksheets(ActiveSheet.Name).Cells(Results
Row + LoopCounter, 1) & _

" <> " &
Workbooks(ExtractBook).Worksheets(ActiveSheet.Name).Cells(RawData
Row + LoopCounter, 1), _

vbExclamation + vbOKOnly, "Samples Don't Match!!! Data Not
Extracted"

    End If

    ptrRow = ptrRow + 1: LoopCounter = LoopCounter + 1
Wend
```

Some degree of explanation is in order for the foregoing routine. First, the function `LastColinRow` is utilized to determine the number of measurements. In the [RawData] section, the number of measurements for each sample will be the number of columns populated for a particular sample's row *minus* two. Two is subtracted because the first two columns in this section hold the sample name and a baseline (denoted B) in the Worksheet.

It is also important to look at the looping structure utilized to access and acquire the data. While "hard limits" could be established using the locations determined by the landmarks, it is far simpler to do the following. Because each section of the report is separated by an empty row, a While loop can be utilized that starts looping at the section of interest ([RawData]) and continues looping until a blank cell is encountered. Notice that `ptrRow` is initially set to equal `RawDataRow + 1` because the samples begin one row below the landmark [RawData]. The variable `LoopCounter` simply keeps track of how many times the loop has cycled, which, when utilized with other landmark locations, can determine data locations to be extracted or written to.

FIGURE 3.13 Sample integrity check message box.

The most important feature of this routine, however, is the integrity check. It is very easy to acquire and place data without regard to its validity. Doing so is an extremely dangerous task; in fact, it could be considered the "unsafe sex" of computation. In the foregoing example, the compound names in each section are run sequentially from top to bottom. Just because the sample names are *supposed to* run sequentially from top to bottom in each section, it does not mean *they will*. If the data that are supposed to match one sample is pulled from another section of the report without verification, erroneous information could be compiled in the report. Therefore, the following (If/Then) statement was added to check and make sure the samples are in fact identical between sections.

```
If
Workbooks(ExtractBook).Worksheets(ActiveSheet.Name).Cells
(ResultsRow + LoopCounter, 1) = _
Workbooks(ExtractBook).Worksheets(ActiveSheet.Name).Cells
(RawDataRow + LoopCounter, 1) Then
```

If the samples are equivalent in each section, then the information (in this case the number of measurements) is added to the results area of the Worksheet. If the samples are not equivalent, then a message box such as that depicted in Figure 3.13 is shown, and the information for that sample pair is omitted from the results area. Integrity checks such as this are critical when preparing automated reports, especially when such reports will be uploaded to a database. Databases on things such as tests for drug-like properties can cost enormous sums of money to build, and once the integrity of such a database is compromised, it is extremely difficult (if not impossible) to fix it. The maintenance of good quality data is everything for those who work in the scientific world.

The information pertaining to the plate position and molecular weight of each sample is simply copied from the [Samples] section to the [Results] section. The code accomplishing this is

```
'Now Populate Row, Column, and Molecular Weight Fields
Workbooks(ExtractBook).Worksheets(ActiveSheet.Name).Cells(Results
Row, 4) = "[Row]"
Workbooks(ExtractBook).Worksheets(ActiveSheet.Name).Cells(Results
Row, 5) = "[Col]"
Workbooks(ExtractBook).Worksheets(ActiveSheet.Name).Cells(Results
Row, 6) = "[MW]"
ptrRow = SamplesRow + 1: LoopCounter = 1
While
Workbooks(ExtractBook).Worksheets(ActiveSheet.Name).Cells
(ptrRow, 1).Value <> ""
'Integrity Check - Only Add Data if Samples are IDENTICAL
```

```
        If
        Workbooks(ExtractBook).Worksheets(ActiveSheet.Name).Cells
        (ResultsRow + LoopCounter, 1) = _
        Workbooks(ExtractBook).Worksheets(ActiveSheet.Name).Cells
        (SamplesRow + LoopCounter, 1) Then
          'For [Row]
  Workbooks(ExtractBook).Worksheets(ActiveSheet.Name).Cells(Results
  Row + LoopCounter, 4) = _

  Workbooks(ExtractBook).Worksheets(ActiveSheet.Name).Cells(Samples
  Row + LoopCounter, 2)

  'For [Col]
  Workbooks(ExtractBook).Worksheets(ActiveSheet.Name).Cells(Results
  Row + LoopCounter, 5) = _

  Workbooks(ExtractBook).Worksheets(ActiveSheet.Name).Cells(Samples
  Row + LoopCounter, 3)

  'For [MW]
  Workbooks(ExtractBook).Worksheets(ActiveSheet.Name).Cells(Results
  Row + LoopCounter, 6) = _

  Workbooks(ExtractBook).Worksheets(ActiveSheet.Name).Cells(Samples
  Row + LoopCounter, 4)

  Else

  MsgBox
  Workbooks(ExtractBook).Worksheets(ActiveSheet.Name).Cells(Results
  Row + LoopCounter, 1) & _

  " <> " &
  Workbooks(ExtractBook).Worksheets(ActiveSheet.Name).Cells(Samples
  Row + LoopCounter, 1), _

    vbExclamation + vbOKOnly, "Samples Don't Match!!! Data Not
    Extracted"
    End If
    ptrRow = ptrRow + 1: LoopCounter = LoopCounter + 1
  Wend
```

Like the earlier routine, this block of code also creates headers for the information to be added to the results section of the report ([Row],[Col], and [MW]). It also utilizes the same loop-type structure and pointer mechanisms as the earlier routine. An integrity-checking mechanism is also present as in the previous routine. In this instance, however, the integrity check could be taken one step further. It would also be prudent to check to make sure the ([Row], [Col], and [MW]) columns in the sample area are where they are supposed to be (columns 2, 3, and 4, respectively). Doing so would not require a great deal more code, and would ensure that not only was the information copied from an identical sample, but that the specific pieces of information to be copied are present at their expected locations.

The sample application can be run by selecting ADA->Chapter 3->Extract Example. The user will be prompted to load a file (Use "Extract.xls" from the CD-ROM). Once the file has been loaded, the manipulations described previously will be performed, and the Worksheet shown in Figure 3.14 will be automatically produced. Notice that the Worksheet now has additional columns in the results section indicating the total number of measurements taken, the plate positions (row and column), and the MW of each sample.

	A	B	C	D	E	F	G	H	I	J	K	L	M	N	O	P
1	[ScreenTitle]															
2	Electron Scanner 01															
3																
4	[Samples]	[Row]	[Col]	[MW]												
5	Sample-01	A	1	250.729												
6	Sample-02	A	3	310.781												
7	Sample-03	A	4	340.81												
8	Sample-04	A	7	363.849												
9	Sample-05	A	8	389.887												
10	Sample-06	A	9	363.849												
11	Sample-07	A	11	324.812												
12	Sample-08	B	1	377.794												
13	Sample-09	B	2	374.872												
14	Sample-10	B	3	353.854												
15																
16	[RawData]	B	1	2	3	4	5	6	7	8	9	10	11	12	13	14
17	Sample-01	0.092	0.102	0.112	0.122											
18	Sample-02	0.092	0.102	0.112	0.122	0.132	0.142	0.152	0.162	0.172	0.182	0.192				
19	Sample-03	0.091	0.101	0.111	0.121	0.131	0.141	0.151								
20	Sample-04	0.091	0.101	0.111	0.121	0.131	0.141	0.151	0.161	0.171	0.181	0.191	0.201	0.211		
21	Sample-05	0.090	0.100	0.110	0.120	0.130	0.140	0.150	0.160	0.170	0.180	0.190	0.200	0.210	0.220	0.230
22	Sample-06	0.090	0.100	0.110	0.120	0.130	0.140	0.150	0.160	0.170						
23	Sample-07	0.091	0.101	0.111	0.121	0.131	0.141	0.151	0.161	0.171	0.181	0.191	0.201			
24	Sample-08	0.090	0.100	0.110	0.120	0.130	0.140	0.150	0.160	0.170	0.180	0.190	0.200	0.210	0.220	0.230
25	Sample-09	0.090	0.100	0.110	0.120	0.130										
26	Sample-10	0.096	0.106	0.116	0.126	0.136	0.146	0.156	0.166							
27																
28	[Results]	[Max]	[No.]	[Row]	[Col]	[MW]										
29	Sample-01	0.122	3	A	1	250.7										
30	Sample-02	0.192	10	A	3	310.8										
31	Sample-03	0.151	6	A	4	340.8										
32	Sample-04	0.211	12	A	7	363.8										
33	Sample-05	0.230	14	A	8	389.9										
34	Sample-06	0.170	8	A	9	363.8										
35	Sample-07	0.201	11	A	11	324.8										
36	Sample-08	0.230	14	B	1	377.8										
37	Sample-09	0.130	4	B	2	374.9										
38	Sample-10	0.166	7	B	3	353.9										

FIGURE 3.14 The extract example final Worksheet.

3.10 SUMMARY

In this section the reader is taught how to manipulate information within the Worksheet environment through the proper use of variables. The reader is shown how to acquire information about Worksheets utilizing VBA code to answer questions such as (1) is a particular Worksheet hidden or visible? (2) how many Worksheets within a particular Workbook contain data? and (3) how many rows or columns of data does a Worksheet, or row, or column within a Worksheet contain?

The proper scope and use of variables are covered along with special variable typing such as static variables, object variables, and arrays. The use of functions is touched on in this chapter but will be covered in depth in Chapter 4. The differentiation between passing parameters by value or by reference is explained along with the possible ramifications of being ignorant as to what the difference is. The use of object variables is covered, which allows the reader to make changes on entire collections of objects by means of looping structures as opposed to coding all changes individually. Looping structures are also covered as they relate to populating arrays and other variables with Worksheet data, or their use in searching for particular discrete pieces of information within a Worksheet. The utilization of landmarks to locate information within data files is also explained. The generation of custom reports is touched on in this chapter, but this topic will be covered in detail in Chapters 6 and 10. At this point, the reader is able to accomplish the following tasks utilizing VBA code: (1) Extract and place information anywhere within a Worksheet/Workbook pair in the Excel environment. (2) Load, save, and manipulate files and acquire file information within the Excel environment. (3) Perform basic manipulations on data within Worksheets such as sorting, copying, and deleting data. (4) Preparing very basic reports. The next few chapters will focus on allowing the reader to perform more complex analyses utilizing custom functions from both within Excel and utilizing third-party calculation tools.

4 Utilizing Functions in Excel

4.1 INTRODUCTION

Functions are the cornerstones of all data analysis. Data analysis requires that calculations take place. Calculations require functions to execute the computations. In Excel there are two kinds of functions. The first type of function is that which resides in the cell of a Worksheet. These types of functions typically carry out mathematical operations utilizing parameters contained within other Worksheet cells. However, Worksheet cell functions are capable of doing far more than just ordinary mathematical operations on Worksheet data, and this chapter will introduce the reader to some techniques for enhancing Worksheet cell functions to accomplish far more than they were ever intended to. The second type of function is a VBA function, which is simply a function written in VBA code. What is not known to many users is that VBA functions can in fact be utilized from within Worksheet cells. The methods for accomplishing this will be covered in this chapter as well.

Functions are wonderful tools for computation, but often not enough thought is given as to what should be done if something goes wrong when a function tries to calculate a value. Trying to calculate values utilizing cells that are empty can wreak havoc with even the best intended functions. For nearly every calculation, there are conditions that will cause it to become unstable and return an invalid result or crash. Sometimes, determining when such conditions will occur proves to be impossible. Because such an eventuality can occur, it is best to prepare for it up front whenever possible. The reader will be shown how to handle errors that can occur when utilizing functions.

4.2 CREATING AND UTILIZING VBA FUNCTIONS IN CODE

Unlike a subroutine, a function is capable of returning a value that can be of any variable type. Most functions return strings, integers, single or double precision values, or Boolean values (True/False). Because functions return a value, they are often called by assigning a variable of the same type equal to them, or substituting a function where a value would normally go. A simple example function:

```
Function DivideBy(ByVal Numerator As Double, ByVal Denominator
As Double) As Double
'Simple Function to divide two numbers
DivideBy=Numerator/Denominator
End Function
```

This function will divide two numbers where the numerator and denominator are passed parameters. The function can be executed from the immediate (or debug) window by typing the following:

```
Debug.Print DivideBy(3,4)
```

which will return this result:

```
0.75
```

Here the function is placed at the end of the Debug.Print statement where some kind of value is expected. Functions can also be called by being assigned to a variable. Typing the following in the immediate window will illustrate this concept:

```
result = DivideBy(3,4): Debug.Print "result = ";result
```

The following will then be printed:

```
result = 0.75
```

Here the returned value of the function (which is a double precision number) is assigned to the variant variable result, which is then printed to the immediate window. When utilizing functions, it is important to understand that the values they return are of a certain type. Therefore, when a variable is set equal to the value of a function, the variables type must be the same as that which will be returned by the function (preferred), or the variable must be of type variant (less desirable method — see Chapter 3).

Functions can also be utilized to *replace* a passed parameter in a subroutine. Consider the following subroutine:

```
Sub TestScore(Score As Single)
  Debug.Print "Your Test Score is: "; Trim(Score * 100); "%"
End Sub
```

Here, a score is required that is a decimal number between 0 and 1. In this instance, such a score would be the number of correct answers divided by the total number of problems in a test. Thus, Score could be replaced by the earlier function DivideBy. Suppose a student were to answer 82 of 100 answers correctly. The subroutine TestScore could be called in this fashion:

```
TestScore(DivideBy(82,100))
Your Test Score is: 82%
```

Notice that the passed parameter Score is only dimensioned as Single. The consequence of this is that the double precision value returned by the DivideBy function will be truncated to a single precision value when the single precision passed parameter Score is assigned the returned value from the double precision function DivideBy.

One final aspect to consider when programming in Excel VBA, which is not an issue when programming in regular Visual Basic, is that the return value of a function can also be assigned to a cell within a Worksheet. Thus statements such as this are valid in Excel VBA.

```
Cells(1,3).Value = DivideBy(500,10)
```

Here a value of 50 (500/10 = 50) is assigned to row 1, col 3 (C) of the active Worksheet.

```
ActiveWorkbook.ActiveSheet.Cells(1,4).Value = DivideBy(500,100)
```

Such a statement can, of course, be explicitly referenced by preceding the Cells method with Workbook and Worksheet references as just previously shown. In this example, the value 5 is placed in row 1, col 4 (D) of the active Worksheet.

4.3 HANDLING ERRORS IN VBA FUNCTIONS

The DivideBy function has performed as expected in the examples given so far. However, consider this case:

```
Debug.Print DivideBy(100,0)
```

Typing the above in the immediate window will cause an error (Figure 4.1). Division by zero, while possible in theory — $\lim_{n \to 0} (\frac{x}{n}) = \infty$ (when x > 0) — is not allowed in Excel. Although few would make such an overt mistake within a function call, suppose the passed parameters to the DivideBy function were calculated by other functions that were extremely complex in nature. It is easy to see how such a call could mistakenly happen. All functions have roots, and depending upon what numbers are plugged in where, a function could return a value of zero. Therefore, when performing

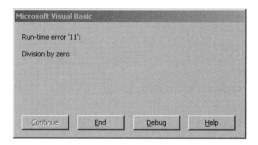

FIGURE 4.1 Division by zero in VBA causes problems.

division, it is good to have an error-handling mechanism. The following is an improved division function that has an error-handling mechanism.

```
Function Division(ByVal Numerator As Double, ByVal Denominator
As Double, _
ByVal ReturnInfiniteSymbol As Boolean) As Variant
'Function to divide two numbers with Error Handling Mechanism
On Error GoTo Fix_It:
Division = Numerator/Denominator
Exit Function

Fix_It:
If Error <> "Division by zero" Then Exit Function
Select Case ReturnInfinite
  Case True
    'Set the Value of the Function Equal to the Infinite Symbol
    Division = "¥"
  Case False
    'Return an insanely Large Number
    Division = 1E+308
End Select
End Function
```

Here, a choice is given on how to handle the division by zero error. If an infinite result occurs, either an infinite symbol can be returned (∞) or the largest possible number in Excel VBA (10^{308}) will be returned. What is returned is dependent upon the setting of the passed Boolean parameter ReturnInfiniteSymbol. The fact that the function is set as type variant allows it to return either a symbol or a number. The infinity symbol (∞) will only show up as such if the Worksheet cell's font is set to type symbol. In the immediate window, the default font is never symbol, so a weird ASCII character (¥) will be printed for the infinity symbol (∞) in the debug window.

```
Debug.Print Division(10,0,True)
¥
Debug.Print Division(10,0,False)
1E+308
```

Notice that the error-handling mechanism checks to make sure the error encountered was in fact a division by zero error. This is very important because, if another type of error should occur, this subsection within the function is not equipped to handle it. Notice in Figure 4.1 that both an error number and an error description are given in the dialog box. Either can be utilized from VBA to determine the type of error that has occurred. The `Err` object will return the error number (in the case of division by zero, this is 11) of the last logged error. The `Error` Object will return the text message associated with the last logged error (in the case of division by zero, this is "Division by zero"). `Error` is also function that will return the text message associated with a particular error number. Thus the statement:

```
Debug.Print Error(11)
Will produce:
Division by zero
```

What good does it do to return an insanely large number if division by zero has occurred? Mathematically, this number (10^{308}) is the best representation of the true result of division by zero (infinity) that is possible in Excel VBA. By catching the error and returning something, the program will not be halted in its tracks. Chances are good that such a large number returned will throw off all subsequent calculations in the report to such a degree that any scientist or engineer looking at the results will immediately be clued in that something went drastically wrong. Such an error-checking mechanism could even be built into a Worksheet, and the process for doing so will be discussed later in this chapter.

4.4 ADDING A FUNCTION TO A WORKSHEET CELL USING VBA CODE

In the previous section, it was shown how to write the value returned from a VBA function into a Worksheet cell. In some instances, this may be acceptable, but often it is preferable to write a function into a Worksheet cell. Consider the following VBA code:

```
Sub TimesTable(ByVal Number As Integer, wkbook, wksheet)
'Create a Simple Times Table
Dim row As Integer
For row = 1 To 12
      Workbooks(wkbook).Worksheets(wksheet).Cells(row, 1) = row
      Workbooks(wkbook).Worksheets(wksheet).Cells(row, 2) = "x"
      Workbooks(wkbook).Worksheets(wksheet).Cells(row, 3) = Number
      Workbooks(wkbook).Worksheets(wksheet).Cells(row, 4) = "="
      Workbooks(wkbook).Worksheets(wksheet).Cells(row, 5) = row
      * Number
Next row
End Sub
```

This subroutine will create a simple times table for the number passed to it in the specified Workbook/Worksheet pair as illustrated in Figure 4.2.

The problem with the Worksheet illustrated here is that if one of the numbers in columns A or C is changed, the answer in column E will remain the same! This is because a static value is written into the Worksheet cell, in this case (`Number * row`). This is a less than ideal situation, especially when creating a custom report in which users may change values that *should* alter subsequent calculations within the Worksheet. Rather than write the static answer into column E, it would be much better to write a formula to each Worksheet cell in column E that multiplies columns A and C. The `TimesTable` subroutine shown previously will be redone to illustrate how to accomplish this.

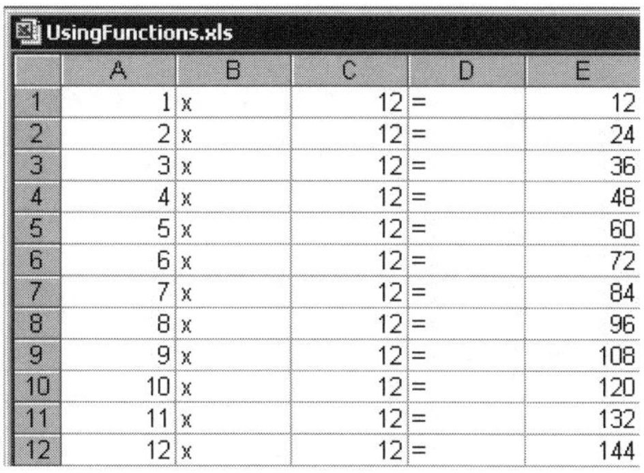

UsingFunctions.xls

	A	B	C	D	E
1	1	x	12	=	12
2	2	x	12	=	24
3	3	x	12	=	36
4	4	x	12	=	48
5	5	x	12	=	60
6	6	x	12	=	72
7	7	x	12	=	84
8	8	x	12	=	96
9	9	x	12	=	108
10	10	x	12	=	120
11	11	x	12	=	132
12	12	x	12	=	144

FIGURE 4.2 A simple times table.

In this particular case, each cell in column E that is to hold an answer would have a formula like this:

```
=A[row]*C[row]
```

where [row] is equal to the row number. For example, the formula in row 1 would be the text string:

=A1*C1

Formulas are in fact text strings and can be written into any particular cell as such. Every formula must begin with the "=" sign!

Note: Functions attached to cells are also referred to as formulas. When a mathematical expression is attached to a Worksheet cell, it can be called either a function or a formula. The terms are utilized interchangeably in this chapter. Oddly enough, when attaching a function to a cell, the property, Formula is utilized. Yet, when the formula wizard appears, and a formula is to be selected, the notation "More Functions …" is utilized to allow the user access to a GUI that shows all the available formulas, but the GUI is titled "Paste Function" (see Figure 4.4). This is very confusing.

The following subroutine will add any formula to any cell. How can one determine what the text string should be for a particular formula? Simply click on the cell that is to utilize the formula and then click on the "=" button on the formula bar as illustrated in Figure 4.3.

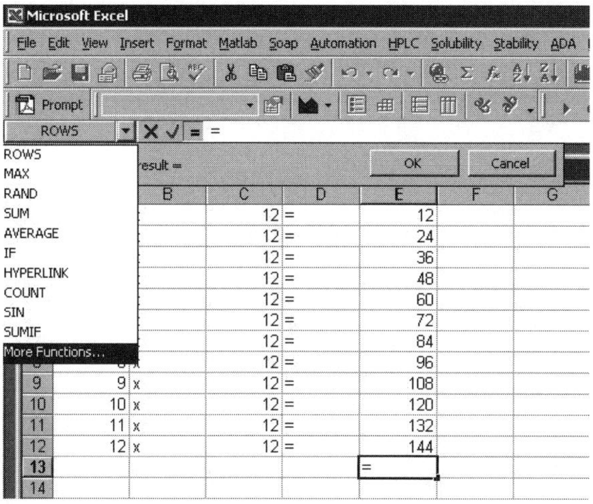

FIGURE 4.3 Determining the proper formula syntax.

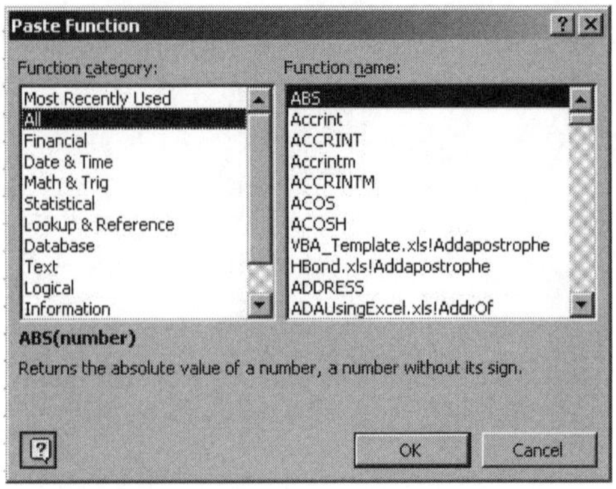

FIGURE 4.4 Choosing a function in Excel.

A dialog will pop up, which shows the most recently used functions. If the user selects "More Functions . . .", then the dialog box in Figure 4.4 will appear, which allows the user to choose any function present in Excel (both built-in functions or VBA functions).

```
Sub AddFormula(wkbook, wksht, row, col, formula$)
'Subroutine adds formula$ to the specified cell in a Worksheet
ActivateWorkbook (wkbook)
Worksheets(wksht).Select
'Ensure formula starts with "=" !
If Left(formula$, 1) <> "=" Then
    formula$ = "=" & formula$
End If
Range(Cells(row, col), Cells(row, col)).Select
ActiveCell.formula = formula$
End Sub
```

This subroutine will add the formula passed to it to the specified location (cell). The subroutine checks to make sure that the passed formula string begins with "=." A common mistake when specifying formulas is to omit the equals sign. The following is the code for the improved version of the subroutine `TimesTable`, which will write the appropriate formula into column E.

```
Sub TimesTable(ByVal Number As Integer, wkbook, wksheet)
'Create a Simple Times Table
Dim row As Integer, formula$
For row = 1 To 12
    Workbooks(wkbook).Worksheets(wksheet).Cells(row, 1) = row
    Workbooks(wkbook).Worksheets(wksheet).Cells(row, 2) = "x"
    Workbooks(wkbook).Worksheets(wksheet).Cells(row, 3) = Number
    Workbooks(wkbook).Worksheets(wksheet).Cells(row, 4) = "="
    'Build Function String
    formula$ = "=A" & Trim(Str(row)) & "*C" & Trim(Str(row))
    Call AddFormula(wkbook, wksheet, row, 5, formula$)
Next row
End Sub
```

Notice that the formula is built on each pass of the loop utilizing a string variable, and that the formula is added to the cell via the AddFormula subroutine. If the AddFormula subroutine were not utilized, adding the formula to the cell would take a form like this:

```
Workbooks(wkbook).Worksheets(wksheet).Cells(row, 5) = "=A1*C1"
```

It is somewhat confusing to have two equals signs next to each other, which is why the proper syntax is pointed out here. If the revised `TimesTable` subroutine is run, and values in columns A or C are changed, the corresponding answer will change in column E. This is a superior way to prepare a report, as it will dynamically change and automatically update the results whenever report parameters are changed. Another added benefit of preparing reports like this is that they allow "what if" scenarios to be run. This can be done by changing certain parameters within the report and viewing how the changes made affect the outcome of the report parameters.

4.5 CREATING ADDITIONAL BUILT-IN FUNCTIONS FOR EXCEL

The functions (or formulas) included with Excel are rather limited in scope. What is not widely documented is that it is possible to create new formulas for use in Excel by using VBA functions. To illustrate how to accomplish this, a formula will be created to solve the quadratic formula. (No such formula is currently built into Excel.) A quadratic equation is of the form

$$ax^2 + bx + c = 0 \qquad (4.1)$$

The roots of such an equation can be solved by means of the quadratic equation:

$$= \frac{-b \pm \sqrt{b^2 - 4ac}}{2a} \qquad (4.2)$$

First, a function needs to be created that will solve the quadratic equation given in (4.2). Notice that, in Equation 4.2, there is a $\pm$ sign. This is because the quadratic equation calculates two roots. The function created must, therefore, be capable of calculating two roots. This poses a problem because, if the function is to be utilized as a formula, it can only return one value to the cell. In the function QuadEq, a mechanism has been added to allow the user to select which root will be returned. This is done by means of the passed parameter RootNumber. If RootNumber is set to 1, then the $\pm$ term is set to (+), and vice versa if RootNumber is set to 2.

```
Function QuadEq(ByVal a As Double, ByVal b As Double, ByVal
c As Double, Optional RootNumber As Integer = 1) As Double
'Solves the Quadratic Equation, User must specify which Root
to Return
'Root 1 = + sqroot term : Root 2 = - sqroot term
If RootNumber <> 1 And RootNumber <> 2 Then Exit Function
Select Case RootNumber
   Case 1
      QuadEq = (-b + Sqr((b ^ 2 - 4 * a * c))) / (2 * a)
   Case 2
      QuadEq = (-b - Sqr((b ^ 2 - 4 * a * c))) / (2 * a)
End Select
End Function
```

Consider the following case:

$$x^2 - 2x - 15 \qquad\qquad (4.3)$$

The roots of this equation are: 3,5
Because (4.3) factors into $(x + 3)(x - 5)$
In Equation 4.3, the parameters for the quadratic equation are a = 1, b = 2, and c = 15.
If the following two commands are typed:

```
Debug.Print "Root 1: "; QuadEq(1, 2, 15, 1)
Debug.Print "Root 2: "; QuadEq(1, 2, 15, 2)
```

The following results are displayed:

Root 1: 5
Root 2: 3

This function works well except in cases where the roots are complex. Consider the case of

$$x^2 + 4 \qquad\qquad (4.4)$$

This function does not factor rationally; its roots are complex. Trying to run the QuadEq function on the function produces an error. The error occurs because, in this case, the function is trying to take the square root of a negative number, which is considered an illegal operation in VBA.

This function can be modified to have the ability to factor a function with complex roots. To do so first requires that the function check if the quadratic equation to be factored has complex roots. This can be easily done by looking at the discriminant of the quadratic Equation 4.2. The discriminant is

$$b^2 - 4ac \qquad\qquad (4.5)$$

which is the portion of the quadratic equation under the radical sign. If the discriminant is negative, then the roots to be solved for are complex, and special actions must be taken to factor the function. In this instance, the roots to be solved for must be split into real and complex portions. Depending upon which root is to be returned, the real and complex portions of the root must be added or subtracted from each other and returned as a single complex number with both real and imaginary portions. To do this, the improved function must return a variant or a string. The function AdvQuadEq is capable of calculating the roots of quadratic equations even when the roots are complex.

```
Function AdvQuadEq(ByVal a As Double, ByVal b As Double, ByVal
c As Double, _
Optional RootNumber As Integer = 1) As Variant
'Solves the Quadratic Equation, User must specify which Root
to Return
'Root 1 = + sqroot term : Root 2 = - sqroot term
Dim discriminant As Double, complex As Boolean
Dim real As Single, cmplx As Single
If RootNumber <> 1 And RootNumber <> 2 Then Exit Function
discriminant = b ^ 2 - 4 * a * c
```

```
If discriminant < 0 Then
    discriminant = Abs(discriminant)
    complex = True
End If
Select Case RootNumber
    Case 1
      Select Case complex
        Case True
            real = -b / (2 * a): cmplx = Sqr(discriminant) / (2 * a)
            AdvQuadEq = real & " + " & cmplx & "i"
      Case False
            AdvQuadEq = (-b + Sqr((discriminant))) / (2 * a)
End Select
    Case 2
      Select Case complex
        Case True
            real = -b / (2 * a): cmplx = Sqr(discriminant) / (2 * a)
            AdvQuadEq = real & " - " & cmplx & "i"
      Case False
            AdvQuadEq = (-b - Sqr((discriminant))) / (2 * a)
  End Select
End Select
End Function
```

If the following function is used to calculate the roots of Equation 4.4, the following will be returned:

Root 1: 0 + 2i
Root 2: 0 − 2i

The resulting roots from this factorization contain only an imaginary component. Notice that i is appended to the end of the complex number to signify that it is the imaginary portion of the complex number. (Some mathematical references utilize j as opposed to i.) Recall that i is

$$i = \sqrt{-1} \tag{4.6}$$

If Equation 4.3 is modified to have complex roots by changing the (15) to (+15), the subroutine returns the following results, which contain both a real and imaginary component:

Root 1: 1 + 3.741657i
Root 2: 1 − 3.741657i

For functions with real roots, a single number is returned as in the earlier example. Now that a function has been developed to calculate the roots of quadratic equations using the quadratic formula, the question is, how can this function be utilized by a cell within a Worksheet as a formula? To utilize an Excel VBA formula as a function, click the "=" sign to edit the formula for that particular cell and then select "More Functions" from the drop-down function dialog box. A GUI such

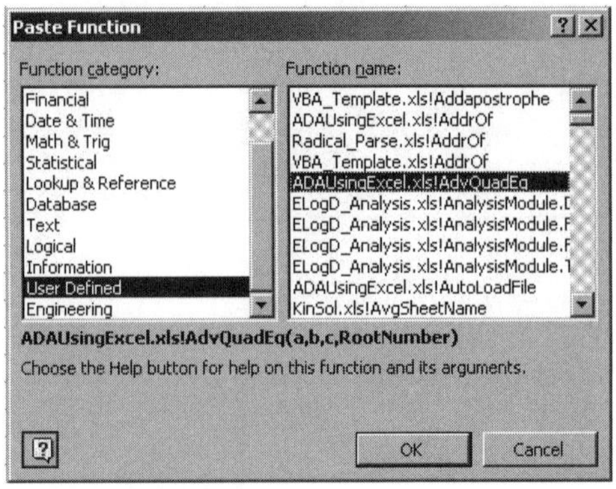

FIGURE 4.5 Choosing a VBA function as a formula.

as that shown in Figure 4.5 will appear. Under the Function category, "User Defined" must be selected. Notice that the VBA functions in the user-defined category are prefixed with the name of the Workbook that the function resides in. To utilize the function just created to factor quadratic equations, the item labeled "ADAUsingExcel.xls!AdvQuadEq" must be selected.

Suppose the roots of Equation 4.4 wish to be computed in a Worksheet. The coefficients of Equation 4.4 can be placed in the Worksheet as shown in Figure 4.6. The coefficients for Equation 4.4

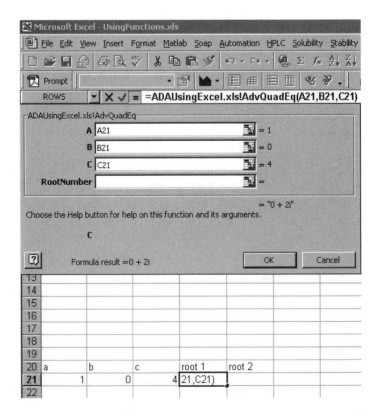

FIGURE 4.6 Calculating roots using the new VBA function.

18					
19					
20	a	b	c	root 1	root 2
21	1	0	4	0 + 2i	0 - 2i
22	1	0	-4	2	-2
23					
24					

FIGURE 4.7 Utilizing the VBA function within a Worksheet.

starting from the highest degreed polynomial are 1,0,4, which are placed in cells A21, B21, and C21, respectively. After choosing the function "ADAUsingExcel.xls!AdvQuadEq" in the formula wizard, the coefficients must be entered in the formula wizard along with the root number (which may be omitted if the default root of 1 is to be calculated) as shown in Figure 4.6.

The results of utilizing this function to calculate the two roots of Equation 4.4 are shown in Figure 4.7. The two roots of Equation 4.4 are shown in cells D21 and E21. In the next row (22), Equation 4.4 is altered to have real roots by changing coefficient c to -4 (i.e., $x^2 - 4$). Here the two real roots of 2 and -2 are returned to cells D22 and E22, respectively.

One last point worth mentioning about custom functions is the lack of a description for their use within the formula dialog box. In Figure 4.6, notice that the custom VBA function "ADAUsingExcel.xls!AdvQuadEq" contains only the following instructions or descriptions for use: "Choose the Help button for help on this function and its arguments." Obviously, this is not very helpful for someone wishing to utilize the function who is unfamiliar with its parameters or use. A custom description for a VBA function can be added to the formula dialog box. This is done by selecting Tools->Macro->Macros. Although functions are *not* listed in the Macros dialog box, Excel is cognizant of their existence. In the macro name text box, the full (function) macro name must be typed, which, in this instance, is "ADAUsingExcel.xls!AdvQuadEq" (see Figure 4.8).

If the function name has been typed with no errors (and should always be prefixed with WorkbookName!), when the Options button is pressed in the Macros dialog box (GUI), the dialog box shown in Figure 4.9 will appear.

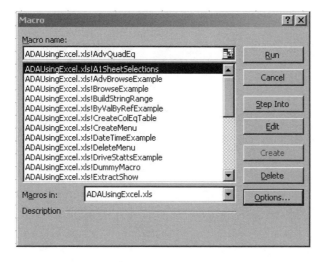

FIGURE 4.8 The macro dialog box for accessing function options.

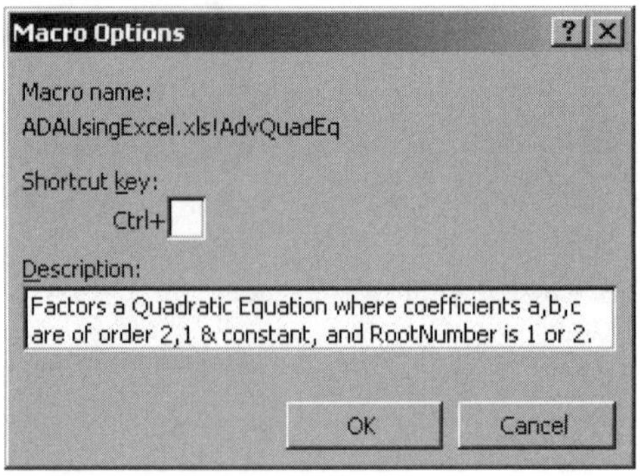

FIGURE 4.9 Setting macro (function) descriptions.

Within this dialog box, both a shortcut key and a description can be set for the chosen macro (in this case a function macro). For the function created in this chapter, the description shown in Figure 4.10 was typed in the text box in the GUI shown in Figure 4.9. This description will now appear every time the function is utilized in Excel's Worksheet formula dialog box. Such descriptions are extremely useful when writing custom functions that will be utilized by numerous members of an organization.

4.6 DYNAMIC FORMATTING OF WORKSHEETS USING FUNCTIONS

One of the focal points of this text is to take information in, process it, and then present it to a user in a manner that is useful for whatever type of project is being worked on. The problem with presenting a user with a report is that the user can change things within the report. In some instances this may be what is desired, especially if a "what if" scenario needs to be analyzed. All too often, however, a parameter within a report may be inadvertently changed to a value that is invalid or outside a reasonable range. In such an instance, it would be useful to automatically

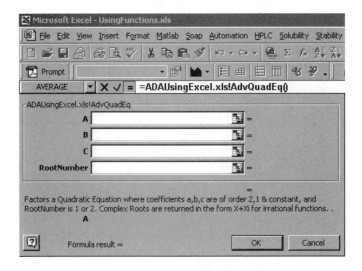

FIGURE 4.10 The formula dialog box now contains a description.

TABLE 4.1
Proposed Voltage Levels for New Cars

Volts (V)	Current (mA)	Resistance (k)
12	0.01	1200
18	0.01	1800
24	0.01	2400
30	0.01	3000
36	0.01	3600
42	0.01	4200
48	0.01	4800

apply formatting to a cell in which an invalid entry has been made. Utilizing the conditional formatting feature, Excel can dynamically change the formatting of cells based upon their contents. For example, if a cell has a value greater than a specified maximum, or is changed to a value greater than a specified maximum, the background of the cell can be made yellow (or any color for that matter), and an outline can be placed around the cell. This type of functionality obviously has a great deal of utility in the real world.

To show how to use the conditional formatting available in Excel, the Worksheet shown in Table 4.1 will be utilized as an example. This table shows proposed voltage, current, and resistance levels for future automobile electrical systems. Here, it is assumed that the current level will remain constant and the system (battery) voltage can be increased. Table 4.1 shows how each 6-V increase in system voltage will affect the resistance or load that can be driven.

Suppose, this table is contained within a Worksheet and it is desired to highlight the row with the current electrical scheme utilized by most new cars, which is 12-V. This can be accomplished by selecting from the menu Format->Conditional Formatting, which will bring up the GUI shown in Figure 4.11.

To highlight the rows that contain a system voltage of 12, the following must be done. First, in the drop-down box, the user must choose between "Formula Is" and "Cell Value Is." In this instance, it is necessary to utilize "Formula Is" option because the cells in a row must be colored based on the value present in another cell (in this case col A, row n), not just the value within their own cell. The following formula will accomplish this:

$$=IF(\$A2=12,TRUE,FALSE)=TRUE \tag{4.7}$$

In this formula, the column is fixed (denoted by the $ in the cell reference $A2). When this formula is painted across the other cells, the row number will change but the column number (A) will remain

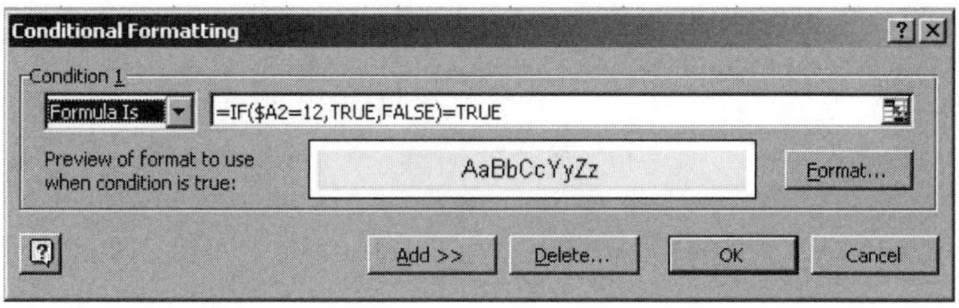

FIGURE 4.11 Setting up conditional formatting.

TABLE 4.2
Fixed Formula Cell References

Fixed Cell Range Designations		
Cell Ref	**Col**	**Row**
$A$1	Fixed	Fixed
A$1	Moving	Fixed
$A1	Fixed	Moving
A1	Moving	Moving

fixed. This allows each individual cell to set a format based on the value of the cell contained in column 1 (*A*) of the current row. (Note: Cell references can be fixed in formulas by pressing the F4 function key. Each push of the F4 function key will cycle through every permutation of formula fixation. See Table 4.2.)

The formula {IF($A2=12,TRUE,FALSE)} in (4.7) checks to see if the value in column *A* of the current cells row is equal to 12. If it is equal to 12, True is returned; else False is returned. When this formula is utilized in the conditional formatting GUI, a value that defines the condition for when the formatting is to be applied must be specified. This is why the "=TRUE" is appended to the end of the formula. The conditional formatting shown in Figure 4.11 is then applied to cell A2 within the Worksheet (row 1 contains the labels). Once this conditional formatting has been applied to one cell, it can be applied to other cells utilizing the Format Painter tool shown in Figure 4.12. Simply click on the cell that has the format desired, click on the format painter button, and then drag the format across the cells desired.

Looking at Table 4.1, column 3 (C) is calculated utilizing column A and column B. The resistance is calculated using Ohm's law:

$$v = i\text{R} \qquad (4.8)$$

which states that the voltage is equal to the current multiplied by the resistance. Rearranging Equation 4.8 yields

$$R = \frac{v}{i} \qquad (4.9)$$

Column C (Resistance) is therefore equal to column A divided by column B. Suppose that a report is produced and, during the process of handling the report, someone accidentally writes a static value to an entry in column C, replacing the formula. If a value in column A or column B were to change, the corresponding value in column C would not change. Such a condition could be very dangerous, especially when "what if" scenarios are to be run in the report. To eliminate

FIGURE 4.12 The Format Painter tool.

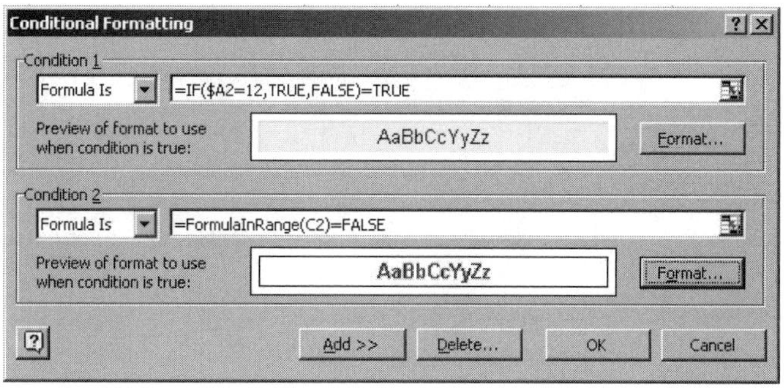

FIGURE 4.13 Setting a second conditional formatting statement.

the possibility of such an occurrence, it would be desirable to color the font of a cell red and place an outline around the cell if any cell in column C of this Worksheet does not contain a formula. This can be done rather easily by adding a second condition of conditional formatting for the cells in column C. However, to accomplish this, first a function must be created that will determine if a cell does or does not contain a formula. The following simple function will return False if every cell in the range passed to the function does not contain a formula.

```
Function FormulaInRange(Cells As Range) As Boolean
'Function will return True if cell or range of cells contains
a formula
FormulaInRange = Cells.HasFormula
End Function
```

Using this function, a second conditional formatting statement can be applied as shown in Figure 4.13.

$$=FormulaInRange(C2)=FALSE \qquad\qquad (4.10)$$

When the FormulaInRange function returns False for a cell, its font will be set to bold red and an outline will be placed around the cell as illustrated in Figure 4.13.

By using the format painter to apply this conditional formatting statement to all the cells utilized in column C, any cell in column C that is inadvertently replaced with a static value instead of a formula will stand out like a sore thumb by having its font turned bold red and being outlined as shown in Figure 4.14.

	A	B	C
1	Volts (V)	Current (mA)	Resistance (kΩ)
2	12	0.01	1200
3	18	0.01	1800
4	24	0.01	2400
5	30	0.01	3000
6	36	0.01	3600
7	42	0.01	**4200**
8	48	0.01	4800

FIGURE 4.14 Conditional formatting in action.

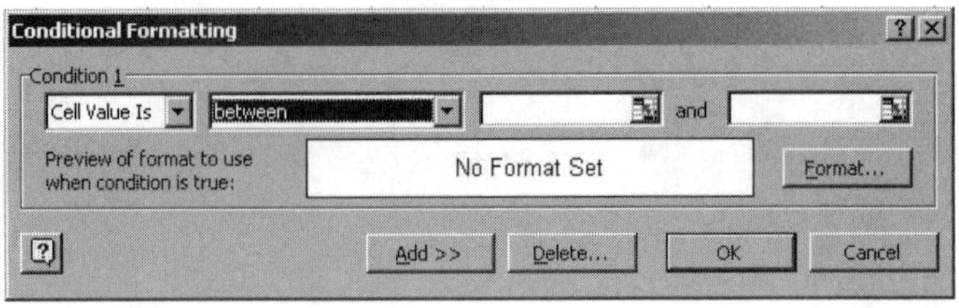

FIGURE 4.15 Conditional formatting using cell values.

In the foregoing example, a static value was placed in cell C7, and the row containing the standard 12-V electrical scheme is highlighted with a yellow background.

One last point worth mentioning when setting up conditional formatting statements manually in Excel: In addition to the "Formula Is" option, there is also a "Cell Value Is" option. In general, the "Cell Value Is" options utility is limited to setting formatting based only on the conditions that occur within the cell to which the formatting conditions are to be attached to. Figure 4.15 shows the options available when utilizing the "Cell Value Is" option for conditional formatting.

4.7   APPLYING DYNAMIC FORMATTING USING VBA

The conditional formatting techniques described in the previous section add tremendous utility to Worksheets in that they allow Worksheets to automatically red-flag entries within them that may be incorrect. Applying such formatting by hand is impractical when a series of reports must be prepared on an ongoing basis. One solution would be to utilize a Workbook template that would have the proper conditional formatting functions built into the Worksheets. Such an approach only works when the cells to utilize the conditional formatting will be confined to a known fixed region for every report. Although this may be the case in some instances, often the region of cells a parameter or group of parameters will occupy will fluctuate due to a number of factors. In such instances, utilizing a template with "hard" or fixed conditional formatting functions will not work. For the ultimate in flexibility, a method of applying conditional formatting dynamically to Worksheets utilizing VBA is required. This section shows how to accomplish this.

The first requirement is having a subroutine that will apply conditional formatting utilizing Excel VBA code. The following subroutine will accomplish this and will add up to three specified conditional formatting conditions, which are specified via the passed parameter ConditionNumber.

```
Sub ApplyCondFormat(wkbook, wksheet, ByVal cellrange As Range,
formula$, Optional backcolor As Integer, _
Optional fontBold As Boolean = False, Optional fontItal As
Boolean = False, Optional fontColor As Integer = 1, _
Optional Outline As Boolean, Optional ConditionNumber As Integer
= 1, Optional DeleteExisting As Boolean = False)

Workbooks(wkbook).Activate
Worksheets(wksheet).Activate 'No Auto Fill ?

'Set Workbook\Worksheet to be ACTED upon
Workbooks(wkbook).Worksheets(wksheet).Activate

'Now Select the Range to be Acted Upon
cellrange.Select
```

```
'Add Formula Here (Conditions under which to apply formatting)
If DeleteExisting = True Then Selection.FormatConditions.Delete
Selection.FormatConditions.Add Type:=xlExpression,
Formula1:=formula$

'If Specified Add Background Color Here
If IsMissing(backcolor) = False Then
    Selection.FormatConditions(ConditionNumber).Interior.Color
    Index = backcolor
End If
'Set Font Attributes Here
With Selection.FormatConditions(ConditionNumber).Font
 .Bold = fontBold
 .Italic = fontItal
 .ColorIndex = fontColor
End With
'For Cell Outlining
With Selection.FormatConditions(ConditionNumber).Borders
 .LineStyle = xlContinuous
 .Weight = xlThin
 .ColorIndex = xlAutomatic
End With
End Sub
```

Conditional Formatting does have some limitations, however, and the biggest limitation is that functions that reside in other Worksheets or Workbooks cannot be utilized as conditional formatting criteria, as shown in Figure 4.16. A sample application has been created to demonstrate the use and limitations of conditional formatting using VBA macro structures. The sample can be run by selecting ADA->Chapter 4->Conditional Formatting from the Excel menu. This will bring up the GUI shown in Figure 4.17.

Pushing the first button titled "Load Example Sheet" will automatically load a template file required for the example. The code and mechanics of how this is accomplished is discussed in detail in Chapter 5. The second button titled "Highlight Default" applies conditional formatting, which

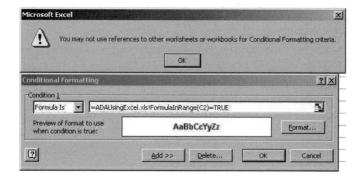

FIGURE 4.16 Conditional formatting limited to native functions.

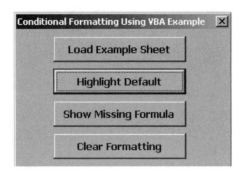

FIGURE 4.17 Conditional formatting example GUI.

highlights the rows that contain information about the default vehicle voltage scheme (12 V). The code that generates this formatting is as follows:

```
Private Sub Button_Default_Click()
Dim row As Integer, col As Integer
Dim cellrange$, formatformula$
 For row = 2 To 8
   For col = 1 To 3
     cellrange$ = Chr(64 + col) & Trim(str(row))
     formatformula$ = "=IF($A" & Trim(str(row)) &
     "=12,TRUE,FALSE)=TRUE"
     Call ApplyCondFormat(FormatBook, ActiveSheet.Name,
     Range(cellrange$), formatformula$, 6)
   Next col
Next row
End Sub
```

Looking at the foregoing snippet of code, notice that a `For/Next` looping structure is utilized to build a string that represents the cell range and cell formatting formula to be utilized when the `ApplyCondFormat` subroutine is called.

The trouble arises when trying to code a VBA routine to apply a conditional formatting routine, which utilizes a custom function. Recall that such functions may not reside in another Workbook or Worksheet and must be native to the Workbook or Worksheet they will be utilized in. One potential workaround for this problem is to construct templates that have the required custom functions contained within them as a code module. This is done with the Workbook template "DynamicFormat.xlt," which contains the function `FormulaInRange`.

The button titled "Show Missing Formula" *should* add a conditional reference using the `FormulaInRange` function to highlight a cell in red to indicate when it contains a static value instead of a formula. Such an indicator is very useful because static values will not automatically update should other parameters in the Worksheet be altered. A bug within the Excel VBA framework, however, prevents this code from executing properly. The code to attempt to execute this formatting is as follows:

```
Private Sub Button_Formula_Click()
Dim row As Integer, col As Integer
Dim cellrange$, formatformula$
```

```
For row = 2 To 8
    cellrange$ = "C" & Trim(str(row))
    formatformula$ = "=FormulaInRange(" & cellrange$ & ")=False"
    Call ApplyCondFormat(FormatBook, ActiveSheet.Name,
    Range(cellrange$), formatformula$, , True, , 3, , 2)
Next row
End Sub
```

The unfortunate thing that happens with this code is that the conditional formatting function appears to be applied ok to cell C2; however, when the code reaches cell C3, it abnormally terminates. Cell C3 contains the conditional formatting function, but no specified formatting will be applied should the condition prove true (Figure 4.18). Cells C4–C8 then have no conditional formatting applied will them. Even more strange is that, although the correct formatting appears to be applied in cell C2, it will not function correctly if the value in cell C2 is changed from a formula to a static value (Figure 4.18).

Finally, the "Clear Formatting" button erases all of the formatting that may have been applied to the region of cells in the Worksheet that contain data. The code to accomplish this is

```
Range("A1:C8").Select
Selection.FormatConditions.Delete
```

This section would not be complete without a method to automate the format painter tool shown in Figure 4.12. Here the conditional formatting statements present in the first range are copied into the second range. For example, here, the conditional formatting statements present in cell A1 will be copied over the range of A1:H30.

```
Call PaintFormats(Range("A1"),Range("A1:H30"))
```

The subroutine which accomplishes this is:

```
Sub PaintFormats(ByVal CellWithFormat As Range, ByVal
FormatArea As Range)
'This subroutine will paint formats from a given cell to a
Range of cells
CellWithFormat.Select
Selection.Copy
FormatArea.Select
Selection.PasteSpecial Paste:=xlFormats, Operation:=xlNone
End Sub
```

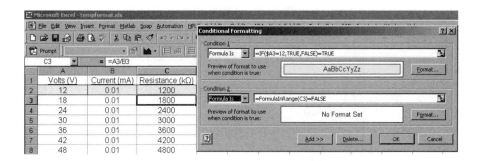

FIGURE 4.18 Abnormal termination in VBA code results in unset formats.

Using this subroutine, it is possible to work around the bug involved in writing conditional formatting statements with functions by writing a conditional formatting statement to a single cell and painting it to the other cells. The following subroutine modification will accomplish this:

```
Private Sub Button_Formula_Click()
Dim row As Integer, col As Integer
Dim cellrange$, formatformula$
  For row = 2 To 8
    cellrange$ = "C" & Trim(str(row))
    formatformula$ = "=FormulaInRange(" & cellrange$ & ")=False"
    Call ApplyCondFormat(FormatBook, ActiveSheet.Name,
    Range(cellrange$), formatformula$, , True, , 3, , 2)
    'WorkAround - After writing formula to a single cell paint
    to others
    Call PaintFormats(Range("C2"), Range("C2:C8"))
    Exit Sub
  Next row
End Sub
```

Looking at Figure 4.19, notice that when a value in column C is replaced with a static value as opposed to a formula, the font in the cell containing the static value is colored red. This alerts the user that a formula is missing from that cell and that the results will not be automatically updated should other values in columns A and B change. In Figure 4.19, cell C7 had its formula replaced with a static value and its contents have turned red.

One last and very important aspect of conditional formatting is to point out that the first condition overrides the second, and the second overrides the third. What is meant by this is that if the first condition is True, then only the formatting associated with the first condition will be implemented. If the second condition is also true, its formatting will be ignored. Looking at Figure 4.20, if cell C2 is changed from a formula to a number, the font in cell C2 should be turned red by the criteria set under conditional formatting rule number 2. However, because conditional formatting rule number 1 is true (cell A2 = 12), the background of cell C2 is highlighted yellow in accordance with conditional formatting rule number 1, and the formatting associated with conditional formatting rule number

	A	B	C
1	Volts (V)	Current (mA)	Resistance (kΩ)
2	12	0.01	1200
3	18	0.01	1800
4	24	0.01	2400
5	30	0.01	3000
6	36	0.01	3600
7	42	0.01	4200
8	48	0.01	4800

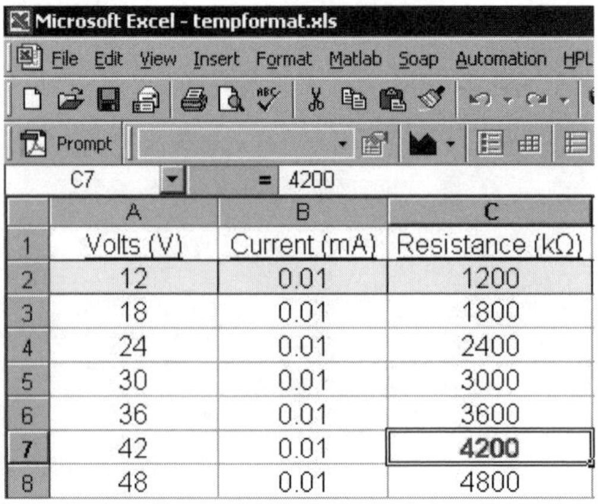

FIGURE 4.19 Conditional format workaround results.

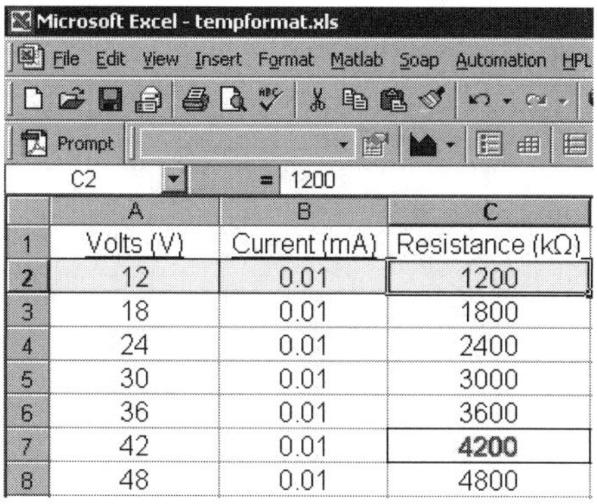

FIGURE 4.20 One condition overriding another.

2 (turning the font red) is ignored. Notice, however, that if the first conditional formatting statement is made false by setting cell A2 to anything but 12, conditional formatting rule 2 will be implemented and the font will turn red in cell A2. (After changing the value in cell A2, it may be necessary to hit F9 to update the Worksheet values depending upon the user's settings.) This is illustrated in Figure 4.21.

4.8   USING THE MACRO RECORDER TO CAPTURE A PROCESS

Although many of the properties and methods available for use in Excel VBA are documented, there are many processes that are not. Ironically, the more complicated (and thus useful) a process is, the less likely it is that it is documented. Fortunately, there is a mechanism built into Excel VBA that allows users to record *actions* taken in Excel as macros. The beauty of this utility is that it not only allows the user to repeat the actions recorded upon demand, but also allows the user to look at the underlying VBA code that causes those actions to occur. For extremely technical and complicated processes, this functionality is extremely useful in that it allows the user to record and capture segments of VBA code that will accomplish a desired task. To demonstrate how this process is accomplished, the macro recorder will be utilized to record macros that perform linear regression on a data set. It will then be shown how to modify the code generated by the macro recorder to create generic subroutines that can be used over and over again.

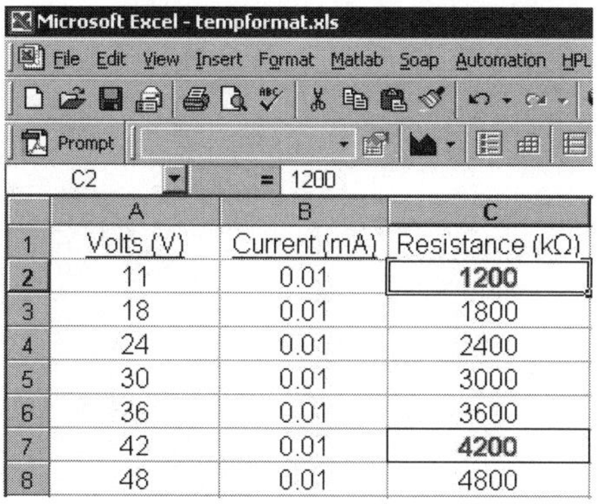

FIGURE 4.21 Conditional formatting using 2nd statement rules.

TABLE 4.3
Linear Regression Sample Data Set

X	Y
1	0.37
2	2.87
3	2.3
4	3.2
5	5.11
6	5.42
7	7.33
8	8.27
9	9.79
10	9.18

For this example, a sample data set must be constructed. This is a relatively easy task to accomplish. Here, a nearly linear line will be constructed, with some degree of error intentionally introduced. The data set constructed for this sample is shown in Table 4.3.

Here the X values are simply created using 1 to 10. The corresponding Y value is equal to the X value ± a value from 0 to 1. This is accomplished rather easily using the formula

$$= A2 + RANDBETWEEN(-100,100)/100$$

where A2 is really A[row] ({i.e., for row 3 = A3}).

Excel has a tool for accomplishing linear regression in a relatively painless manner with the Analysis Toolpak. In order to utilize the regression tool, the analysis toolpak must be installed. To do so, choose from the menu Tools->Add-Ins and make sure the Analysis Toolpak and Analysis Toolpak VBA is checked as shown in Figure 4.22.

The Analysis Toolpak not only has tools built into Excel, but contains VBA routines that can be utilized to execute all the mechanisms built into the Analysis Toolpak. Unfortunately, use of the Analysis Toolpack VBA routines is not very well documented. The macro recorder, however, can

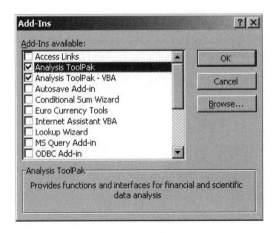

FIGURE 4.22 Enabling use of the Analysis Toolpak.

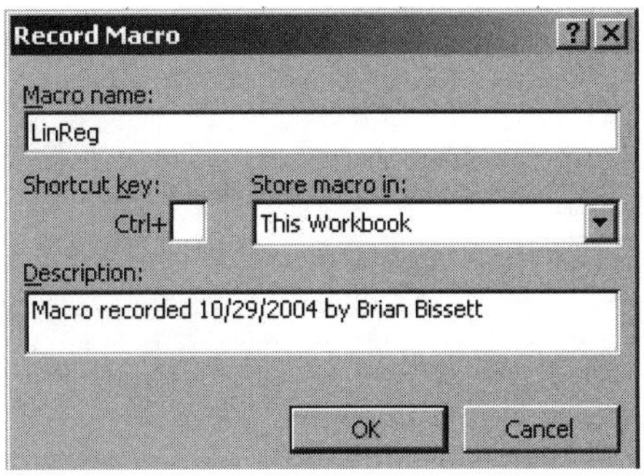

FIGURE 4.23 Recording a new macro named "LinReg."

be used to capture the VBA commands utilized to accomplish a process. In this example, the macro recorder will capture the commands involved in utilizing the linear regression tool. The macro recorder can be started by selecting Tools->Macro->Record New Macro from the menu. This will then bring up the dialog box shown in Figure 4.23.

The macro recording process will start after the user fills out the macro name to be recorded (in this case "LinReg") and presses the OK button. When the recording process starts, a GUI like that shown in Figure 4.24 will appear. Excel will continue recording the VBA commands required to automate the processes the user is carrying out until the square stop button is pressed (much like a tape recorder).

To record the linear regression process, choose from the menu Tools->Data Analysis and select Regression in the dialog box as shown in Figure 4.25.

The sample linear regression data set is stored in the Workbook "linearfitexample.xls." If the user fills out the GUI as shown in Figure 4.26 and then presses the OK button, the following macro will be recorded:

```
Sub LinReg()
' LinReg Macro
' Macro recorded 11/01/2004 by Brian Bissett
    Application.Run "ATPVBAEN.XLA!Regress",
    ActiveSheet.Range("$B$1:$B$11"), _
    ActiveSheet.Range("$A$1:$A$11"), False, True, ,
    "FitResults", False, _
    False, False, True, , False
End Sub
```

Looking at this code, recording the linear regression process tells the user several things. First, the regression is run in the template "ATPVBAEN.XLA," and the subroutine name is Regress. This subroutine has numerous parameters, and it is not immediately apparent looking at the code what

FIGURE 4.24 When recording a new macro, this GUI will appear.

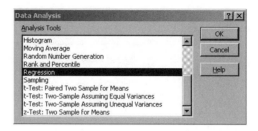

FIGURE 4.25 Launching the linear regression tool.

each parameter does. Even using the autocomplete feature as shown in Figure 4.27 provides little insight into the function of each parameter.

Of the 30 arguments passed to the Regress subroutine, only the X data range, Y data range, and new Worksheet name are readily apparent by looking at the recorded macro code. One way to discern what all the other parameters mean is to execute the code and randomly vary the Boolean results and observe the outcome. A more practical way is to conduct a search for the macro "ATPVBAEN.XLA!Regress" on the Internet and see if any practical information can be located.

When the Regress subroutine in the Analysis Toolpak is run, a line is fitted to the specified data of the form

$$y = mx + b \tag{4.11}$$

where m = slope, and b = y intercept.

On the analysis sheet created by the Regress macro, the slope is for some reason labeled X (instead of m) (cell B18, if the results are placed into a new Worksheet), and the intercept is appropriately labeled intercept (cell B17, if the results are placed into a new Worksheet). For the

	A	B	C	D	E	F	G	H
1	X	Y	**Regression**					? X
2	1	0.37						
3	2	2.87	Input					OK
4	3	2.3	Input Y Range:		$B$1:$B$11			
5	4	3.2	Input X Range:		$A$1:$A$11			Cancel
6	5	5.11						
7	6	5.42	☑ Labels		☐ Constant is Zero			Help
8	7	7.33	☐ Confidence Level	95	%			
9	8	8.27						
10	9	9.79	Output options					
11	10	9.18	○ Output Range:					
12			● New Worksheet Ply:	FitResults				
13			○ New Workbook					
14								
15			Residuals					
16			☐ Residuals		☐ Residual Plots			
17			☐ Standardized Residuals		☑ Line Fit Plots			
18								
19			Normal Probability					
20			☐ Normal Probability Plots					
21								

FIGURE 4.26 Executing the regression tool on the sample data set.

```
Sub LinReg()
'
' LinReg Macro
' Macro recorded 11/01/2004 by Brian Bissett
'

'
    Application.Run "ATPVBAEN.XLA!Regress", ActiveSheet.Range("$B$1:$B$11"), _
        ActiveSheet.Range("$A$1:$A$11"), False, True, , "FitResults", False, _
        False, False, True, , False
End Sub      Run([Macro], [Arg1], [Arg2], [Arg3], [Arg4], [Arg5], [Arg6], [Arg7], [Arg8], [Arg9], [Arg10], [Arg11], [Arg12], [Arg13], [Arg14], [Arg15], [Arg16], [Arg17], [Arg18], [Arg19], [Arg20],
             [Arg21], [Arg22], [Arg23], [Arg24], [Arg25], [Arg26], [Arg27], [Arg28], [Arg29], [Arg30])
```

FIGURE 4.27 The autocomplete function shows all the arguments but gives few clues as to their use.

sample data set in this example, m = 1.032 and b = 0.292. A graph of the original and predicted (or fitted) data sets is shown in Figure 4.28.

Directly below the graph is a table of the original X/Y data set plus the Y predicted (or fitted) values for the data set using the aforementioned slope and intercept. The final column $Y - Y_{Predicted}$ are also referred to as the residuals. In this instance, if a residual is positive, the fitted point lies below the actual data, and if a residual is negative, it lies above the actual data. The r^2 value (also known as the coefficient of determination value or COD) is located in cell B5 if the results are placed into a new Worksheet. The r^2 value is a measurement that describes just how well the curve is fitted to the data, where a 1.0 is a perfect fit and a 0 is indicative of zero correlation. In this instance, $r^2 = 0.957$, which is considered a fair but less than optimal fit to the data. In general, when $r^2 > 0.98$, the fit is considered good, and when $r^2 < 0.90$, it may be indicative that no meaningful correlation exists. The fit in this example, therefore, provides a useful but not necessarily precise prediction of any particular data point.

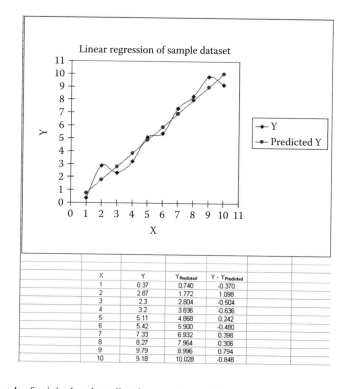

X	Y	$Y_{Predicted}$	$Y - Y_{Predicted}$
1	0.37	0.740	-0.370
2	2.87	1.772	1.098
3	2.3	2.804	-0.504
4	3.2	3.836	-0.636
5	5.11	4.868	0.242
6	5.42	5.900	-0.480
7	7.33	6.932	0.398
8	8.27	7.964	0.306
9	9.79	8.996	0.794
10	9.18	10.028	-0.848

FIGURE 4.28 Graph of original and predicted regression results.

4.9 CREATING A LINEAR REGRESSION TOOL USING THE VBA ANALYSIS TOOLPAK

The creation of a Linear Regression Tool would be a useful addition to Excel. Ideally, such a tool would have a GUI that would allow the user to specify the dependant (y) and independent (x) variables. The tool should produce a line graph that contains both the original data and the linearly fitted data as shown in Figure 4.28. The fitted equation and r^2 value should be placed on the graph, as well as labels for the title and axes. Finally, a data set should be generated for the fitted line with the number of points in the generated data set specified by the user.

The first step toward the construction of such a regression tool would be having a complete understanding of the "ATPVBAEN.XLA!Regress" macro. As discussed in the previous section and shown in Figure 4.27, it is nearly impossible to discern what the arguments are that are passed to this macro that resides in the analysis tool pack. There are only three means of determining what the arguments are for a black box type of function. In some instances, the function may be defined within Excel's help files; in this instance, it is not. If the developer is lucky and searches the Web or the Microsoft Knowledge Base, there may be an article someone has written or contributed that describes using the function within the VBA framework and identifies what its passed parameters control. In this instance, some articles can be found on the Web that describe *some* but not *all* of the passed parameters in the "ATPVBAEN.XLA!Regress" macro. The final method of argument identification within a black box function is the brute force method. This involves altering a single parameter at a time and observing the affects such an alteration produces on the macros output. Although such an approach may seem somewhat primitive, in some instances it may be the only means possible by which a developer can identify a black box function's passed parameters. If a black box function has a GUI associated with it, as the "ATPVBAEN.XLA!Regress" macro does (see Figure 4.26), this will lend some clues as to what the various passed parameters could be utilized for within the macro. The author utilized all of the above methods to identify the following useful passed parameters in the "ATPVBAEN.XLA!Regress" macro. The following subroutine allows custom calls to the "ATPVBAEN.XLA!Regress" macro and is self explanatory.

```
Sub MLR(wkbook$, wksheet$, RangeYs, RangeXs, Labels As Boolean, _
ConstantZero As Boolean, ConfidenceLevel%, _
Residuals As Boolean, STDResiduals As Boolean, ResidualPlots
As Boolean, _
LineFitPlots As Boolean, ProbOutput As Boolean, NormalProbPlots
As Boolean, _
Optional newsheetname As Variant = "")
Application.Run "ATPVBAEN.XLA!Regress", _
Workbooks(wkbook$).Worksheets(wksheet$).Range(RangeYs), _
Workbooks(wkbook$).Worksheets(wksheet$).Range(RangeXs), _
Labels, ConstantZero, ConfidenceLevel%, _
newsheetname, Residuals, STDResiduals, ResidualPlots, _
LineFitPlots, False, ProbOutput, NormalProbPlots
End Sub
```

Some thought needs to be given as to what the functionality of this new tool should be. The tool should allow multiple fitting sessions. When multiple fitting sessions occur, the graphs and generated data sets should appear on separate and uniquely named Worksheets and charts. The user should be able to specify the following information for the fitting session:

1. The Ranges of the independent X and dependant Y data to be fitted
2. The total number of points in the data set to be generated using the fitted equation

3. The number of decimal places the fitted data set should utilize
4. The Option to specify the names of the new sheets and charts
5. The Option to specify labels for the axes and title on the graph
6. The Option to smooth the Original data on the graph
7. The Option to have the data displayed on the graph

With these requirements in mind, a subroutine was written with the following passed parameters to fulfill the above specifications.

```
Sub LinFitTool(XRange As String, YRange As String, TotalPoints
As Integer, _
decplaces As Integer, _
Optional Labels As Boolean = False, Optional newsheetname As
String = "", _
Optional SmoothOrigData As Boolean = True, Optional LabelX As
String = "X - Axis Label", _
Optional LabelY As String = "Y - Axis Label", Optional
LabelTitle As String = "Title", _
Optional ChartWithData As Boolean = True)
```

The subroutine is too long to list in its entirety; however, the key and critical components of the subroutine will be listed and explained in this section. The entire subroutine is contained in the accompanying CD-ROM in the "Chapter4" Module of the "ADAUsingExcel.xls" Workbook. This application also requires the construction of a GUI that allows the above options to be selected or specified. A GUI was constructed as depicted in Figure 4.29.

Looking at Figure 4.29, the X (independent) and Y (dependant) data are selected by means of RefEdit controls at the top of the GUI. A series of checkboxes allow the user to specify if the data contains a header, if the original line should be smoothed, if a graph should be included on the Worksheet containing

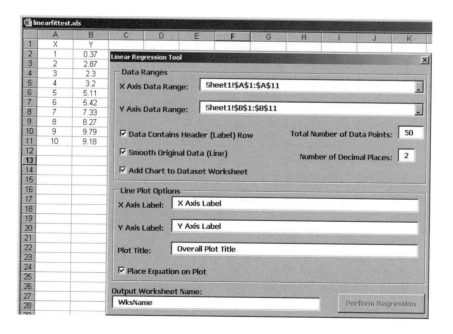

FIGURE 4.29 Linear regression tool GUI.

the generated data set, or if the fitted equation should be placed on the Graph. A series of textboxes let the user specify the number of data points to be utilized in the fitted data set, the number of decimal places to be utilized on the spreadsheet containing the fitted data set, the axis and title labels, and the new Worksheet name. All of the GUI settings are stored in the Windows Registry as shown in Chapter 2 so that they will be "remembered" from one macro execution to the next.

The macro is executed when the "Perform Regression" button is pressed. Upon execution, the macro must ensure that the Workbook\Worksheet pair pointed to by the RefEdit boxes that specify the data range is the active Workbook\Worksheet pair. This can be accomplished with the following single line of code.

```
'Make Sure Sheet in Ref Edit is Active Sheet!
Workbooks(ActiveWorkbook.Name).Worksheets(SheetInRange(XRange))
.Select
```

Next a unique Workbook must be created to hold the generated dataset of the linear regression. If a new sheet name is specified, that name will be utilized.

```
OldSheetName = ActiveSheet.Name
If newsheetname = "" Then
    DataSheet$ = CreateSheet(ActiveWorkbook.Name, "Dataset")
    Else
    DataSheet$ = CreateSheet(ActiveWorkbook.Name, newsheetname)
End If
Sheets(OldSheetName).Activate
```

It is the CreateSheet function that actually creates a new sheet. This function checks to see if the specified sheet name already exists. If the sheet does not exist, the function creates it. If the sheet does exist, then the function appends a number to the sheet name and tries to create it until such time as a unique sheet name is generated. The function returns the name of the created Worksheet.

```
Function CreateSheet(wkbook$, sheetname$) As String
'Creates a sheet Named sheetname$ in wkbook$, iff sheet exists
append number suffix and retry
'Returns the Name of the Created Sheet
Dim newsheetname$, ii As Integer
If SheetExists(wkbook$, sheetname$, True) = False Then
    CreateSheet = sheetname$
    Exit Function
End If
For ii = 1 To 1000
    newsheetname$ = sheetname$ & Trim(str(ii))
    If SheetExists(wkbook$, newsheetname$, True) = False Then
        CreateSheet = newsheetname$
        Exit Function
    End If
Next ii
End Function
```

Notice that the CreateSheet function relies upon a separate function named SheetExists to determine if the new sheet name is already in existence.

```
Function SheetExists(wkbook$, sheetname$, Optional addsheetname
As Boolean = False) As Boolean
'This function will determine if sheetname$ exists in Workbook$
'If the sheet exists returns TRUE, Returns False if sheet
added or Not Present
Dim i As Integer, Count As Integer, DataBook$
Call ActivateWorkbook(wkbook$)
'rotate through all the Worksheets by getting the count of
Worksheets
For i = 1 To Worksheets.Count
  ActiveWorkbook.Worksheets(i).Activate
  If Worksheets(i).Name = sheetname$ Then
     'Sheet Exists
     SheetExists = True
     Exit Function
     End If
Next i
'If we get this far sheetname$ is NOT in wkbook$
SheetExists = False
If addsheetname = False Then
     Exit Function
   Else
     'Add and Name Sheet at the same time!
     Sheets.Add.Name = sheetname$
     SheetExists = False
     Exit Function
End If
End Function
```

With a new sheet created to hold the fitted data set, the regression can begin. This is done with the following call to the "ATPVBAEN.XLA!Regress" macro.

```
Application.Run "ATPVBAEN.XLA!Regress",
ActiveSheet.Range(YRange), _
ActiveSheet.Range(XRange), False, Labels, , "FitResults",
False, _
False, False, True, , False
```

With the linear regression complete, the fitted line parameters COD (r^2), Y-Intercept, and slope can be extracted from the results using the FindPos, FindRow, and Find Col functions. The COD and Y-Intercept can be determined rather easily by means of the following code snippets:

```
COD =
Workbooks(ActiveWorkbook.Name).Worksheets("FitResults").Cells
(FindRow(FindPos(ActiveWorkbook.Name, "FitResults", "R
Square", False, True)), FindCol(FindPos(ActiveWorkbook.Name,
"FitResults", "R Square", False, True)) + 1)
```

```
Intercept =
Workbooks(ActiveWorkbook.Name).Worksheets("FitResults").Cells(
FindRow(FindPos(ActiveWorkbook.Name, "FitResults",
"Intercept", False, True)),
FindCol(FindPos(ActiveWorkbook.Name, "FitResults",
"Intercept", False, True)) + 1)
```

The slope is somewhat more problematic to locate because the label that identifies it will vary depending upon if the data has a header row of labels. If the data has a header row of labels, the slope will be identified using that label. If no labels are present, the slope will be identified with heading "X Variable 1." The following code snippet compensates for this irregularity. (The reason for this irregularity is unknown.)

```
'Finding Slope will depend upon if Labels Exist
If Labels = True Then
    'Data has label, use value in first X cell
    SlopeSearch$ =
    Workbooks(ActiveWorkbook.Name).Worksheets(SheetInRange(XRange)).
    Cells(FirstRowinRange(XRange),    ColumninRange(XRange))
    Slope =
    Workbooks(ActiveWorkbook.Name).Worksheets("FitResults").Cells
    (FindRow(FindPos(ActiveWorkbook.Name, "FitResults",
    SlopeSearch$, False, True)),
    FindCol(FindPos(ActiveWorkbook.Name, "FitResults",
    SlopeSearch$, False, True)) + 1)
    Else
    'No labels, search for "X Variable 1"
    Slope =
    Workbooks(ActiveWorkbook.Name).Worksheets("FitResults").Cells
    (FindRow(FindPos(ActiveWorkbook.Name, "FitResults", "X
    Variable 1", False, True)),
    FindCol(FindPos(ActiveWorkbook.Name, "FitResults", "X
    Variable 1", False, True)) + 1)
End If
```

With the regression parameters extracted, a string (Plot$) can be created that contains the fitted equation and the r^2 value. This string can then be pasted onto the graph if so desired.

```
Plot$ = "y = " & (Format(Slope, "0.000")) & "*x + " &
(Format(Intercept, "0.000")) & _
Chr(10) & "R2 = " & (Format(COD, "0.000"))
```

Next, a series of statements are issued that govern the creation of the graph. These statements will not be repeated here for the sake of brevity but are present in the source code included with the text on the accompanying CD-ROM. A data set must now be constructed that represents the linear regression equation to the line. The number of points contained in this data set is specified by the user in the GUI. The first step toward creating this data set is to determine the minimum and maximum values to be contained within the data set. The following snippet of code accomplishes this:

```
If Labels = False Then
  Xmin =
  Workbooks(ActiveWorkbook.Name).Worksheets(SheetInRange(XRange)).
  Cells(FirstRowinRange(XRange), ColumninRange(XRange))
```

```
Else
  Xmin =
  Workbooks(ActiveWorkbook.Name).Worksheets(SheetInRange(XRange)).
  Cells(FirstRowinRange(XRange) + 1, ColumninRange(XRange))
End If
  Xmax =
  Workbooks(ActiveWorkbook.Name).Worksheets(SheetInRange(XRange)).
  Cells(LastRowinRange (XRange), ColumninRange(XRange))
```

To determine the location of the minimum and maximum values in the dataset, the use of four custom functions is required: `SheetInRange`, `LastRowinRange`, `FirstRowinRange`, and `ColumninRange`.

```
Function SheetInRange(RefEditTxt) As String
'Returns the SheetName in the RefEdit Controls Range
SheetInRange = Mid(RefEditTxt, 1, InStr(RefEditTxt, "!") - 1)
End Function

Function FirstRowinRange(ByVal RangeAddress) As Integer
'Returns the First Row in a Range
Dim topleft$, bottright$
Dim startrow As Integer, endrow As Integer
Dim CtrlRange As Range, ii As Integer
If InStr(RangeAddress, "$") <> 0 Then
    'Convert Range Address to R1C1 Format
    Set CtrlRange = Range(RangeAddress)
    RangeAddress = CtrlRange.Address(, , xlR1C1)
End If
If InStr(RangeAddress, ":") = 0 Then
    FirstRowinRange = 1
    Exit Function
    Else
    topleft$ = Mid(RangeAddress, 1, InStr(RangeAddress, ":") - 1)
    bottright$ = Mid(RangeAddress, InStr(RangeAddress, ":") + 1,
    Len(RangeAddress))
End If
'Determine Starting Row
For ii = 1 To Len(topleft$)
  If Mid(topleft$, ii, 1) = "C" Then
    FirstRowinRange = Val(Mid(topleft$, 2, ii - 1))
    Exit Function
  End If
Next ii
End Function

Function LastRowinRange(ByVal RangeAddress) As Integer
'Returns the Last Row in a Range
```

```
Dim topleft$, bottright$
Dim startrow As Integer, endrow As Integer
Dim CtrlRange As Range, ii As Integer
If InStr(RangeAddress, "$") <> 0 Then
    'Convert Range Address to R1C1 Format
    Set CtrlRange = Range(RangeAddress)
    RangeAddress = CtrlRange.Address(, , xlR1C1)
End If
If InStr(RangeAddress, ":") = 0 Then
    LastRowinRange = 1
    Exit Function
    Else
    topleft$ = Mid(RangeAddress, 1, InStr(RangeAddress, ":") - 1)
    bottright$ = Mid(RangeAddress, InStr(RangeAddress, ":") + 1,
Len(RangeAddress))
End If
'Determine Ending Row
For ii = 1 To Len(bottright$)
 If Mid(bottright$, ii, 1) = "C" Then
    LastRowinRange = Val(Mid(bottright$, 2, ii - 1))
    Exit Function
End If
Next ii
End Function

Function ColumninRange(ByVal RangeAddress) As Integer
'Returns the Column Number in a Range, (if more than one
column in Range returns 0)
Dim topleft$, bottright$
Dim startcol As Integer, endcol As Integer
Dim CtrlRange As Range, ii As Integer
If InStr(RangeAddress, "$") <> 0 Then
    'Set CtrlRange Range object to the range address passed
    Set CtrlRange = Range(RangeAddress)
    'Convert Range Address to R1C1 Format
    RangeAddress = CtrlRange.Address(, , xlR1C1)
End If
If InStr(RangeAddress, ":") = 0 Then
    'ColsinRange = 1
    'Exit Function
    Else
topleft$ = Mid(RangeAddress, 1, InStr(RangeAddress, ":") - 1)
bottright$ = Mid(RangeAddress, InStr(RangeAddress, ":") + 1,
Len(RangeAddress))
End If
```

```
'Determine Starting Column
For ii = 1 To Len(topleft$)
  If Mid(topleft$, ii, 1) = "C" Then
     startcol = Val(Mid(topleft$, ii + 1, Len(topleft$)))
     Exit For
  End If
Next ii
'Determine Ending Column
For ii = 1 To Len(bottright$)
  If Mid(bottright$, ii, 1) = "C" Then
     endcol = Val(Mid(bottright$, ii + 1, Len(bottright$)))
     Exit For
End If
Next ii
If startcol = endcol Then
    ColumninRange = startcol
    Else
    ColumninRange = 0
End If
End Function
```

With the minimum and maximum values now known, it is possible to create a data set with equidistant members with the following snippet of code.

```
For ii = 1 To TotalPoints - 1
  'Create a spaced set of points (X's)
  XVal = Xmin + (ii - 1) * ((Xmax - Xmin) / (TotalPoints - 1))
  'Solve for y of current Xval
  Yval = Slope * XVal + Intercept
  'Write XY pair to Worksheet
  Workbooks(ActiveWorkbook.Name).Worksheets(DataSheet$).Cells
  (i i, 1) = XVal
  Workbooks(ActiveWorkbook.Name).Worksheets(DataSheet$).Cells
  (ii, 2) = Yval
Next ii
'Last Point
XVal = Xmax
'Solve for y of current Xval
Yval = Slope * XVal + Intercept
Workbooks(ActiveWorkbook.Name).Worksheets(DataSheet$).Cells
(ii, 1) = XVal
Workbooks(ActiveWorkbook.Name).Worksheets(DataSheet$).Cells
(ii, 2) = Yval
```

The desired precision of the generated data set can then be set with the `SetColPrecision` subroutine as follows.

```
'Set Column Precision for Linear Fitted Dataset
Call SetColPrecision(ActiveWorkbook.Name, DataSheet$, 1,
decplaces)
Call SetColPrecision(ActiveWorkbook.Name, DataSheet$, 2,
decplaces)

Sub SetColPrecision(ByVal wkbook, ByVal wksht, ByVal ColNumber,
ByVal decplaces)
'Set the number of Decimal Places for a particular Column in
a Worksheet
Dim numforstr As String, ii As Integer
ActiveWorkbook (wkbook)
Sheets(wksht).Activate
Columns(ColNumber).Select
'This used to work?
'Selection.NumberFormat = decplaces
numforstr = "0."
For ii = 1 To decplaces
  numforstr = numforstr & "0"
Next ii
Selection.NumberFormat = numforstr
End Sub
```

The last routine in this sample application is worth mentioning because an error-handling mechanism should be in place in the event that a user does not have the analysis toolpack installed on their machine. The following line of code should be placed at the very beginning of the subroutine.

```
On Error GoTo Add_ToolPak
```

At the very end of the subroutine, this snippet of code should then be placed.

```
'If the Analysis ToolPak is not installed tell the user how
to install it!
Add_ToolPak:
If InStr(Error, "'ATPVBAEN.XLA' could not be found.") <> 0 Then
  MsgBox "Choose from Menu: Tools->Add-Ins and Check Analysis
  ToolPak and Analysis ToolPak VBA - Then Rerun Macro.", _
  vbCritical + vbOKOnly, "Macro Terminating due to Error:
  Analysis ToolPak NOT Installed!"
End
End If
```

One word of caution on the foregoing snippet of code. It assumes that any error that occurs is because the Analysis Toolpak was not installed, which may or may not be the case. If an application is to be distributed, more comprehensive methods of error handling should be constructed. However, even when specific actions are delegated for specific errors, the programmer should be cognizant of the fact that an error may be triggered by a variety of causes.

The completed application can be run by selecting ADA->Chapter 4->Linear Regression Tool. This will bring up the GUI shown in Figure 4.29. The sample application can be tested utilizing the included file "linearfittest.xls." Doing so will produce the graph shown in Figure 4.30.

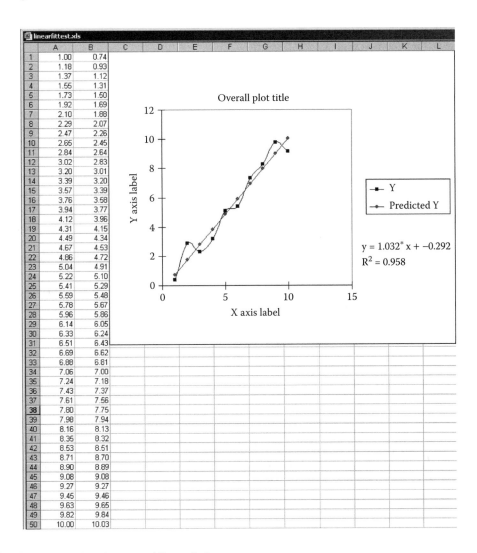

FIGURE 4.30 Graph and generated linear fit data set.

Notice that the graph contains both the actual data (blue line) and the linearly fitted prediction (red straight line). Immediately adjacent to the graph is an equidistant 50-point data set generated by using the linearly fitted equation. The equation and the (r^2) Value are placed on the graph, as well as labels for the axes and a title. Most importantly, notice if the application is run more than once; unique Worksheet and chart names are generated to hold the most recently fitted the data sets and accompanying plots.

4.10 CREATING A POLYNOMIAL REGRESSION TOOL USING THE VBA ANALYSIS TOOLPAK

Linear treatment of data is often preferable because of its simplicity. A linear regression is, in fact, simply the best straight line that can be drawn through a set of plotted points. Nonlinear data are often treated as linear data for modeling purposes just because it is so much easier to do so. However, the sharpness and frequency of curves a plot has in it will be factors that limit if a linear regression will provide a reasonably accurate model of the data set.

In such cases where data have many curves or a single sharp curvature, it is more readily modeled to a polynomial function of the form

$$f(x) = a_0 + a_1 x + a_2 x^2 + a_3 x^3 + , \cdots a_n x^n \qquad (4.12)$$

In Equation 4.12 the degree of the polynomial is represented by n. The degree of the polynomial is the highest power of x utilized in the function. Typically, a higher degree polynomial will provide a more accurate estimate *at the location of the data points*. A polynomial of higher degree, however, does not always lead to a better overall approximation of the data. This concept is illustrated in Figure 4.31.

Figure 4.31 contains some data (black dots) that are fitted by both a third-degree polynomial (red line) and a seventh-degree polynomial (blue line). The curve of the third-degree polynomial runs almost directly down the center of all the data points. Although the curve of the seventh-degree polynomial *is closer to the data points*, it is not necessarily a better representation of the behavior of the data set. This is because, as the degree of a polynomial increases, its tendency to wildly fluctuate above and below the values of the data points also increases. Notice in Figure 4.31 that, immediately before $x = 2$ and after $x = 9$, the seventh-degree polynomial oscillates outside the corridor, that one would expect a reasonable value to fall. If this same data set were fit to successively higher-degree polynomials, the effect would become even more pronounced and hence the fit less representative of the true nature of the function. A general rule of thumb is to use the lowest-degree polynomial possible to obtain a "reasonable" curve fit.

The fact that regression by polynomials returns multiple coefficients adds another level of complexity to creating a fitting tool. The GUI utilized in the linear regression tool can be reused, but a textbox must be added to the GUI that specifies the degree of the polynomial (see Figure 4.32). As opposed to the linear case that utilized the "ATPVBAEN.XLA!Regress" macro, this tool will utilize the LINEST() and TREND() spreadsheet functions.

The first step toward creating a polynomial regression tool is to determine the coefficients of a fitted curve to the given data set. This can be accomplished using the LINEST() function, but a spreadsheet must be constructed in a certain manner to utilize the LINEST() function.

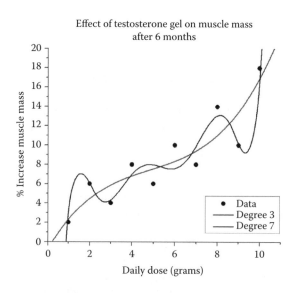

FIGURE 4.31 Higher degree polynomials are not necessarily better approximations.

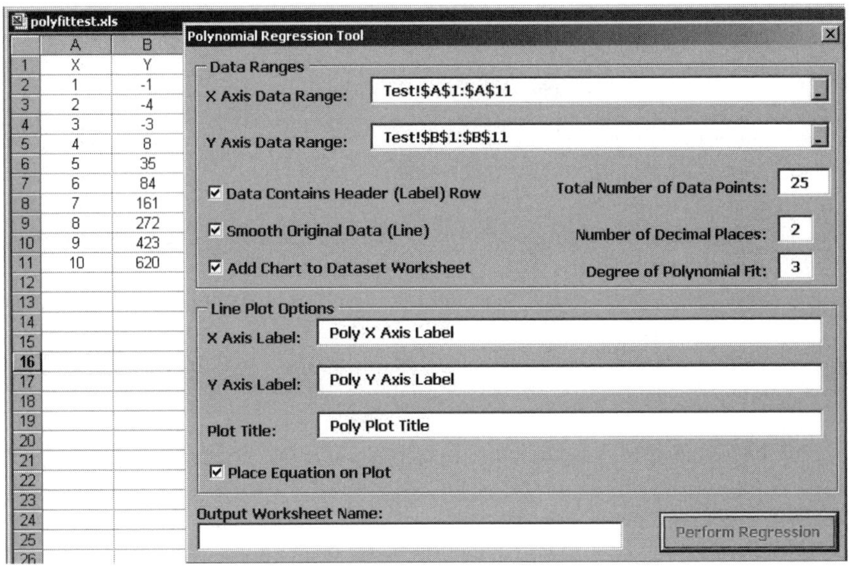

FIGURE 4.32 Polynomial regression tool GUI.

Table 4.4 shows how a spreadsheet must be set up to utilize the LINEST function to calculate the coefficients for a third-degree polynomial fit. The first column must contain the original x data. The subsequent columns must have x raised to the power n up to the nth power or degree of the polynomial to be fit. The last column will contain the y data.

The LINEST function calculates the statistics for a line by using the "least squares" method to calculate a straight line that best fits the given data and returns an array that describes the line. Because this function returns an array of values, it must be entered as an *array formula*.

The equation for the line is:

$$y = mx + b \quad \text{(linear case)} \tag{4.13}$$

$$y = m_1 x + m_2 x^2 + \ldots + b \quad \text{(polynomial case)} \tag{4.14}$$

TABLE 4.4
LINEST Function Spreadsheet Format

X	X2	X3	X4	Y	adj	Y_{adj}
1	1	1	1	0	0	0
2	4	8	16	3	0.3	3.3
3	9	27	81	2	0.2	1.8
4	16	64	256	9	0.6	8.4
5	25	125	625	36	3.3	39.3
6	36	216	1296	85	6.4	78.6
7	49	343	2401	162	2.6	159.4
8	64	512	4096	273	14.2	287.2
9	81	729	6561	424	37.4	386.6
10	100	1000	10000	621	48.4	669.4

---

where the dependent y-value is a function of the independent x-values. The m-values are coefficients corresponding to each x-value, and b is a constant value.

```
LINEST(Y Values,X Values,Const,Statts)
```

Y Values is the set (Column) of known y-values.
X Values is the set (Column) of known x-values.

Const is a logical value specifying whether to force the constant b to equal 0. If const is TRUE or omitted, b is calculated normally.

If Const is FALSE, b is set equal to 0 and the m-values are adjusted to fit the equation:

$$y = m_1 x + m_2 x^2 + \cdots + m_n x^n$$

Stats is a logical value specifying whether to return additional regression statistics. If Stats is TRUE, LINEST returns the additional regression statistics. If Stats is FALSE or omitted, LINEST returns only the m-coefficients and the constant b.

Figure 4.33 shows how the LINEST function is set up within a Worksheet. Notice the green box around the powers of x and the blue box around the vector (column) of y solutions for a given series of x. The LINEST function syntax can be seen in the formula editor in Excel. A solution matrix starts at column H. The solution to the coefficients (m-values) are found in row 1 and colored red. The solution to this data set is the following polynomial:

$$y = 2x - 4x^2 + x^3 \tag{4.15}$$

Here, the x-coefficient (2) is found in location K1, the x^2-coefficient (4) is found in location J1, and the x^3 coefficient (1) is found in location I1. In this case, the b term is essentially zero and is located at location L1 (and is colored blue). Finally, the r^2 (or COD) term is perfect at 1.0 and is located at location H3 (and is colored red).

Notice in Figure 4.33 that column F contains some numbers labeled poly. This column contains the curve fitted to the polynomial specified by the polynomial (4.15) in columns A–E. The curve is calculated utilizing the Worksheet TREND() function.

```
TREND(known_Y,known_X,new_X,const)
```

Known_Y is the set of known y-values.
Known_X is the set of known x-values.

PI	X ✓ =	=LINEST(E2:E11,A2:D11,,TRUE)									
	polyfittest.xls										
Table 04x04.xls											
	A	B	C	D	E	F	G	H	I	J	K	L
1	X	X2	X3	X4	Y	poly		=LINEST(E2:E	1	-4	2	6.49282E-11
2	1	1	1	1	-1	-1		7.3691E-14	1.63E-12	1.2199E-11	3.52E-11	3.12503E-11
3	2	4	8	16	-4	-4		1	9.46E-12	#N/A	#N/A	#N/A
4	3	9	27	81	-3	-3		1.16592E+27	5	#N/A	#N/A	#N/A
5	4	16	64	256	8	8		417202.5	4.47E-22	#N/A	#N/A	#N/A
6	5	25	125	625	35	35		#N/A	#N/A	#N/A	#N/A	#N/A
7	6	36	216	1296	84	84		#N/A	#N/A	#N/A	#N/A	#N/A
8	7	49	343	2401	161	161		#N/A	#N/A	#N/A	#N/A	#N/A
9	8	64	512	4096	272	272		#N/A	#N/A	#N/A	#N/A	#N/A
10	9	81	729	6561	423	423		#N/A	#N/A	#N/A	#N/A	#N/A
11	10	100	1000	10000	620	620		#N/A	#N/A	#N/A	#N/A	#N/A

FIGURE 4.33 Using the LINEST function in a Worksheet.

=TREND(E2:E11,A2:D11)

polyfittest.xls

	A	B	C	D	E	F
1	X	X2	X3	X4	Y	poly
2	1	1	1	1	-1	=TREND(E
3	2	4	8	16	-4	-4
4	3	9	27	81	-3	-3
5	4	16	64	256	8	8
6	5	25	125	625	35	35
7	6	36	216	1296	84	84
8	7	49	343	2401	161	161
9	8	64	512	4096	272	272
10	9	81	729	6561	423	423
11	10	100	1000	10000	620	620

FIGURE 4.34 Using the TREND() Worksheet Function.

New_X (optional) are new x-values for which the TREND function should return corresponding y-values. If New_X is omitted, it is assumed to be the same as Known_X. If Const is TRUE or omitted, b is calculated normally. If const is FALSE, b is set equal to 0 (zero), and the m-values are adjusted so that $y = m_1 x + m_2 x^2 + \cdots + m_n x^n$.

Figure 4.34 shows how the LINEST function is set up within a Worksheet. Notice the green box around the powers of x, and the blue box around the vector (column) of y solutions for a given series of x. The TREND function syntax can be seen in the formula editor in Excel. The fitted curve of solutions to the given polynomial is returned in column F.

With an understanding of how the mechanism works to accomplish polynomial curve fitting in Excel, a tool can now be constructed to automate the process using the GUI shown in Figure 4.32. The first step in this process would be to generate a Worksheet in such a format that the LINEST() and TREND() functions can calculate the regression parameters and return a fitted curve (such as that shown in Figure 4.35).

The code that will create a neatly formatted sheet as shown in Figure 4.35 follows. Note that the degree of the polynomial to be fitted (`PolyDeg`) is a passed parameter to the subroutine in which this code resides, as are `XRange` and `YRange`.

```
'Create Polynomial Fitting Sheet
PolySheet$ = CreateSheet(ActiveWorkbook.Name, "PolyFit")
```

Table 04x04.xls

	A	B	C	D
1	X	X^2	X^3	Y
2	1	1	1	-1
3	2	4	8	-4
4	3	9	27	-3
5	4	16	64	8
6	5	25	125	35
7	6	36	216	84
8	7	49	343	161
9	8	64	512	272
10	9	81	729	423
11	10	100	1000	620

FIGURE 4.35 An example Worksheet for using TREND() and LINEST() Functions.

```
If Labels = False Then
 startrow = FirstRowinRange(XRange)
 Else
 startrow = FirstRowinRange(XRange) + 1
End If

For ii = startrow To LastRowinRange(XRange)
 'X Value Labels
 If ii = startrow Then
    Range(Cells(1, 1), Cells(1, 1)).Select
    ActiveCell.FormulaR1C1 = "X"
    ActiveCell.HorizontalAlignment = xlCenter
End If
 'Write in X Values
 Workbooks(ActiveWorkbook.Name).Worksheets(PolySheet$).Cells
 (ii, 1) = _
 Workbooks(ActiveWorkbook.Name).Worksheets(SheetInRange(XRange)).
 Cells(ii, ColumninRange(XRange))
 For jj = 2 To PolyDeg
 'X Power Labels
 If ii = startrow Then
   Range(Cells(1, jj), Cells(1, jj)).Select
   ActiveCell.FormulaR1C1 = "X" & Trim(str(jj))
   ActiveCell.HorizontalAlignment = xlCenter
   With ActiveCell.Characters(Start:=2, Length:=1).Font
     .Superscript = True
   End With
 End If
 'Write in Polynomial Powers of X
 PolyForm$ = "=A" & Trim(str(ii)) & "^" & Trim(str(jj))
 Call AddFormula(ActiveWorkbook.Name, PolySheet$, ii, jj,
 PolyForm$)
 Next jj
 'Y Value Label
 If ii = startrow Then
    Range(Cells(1, PolyDeg + 1), Cells(1, PolyDeg + 1)).Select
    ActiveCell.FormulaR1C1 = "Y"
    ActiveCell.HorizontalAlignment = xlCenter
 End If
 'Write in Y Values
 Workbooks(ActiveWorkbook.Name).Worksheets(PolySheet$).Cells
 (ii, PolyDeg + 1) = _
 Workbooks(ActiveWorkbook.Name).Worksheets(SheetInRange(YRange)).
 Cells(ii, ColumninRange(YRange))
 Next ii
```

The next step is to create a sheet to hold results calculated by the polynomial fitting session using the LINEST() and TREND() Worksheet functions. This can be accomplished using the following code fragment.

```
'Create Polynomial Fit Results Sheet
ResultsSheet$ = CreateSheet(ActiveWorkbook.Name,
"PolyFitResults")

'Perform Polynomial Fitting
'Create Range for Results based on the Degree of the Polynomial
ResultRange$ = "A1:" & Chr(65 + PolyDeg) & "5"
Sheets(ResultsSheet$).Select
Range(ResultRange$).Select
'Create Formula for Fitting
PolyForm$ = "=LINEST(" & PolySheet$ & "!" & Chr(65 + PolyDeg)
& "2:" & Chr(65 + PolyDeg) & _ Trim(str(LastRowinRange(XRange)))
& _
"," & PolySheet$ & "!A2:" & Chr(64 + PolyDeg) &
Trim(str(LastRowinRange(XRange))) & ",TRUE,TRUE)"
'Do the Fitting
Selection.FormulaArray = PolyForm$
```

Notice that the LINEST() function is utilized as a `formula array`, meaning the function returns an array, which in this case is returned to an appropriate-sized range of cells within a Workbook. As such, the coding within VBA is slightly different.

As with the previous curve-fitting tool, unique names are generated for all Worksheets utilized to hold information about a particular fitting session. Using this methodology, multiple polynomial regressions can occur utilizing a single Workbook without error, and the results of each fitting session will be retained in a unique set of Worksheet names.

The last major step in this process is to create a data set with the number of points specified in the GUI. Although the TREND() function could be utilized to do this, it is easier to reuse the code from the earlier application.

```
'Now Create Sheet to hold polynomial fitted Dataset
DataSheet$ = CreateSheet(ActiveWorkbook.Name, "PolyDataSet")

If Labels = False Then
    Xmin =
    Workbooks(ActiveWorkbook.Name).Worksheets(SheetInRange
    (XRange)).Cells(FirstRowinRange(XRange),
    ColumninRange(XRange))
    Else
    Xmin =
    Workbooks(ActiveWorkbook.Name).Worksheets(SheetInRange
    (XRange)).Cells(FirstRowinRange(XRange) + 1,
    ColumninRange(XRange))

End If
    Xmax =
    Workbooks(ActiveWorkbook.Name).Worksheets(SheetInRange
    (XRange)).Cells(LastRowinRange(XRange),
    ColumninRange(XRange))
```

```
'Write XY labels to Worksheet
Workbooks(ActiveWorkbook.Name).Worksheets(DataSheet$).Cells
(1, 1) = "X"
Workbooks(ActiveWorkbook.Name).Worksheets(DataSheet$).Cells
(1, 1).HorizontalAlignment = xlCenter

Workbooks(ActiveWorkbook.Name).Worksheets(DataSheet$).Cells
(1, 2) = "Ypolyfit"

Workbooks(ActiveWorkbook.Name).Worksheets(DataSheet$).Cells
(1, 2).HorizontalAlignment = xlCenter

For ii = 1 To TotalPoints - 1
 'Create a spaced set of points (X's)
 XVal = Xmin + (ii - 1) * ((Xmax - Xmin) / (TotalPoints - 1))
 'Solve for y of current Xval
 Yval = CalcYValue(XVal, PolyCoeff())
 'Write XY pair to Worksheet
 Workbooks(ActiveWorkbook.Name).Worksheets(DataSheet$).Cells
 (ii + 1, 1) = XVal

 Workbooks(ActiveWorkbook.Name).Worksheets(DataSheet$).Cel
 ls(ii + 1, 2) = Yval

 Next ii
 'Last Point
 XVal = Xmax
 'Solve for y of current Xval
 Yval = CalcYValue(XVal, PolyCoeff())
 Workbooks(ActiveWorkbook.Name).Worksheets(DataSheet$).Cells
 (ii + 1, 1) = XVal
 Workbooks(ActiveWorkbook.Name).Worksheets(DataSheet$).Cells
 (ii + 1, 2) = Yval

 'Set Column Precision for Linear Fitted Dataset
 Call SetColPrecision(ActiveWorkbook.Name, DataSheet$, 1,
 decplaces)
 Call SetColPrecision(ActiveWorkbook.Name, DataSheet$, 2,
 decplaces)
```

The last snippet of code that will be explained in this sample application is the placing of the equation on the plot. This process falls more into the category of arduous than difficult, but there are a few idiosyncrasies that make repeating the code here worthwhile.

```
'Add y=a + bx + bx^2 ... & cod Label here
If frm4PolyRegTool.CheckBox_Eq = True Then
   ActiveChart.Shapes.AddLabel(msoTextOrientationHorizontal,
   110#, 55#, 0#, 0#).Select
   Selection.ShapeRange(1).TextFrame.AutoSize = msoTrue
   'Remove "^"
   For ii = 1 To Len(Plot$)
```

```
  If Mid(Plot$, ii, 1) = "^" Then
    Plot$ = Mid(Plot$, 1, ii - 1) & Mid(Plot$, ii + 1,
    Len(Plot$) - ii)
  End If
Next ii
Selection.Characters.Text = Plot$
Selection.Characters.Font.Size = 16
Selection.Characters.Font.Bold = True
For ii = 1 To Len(Plot$)
  If (Mid(Plot$, ii, 1) = "X" Or Mid(Plot$, ii, 1) = "R") And
  IsNumeric(Mid(Plot$, ii + 1, 1)) Then
    With Selection.Characters(Start:=ii + 1, Length:=1).Font
    .Superscript = True
    End With
  End If
Next ii
End If
```

The completed application can be run by selecting <u>A</u>DA->Chapter <u>4</u>->Polynomial Regression Tool. This will bring up the GUI shown in Figure 4.32. The sample application can be tested utilizing the included file "polyfittest.xls." Doing so will produce the graph shown in Figure 4.36.

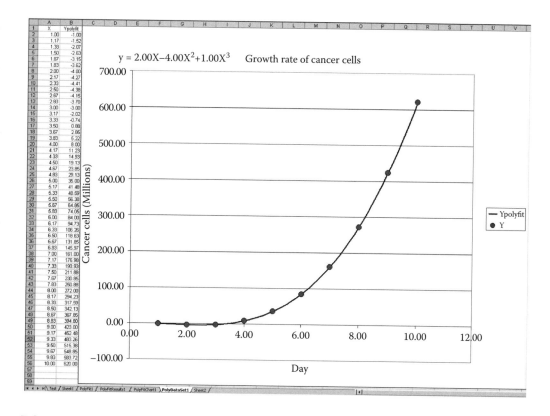

FIGURE 4.36 Result of polynomial fitting of data in "polyfittest.xls."

SUMMARY

Functions are the building blocks of modern analyses. This chapter covers utilizing both Worksheet functions and VBA code functions. It also delves into specialized use of functions, such as calling VBA coded functions from Worksheet cells and populating Worksheet cells with custom functions using VBA. The reader is also shown how to add additional "built-in" functions to Excel. It is functions that allow specialized dynamic formatting of Worksheets, which add a new level of complexity and flexibility when creating custom reports. The use of the Macro Recorder is also covered. The Macro Recorder allows the user to capture the code required to perform a specific process, provided the user knows how to carry out this process manually. Once the code required for a certain task is known, the user is then free to create a generic function to accomplish the task at will within VBA. The chapter concludes with two sample applications that automate two of the more common analyses: linear regression and polynomial regression. Not only are functions utilized to perform the regressions but the user is shown how to extract the parameters of interest generated from the function calls that perform the regressions. Tools such as those shown can then be constructed by the user to accomplish a variety of custom analyses quickly and efficiently.

5 Data Mining in Excel

5.1 INTRODUCTION

Having information is one thing, finding it is another. This chapter is dedicated to helping developers create tools that allow users to rapidly find the information they need. Having a Worksheet that contains all the data related to a particular project, experiment, or method is only the first step toward useful utilization of that information. It can be a daunting task to manually look through a large Worksheet and extricate entries of data whose parameters fall within a certain set of rules.

Table 5.1 shows parameters utilized when testing electronically actuated valves. The trigger voltage is the voltage that is first applied to a valve to actuate it. After a valve has been activated, the trigger voltage will be lowered to a hold in voltage after a specified period of time, usually in milliseconds (ms). The hold in voltage is usually some fraction of the trigger voltage (usually about a third). The purpose of the hold in voltage is to continue to keep the valve actuated, but to do so drawing less current. Although the valve could be driven at the trigger voltage for the entire period of its activation, doing so would be wasteful of energy. In addition, oftentimes, valves will get extremely hot if driven at their trigger voltage for an extended period of time. This will decrease the life of the valve, which is an important consideration, especially if the valve is located in an area that makes its replacement difficult. The applied vacuum is the maximum amount of vacuum present in the lines connected to the valve. The higher the vacuum, the more difficulty the valve will have in either opening or closing because it will have to exert force against the suction pressure that exists in the lines. The valve state is simply whether the valve was determined to be open or closed when the following trigger voltage, hold in voltage, and vacuum were applied to the valve for testing purposes.

Table 5.2 shows the ranges between which each valve actuation parameter may vary. Now, suppose the user wished to know when test conditions occurred that the valve stayed open (did not actuate) and the following conditions occurred:

1. Trigger Voltage > 16.5 V
2. Hold in Voltage > 6.5 V
3. Applied Vacuum < 60 psi

For the information in Table 5.1, this is not difficult to determine manually. There are only 10 items in the table. What happens when there are *thousands* of items in a table? It is impossible to do this by hand. If this information is already in a database, then that database can be queried utilizing the aforesaid conditions. It may not be prudent to put this information into a database unless it meets or does not meet certain conditions. In this chapter, the reader will be shown how to create tools to solve problems just like the one shown.

Pulling parameters out of a pool of information that meet a certain set of specified criteria is called "data mining." A number of tools exist to perform data-mining operations. Often, however, users need some quick and easy methods to extract some information and omit other pieces of information. Having the ability to perform such operations within Excel by utilizing macros gives the user the ability to limit the amount of extraneous information that will be archived or included within a report.

Once information can be qualified as to meeting or not meeting a certain set of selection criteria, it must be decided what should be done with the information that falls within the specified criteria. The information can be treated in one of two ways. The information can be isolated from the pool it resides in, that is, the information can be extracted and written into a Worksheet, Workbook, or file. The other option is to leave the information within the pool it resides in and call attention to it in some manner,

TABLE 5.1
Valve Actuation Parameters

Trigger Voltage	Hold in Voltage	Applied Vacuum	Valve State
16.204	7.922	38.870	Closed
16.873	5.574	73.261	Open
16.107	5.230	68.921	Closed
17.065	6.583	87.497	Open
17.204	7.970	45.522	Open
16.826	5.992	50.624	Closed
16.127	5.829	75.792	Open
17.264	5.828	51.060	Closed
17.788	6.030	40.945	Closed

usually by coloring it to stand out within the Worksheet. This chapter will begin with creating tools to highlight information within Worksheets and then progress to more advanced methods of actually extracting data from a preexisting source. The chapter will conclude with advanced methods for data mining, including extracting information from multiple source files located in different locations.

5.2 THE TERRIBLE TRUTH ABOUT COLORS IN VBA

Excel contains 56 unique colors that can be utilized in Excel VBA by means of the ColorIndex property. The default 56 colors are shown in Figure 5.1. Of these 56 colors, 40 are shown in Excel's Workbook color palette. These colors can be modified to reflect any type of custom color the user desires by selecting from the menu Tools->Options and then clicking on the Color tab. Doing so will display the color palette shown in Figure 5.2.

Any color in the color palette can be modified by simply selecting the color and clicking on the Modify button, which will produce the GUI shown in Figure 5.3.

By utilizing this GUI, the user can make any color possible by simply dragging the crosshairs on the color map to the color shade desired, or even more accurately, by specifying the RGB (Red–Green–Blue) values of the particular color desired.

With such a great deal of functionality built into color selection in Excel, it would seem that color selection using VBA would be a snap. Unfortunately, it is not. Although there are 56 custom colors available on Excel's color palette, there are no system constants in Excel VBA for the standard Excel color palette colors. Although the color palette colors can be utilized in Excel VBA by means of the ColorIndex property, the colors must be referred to by an arbitrary number that gives no sense as to their true color and only refers to their location within the color palette! Worse yet is the fact that if the system colors have been modified, the number assigned to the default color palette will now represent a different color entirely. What is really needed is a set of system constants that will *always* represent the color specified even if the color palette has been modified.

TABLE 5.2
Valve Parameter Value Ranges

Trigger Voltage	Hold in Voltage	Applied Vacuum	Valve State
15–18	5–8	20–100 psi	Open/Closed

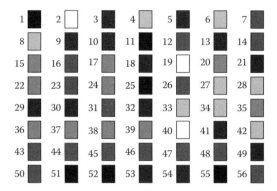

FIGURE 5.1 The Default colors available in Excel.

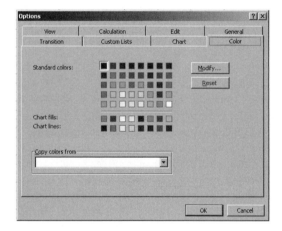

FIGURE 5.2 The Workbook color palette.

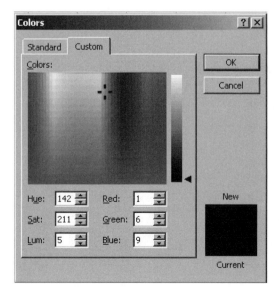

FIGURE 5.3 Modification of default Workbook colors.

TABLE 5.3
VB Color Constants

| Constant | Value | | Description |
	Hex	Dec	
vbBlack	&h00	0	Black
vbRed	&hFF	255	Red
vbGreen	&hFF00	65280	Green
vbYellow	&hFFFF	65535	Yellow
vbBlue	&hFF0000	16711680	Blue
vbMagenta	&hFF00FF	16711935	Magenta
vbCyan	&hFFFF00	16776960	Cyan
vbWhite	&hFFFFFF	16777215	White

Such a set of system constants does exist, but they are VB system constants, not Excel VBA constants. These VB system constants cannot be utilized with the `ColorIndex` property. They must only be used with the `Color` property.

Unfortunately VB only has eight color constants which Excel's Color property will recognize (Table 5.3). In the default color palette, names for Colors 1 through 8 (Black, White, Red, Green, Blue, Yellow, Magenta, and Cyan) have been assigned constants. The value for these constants is shown in Table 5.3. The constants are named accordingly: vbBlack, vbRed, vbGreen, vbYellow, vbBlue, vbMagenta, vbCyan, and vbWhite. Notice that the constants are values from 0 to 16777215. Further observe that the values utilized as color indexes (1–56) are not utilized as VB color constants. This sets the stage for the creation of generic subroutines capable of utilizing both the VB color constants and the `ColorIndex` property settings.

Prior to writing some color setting subroutines, it would be useful to have the ability to generate a color map at will that would show the current system colors in the color palette, their RGB (Red–Green–Blue) component mix, and the hexadecimal numeric representation of each color.

The color map shown in Figure 5.4 can be generated utilizing the subroutine `DisplayColor Palette` that follows. It can be run from the menu ADA->Chapter 5->Show Color Map.

```
Sub DisplayColorPalette(wkbook, wksheet)
'Display All the Current Colors in the Color Palette
Workbooks(wkbook).Worksheets(wksheet).Activate
Dim Color_Index As Integer
Dim RGB As String, str As String
'Create Headers
Cells(1, 1) = "Backgnd"
Cells(1, 2) = "Color"
Cells(1, 3) = "Hex"
Cells(1, 4) = "Red"
Cells(1, 5) = "Grn"
Cells(1, 6) = "Blu"

For Color_Index = 1 To 56
  Cells(Color_Index + 1, 1).Interior.ColorIndex = Color_Index
  Cells(Color_Index + 1, 1).Value = "[Color " & Color_Index & "]"
  RGB = Right("000000" & hex(Cells(Color_Index + 1,
  1).Interior.Color), 6)
```

	A	B	C	D	E	F
1	Backgnd	Color	Hex	Red	Grn	Blu
2		[Color 1]	0	0	0	0
3	[Color 2]	[Color 2]	FFFFFF	255	255	255
4	[Color 3]	[Color 3]	FF0000	255	0	0
5	[Color 4]	[Color 4]	00FF00	0	255	0
6	[Color 5]	[Color 5]	0000FF	0	0	255
7	[Color 6]	[Color 6]	FFFF00	255	255	0
8	[Color 7]	[Color 7]	FF00FF	255	0	255
9	[Color 8]	[Color 8]	00FFFF	0	255	255
10	[Color 9]	[Color 9]	800000	128	0	0
11	[Color 10]	[Color 10]	8000	0	128	0
12	[Color 11]	[Color 11]	80	0	0	128
13	[Color 12]	[Color 12]	808000	128	128	0
14	[Color 13]	[Color 13]	800080	128	0	128
15	[Color 14]	[Color 14]	8080	0	128	128
16	[Color 15]	[Color 15]	C0C0C0	192	192	192
17	[Color 16]	[Color 16]	808080	128	128	128
18	[Color 17]	[Color 17]	9999FF	153	153	255
19	[Color 18]	[Color 18]	993366	153	51	102
20	[Color 19]	[Color 19]	FFFFCC	255	255	204
21	[Color 20]	[Color 20]	CCFFFF	204	255	255
22		[Color 21]	660066	102	0	102
23	[Color 22]	[Color 22]	FF8080	255	128	128
24	[Color 23]	[Color 23]	0066CC	0	102	204
25	[Color 24]	[Color 24]	CCCCFF	204	204	255
26		[Color 25]	80	0	0	128
27	[Color 26]	[Color 26]	FF00FF	255	0	255
28	[Color 27]	[Color 27]	FFFF00	255	255	0
29	[Color 28]	[Color 28]	00FFFF	0	255	255
30		[Color 29]	800080	128	0	128
31		[Color 30]	800000	128	0	0
32	[Color 31]	[Color 31]	8080	0	128	128
33		[Color 32]	0000FF	0	0	255
34	[Color 33]	[Color 33]	00CCFF	0	204	255
35	[Color 34]	[Color 34]	CCFFFF	204	255	255
36	[Color 35]	[Color 35]	CCFFCC	204	255	204
37	[Color 36]	[Color 36]	FFFF99	255	255	153
38	[Color 37]	[Color 37]	99CCFF	153	204	255
39	[Color 38]	[Color 38]	FF99CC	255	153	204
40	[Color 39]	[Color 39]	CC99FF	204	153	255
41	[Color 40]	[Color 40]	FFCC99	255	204	153
42	[Color 41]	[Color 41]	3366FF	51	102	255
43	[Color 42]	[Color 42]	33CCCC	51	204	204
44	[Color 43]	[Color 43]	99CC00	153	204	0
45	[Color 44]	[Color 44]	FFCC00	255	204	0
46	[Color 45]	[Color 45]	FF9900	255	153	0
47	[Color 46]	[Color 46]	FF6600	255	102	0
48	[Color 47]	[Color 47]	666699	102	102	153
49	[Color 48]	[Color 48]	969696	150	150	150
50		[Color 49]	3366	0	51	102
51	[Color 50]	[Color 50]	339966	51	153	102
52		[Color 51]	3300	0	51	0
53		[Color 52]	333300	51	51	0
54	[Color 53]	[Color 53]	993300	153	51	0
55	[Color 54]	[Color 54]	993366	153	51	102
56	[Color 55]	[Color 55]	333399	51	51	153
57	[Color 56]	[Color 56]	333333	51	51	51

FIGURE 5.4 Macro-generated table of system colors.

```
Cells(Color_Index + 1, 2).Font.ColorIndex = Color_Index
Cells(Color_Index + 1, 2).Value = "[Color " & Color_Index & "]"
If Hex2Dec(Right(RGB, 2)) + Hex2Dec(Mid(RGB, 3, 2)) + Hex2Dec
( Left(RGB, 2)) >= 510 Then
  'Color is Extremely Light, Darken Cell Background
  Cells(Color_Index + 1, 2).Interior.Color = vbBlack
End If
'Excel shows nibbles in reverse order so make it as RGB
```

```
    str = Right(RGB, 2) & Mid(RGB, 3, 2) & Left(RGB, 2)
    'generating 2 columns in the HTML table
    Cells(Color_Index + 1, 3) = str
    Cells(Color_Index + 1, 4).formula = "=Hex2dec(""" & Right(RGB,
    2) & """)"
        Cells(Color_Index + 1, 5).formula = "=Hex2dec(""" &
    Mid(RGB, 3, 2) & """)"
        Cells(Color_Index + 1, 6).formula = "=Hex2dec(""" &
    Left(RGB, 2) & """)"
    Next Color_Index
    End Sub
```

   The first thing to notice is that Excel Worksheets have a built-in function called `Hex2Dec`, which will convert a hexadecimal number to a decimal number. Hexadecimal numbers are commonly utilized for representations of color codes, constants, and addresses in many programming languages. Although VBA has a function to convert strings to their hexadecimal equivalents (Hex), no such function exists to convert hexadecimal numbers to their decimal equivalents. For example:

```
Debug.Print Hex(10)
A
```

One would think a reverse function would exist such as:

```
Debug.Print Dec("A")
```

Unfortunately, no such function exists, and typing the one just cited will produce an error. The `Val` function will, however, convert a hexadecimal number to its decimal equivalent. The string representation of the number must be preceded by "&h" or "&H," which denotes a hexadecimal number. The problem is that sometimes hexadecimal numbers are not preceded by the identifying indicator, or sometimes are only preceded by an "h." To utilize the Val function to convert such hexadecimal numbers to their decimal equivalents, the number must be reformatted to have the proper prefix (&H or &h). The following VBA function will convert a hexadecimal number to its decimal equivalent and will ensure that the hexadecimal number passed for conversion is formatted with the proper prefix.

```
Function Hex2Dec(hex) As Long
If UCase(Left(hex, 2)) = "&H" Then
    Hex2Dec = Val(hex)
    Exit Function
ElseIf UCase(Left(hex, 1)) = "H" Then
    hex = "&" & hex
    Hex2Dec = Val(hex)
    Exit Function
Else
    hex = "&h" & hex
    Hex2Dec = Val(hex)
    Exit Function
End If
```

Thus, using the above function:

```
Debug.Print Hex2Dec("A")
```

```
   10
Debug.Print Hex2Dec("hA")
   10
Debug.Print Hex2Dec("&hA")
   10
```

As can be seen from the examples, the formatting issue is taken care of by the function, and the three cases all produce the same conversion result without error.

Notice also that, if the chosen color is very light (where very light has been arbitrarily determined to be a color whose RGB value totals 510 or greater), then the cell holding that color has its background changed to black using the VB color constant vbBlack. To do this, the .Color property is utilized. The reason the .Color property is utilized as opposed to the .ColorIndex property is that the .ColorIndex property for black (which is 1) could be changed to a different color should a user set a custom-defined color for that .ColorIndex. The VB color constant vbBlack will always reflect the color black, however. The advantage of using the VB color constants is that they are fixed and will never change.

Looking at Figure 5.4, some explanation is in order about the six columns created. The first column shows what a cell will look like if its background is set to that color. The second column shows what the text will look like if it is set to that color. As mentioned earlier, if the color is so light that it will be difficult to see against a white background, the cells' background is changed to black so that the true color of the font can be readily observed. The third column named hex is a 6-digit hexadecimal number representative of all the RGB components within that color.

Example: Red–RGB Components (255–0–0)
Hex: FF0000 <-> [FF][00][00] <-> [Red][Grn][Blu]

In this example, the color red has a red component of 255 (the maximum value), a green component of 0, and a blue component of 0. The decimal 255 is the hexadecimal number FF. Therefore, the RGB components for the color red in hex are expressed as FF0000.

A similar table will now be constructed for the VB color constants that, unlike the ColorIndex constants, are guaranteed to always remain fixed and cannot be changed. Creating a subroutine to construct a color index map for the VB color constants is far more problematic than in the previous example. Table 5.4 shows the parameters of interest related to VB's color constants.

Notice that the decimal values corresponding to each color do not follow an *obvious* pattern. However, whenever binary number systems are utilized, a pattern is usually present even if it is not self evident. Notice that the hexadecimal pattern for each color has only two components: "FF"

TABLE 5.4
VB Color Constant Parameters

n	Odd Multiple (oddmult)	Constant	Value Hex	Value Dec	Description	Components 1	2	3	Formula
1	n/a	vbBlack	&h00	0	Black	1	0	0	0
2	n/a	vbRed	&hFF	255	Red	2	0	0	$n + 255$
3	0	vbGreen	&hFF00	65280	Green	2	1	0	$(n - 1) * 256$
4	n/a	vbYellow	&hFFFF	65535	Yellow	2	2	0	$(n - 1) + 255$
5	1	vbBlue	&hFF0000	16711680	Blue	2	1	1	$(n - 2) * 256$
6	n/a	vbMagenta	&hFF00FF	16711935	Magenta	2	1	2	$(n - 1) + 255$
7	2	vbCyan	&hFFFF00	16776960	Cyan	2	2	1	$(n - 4) * 257$
8	n/a	vbWhite	&hFFFFFF	16777215	White	2	2	2	$(n - 1) + 255$

or "00." A color may contain only one component (as in the case of vbBlack) or as many as three components (as in the case of vbWhite). The components area of Table 5.4 contains 3 possible numbers that correspond to the following possibilities:

0 = Component not present.
1 = Component is "00."
2 = Component is "FF."

The shading in Table 5.4 shows evidence of a possible pattern, but it is not really discernable until one looks at the decimal representation of each VB color constant. Here, it is readily observed that, after each group of two colors, there is a sharp increase in the decimal value of the color constants. Upon further inspection it can be observed that the number of significance here is hexadecimal "FF" or 255 decimal. It is now possible to write a formula to predict the value of the next color constant based on the previous color constants' values. This is done in the formula column of Table 5.4.

Notice that for even values of (n), the formula is consistent:

$$evens = (n - 1) + 255 \tag{5.1}$$

Example: (From Table 5.4) Predict Decimal vb Constant for $n = 6$.
= {Decimal Value @ (6.1)} + 255
= {Decimal Value @ $n = (5)$} + 255
= 16711680 + 255
= 16711935

Odd values of (n), however, become more problematic. The key digit constant here $(255)_d$ must be adjusted based on which odd multiple is used. This gives rise to the oddmult column in Table 5.4. Also note that the previous value to be utilized must also be adjusted depending upon the odd multiple: $(n-1)$, $(n-2)$, and $(n-4)$. This pattern can be computed by means of the following formula:

$$odds = (n - 2^{oddmult})*255 + 2^{oddmult} - oddmult \tag{5.2}$$

Example: (From Table 5.4) Predict Decimal vb Constant for $n = 7$.
= {Decimal Value @ (7.2^2)} * 255 + 2^2 – 2
= {Decimal Value @ (7.4)} * 255 + 4 – 2
= {Decimal Value @ (3)} * 255 + 2
= 65280 * 257
= 16776960

Now, knowing the pattern to prediction, it is possible to construct a subroutine to create a VB constant color map. Before this can be done, two simple functions must be created that can determine if a number is odd or even. (The functions `IsOdd` and `IsEven` are available for use in Worksheets, but unbelievably are not built into Excel VBA!)

```
Function IsEven(ByVal lngNumber As Long) As Boolean
  IsEven = (lngNumber \ 2 = lngNumber / 2)
End Function

Function IsOdd(ByVal lngNumber As Long) As Boolean
  IsOdd = (lngNumber \ 2 <> lngNumber / 2)
End Function
```

These two functions work quite simply by taking the number passed to them and performing a regular (floating point) division (/) and an integer-only division (\) by the numeral 2. If the two divisions equal each other (in other words, no remainder is present), then the number passed is even; else it is odd.

The following subroutine will create a color map of the 8 VB color constants. It can be run by selecting <u>A</u>DA->Chapter <u>5</u>->Show <u>V</u>B Color Map from the menu.

```vb
Sub DisplayVBColorCodes(wkbook, wksheet)
'Display All the Current Colors in the Color Palette
Workbooks(wkbook).Worksheets(wksheet).Activate
Dim vbColor As Integer, ColorNames As String
Dim Colors() As String, vbColorValue() As Long
Dim oddmult As Integer

ColorNames =
"vbBlack,vbRed,vbGreen,vbYellow,vbBlue,vbMagenta,vbCyan,vbWhite"
Colors() = Split(ColorNames, ",")

'Create Headers
Cells(1, 1) = "Backgnd"
Cells(1, 2) = "Color"
Cells(1, 3) = "Bin"
Cells(1, 4) = "Hex"

For vbColor = 1 To 8
  ReDim Preserve vbColorValue(vbColor)
  Select Case vbColor
    Case 1
      vbColorValue(vbColor) = 0
    Case 2
      vbColorValue(vbColor) = 255
    Case Else
      If IsOdd (vbColor) = True Then
          'Odd Reliant Upon Index
          vbColorValue(vbColor) = vbColorValue(vbColor - 2 ^
          oddmult) * (255 + ((2 ^ oddmult) - oddmult))
          oddmult = oddmult + 1
  Else
          'Even Always Same Formula
          vbColorValue(vbColor) = vbColorValue(vbColor - 1) + 255
          End If
    End Select
    Debug.Print vbColor, Colors(vbColor - 1),
    vbColorValue(vbColor)
    'Background
    Cells(vbColor + 1, 1).Interior.Color =
    vbColorValue(vbColor)
    Cells(vbColor + 1, 1).Value = Colors(vbColor - 1)
    'Text Color
    Cells(vbColor + 1, 2).Font.Color = vbColorValue(vbColor)
    Cells(vbColor + 1, 2).Value = "[" & Colors(vbColor - 1) & "]"
```

	A	B	C	D
1	Backgnd	Color	Bin	Hex
2		[vbBlack]	0	0
3	vbRed	[vbRed]	255	FF
4	vbGreen	[vbGreen]	65280	FF00
5	vbYellow	[vbYellow]	65535	FFFF
6	vbBlue	[vbBlue]	16711680	FF0000
7	vbMagenta	[vbMagenta]	16711935	FF00FF
8	vbCyan	[vbCyan]	16776960	FFFF00
9	vbWhite		16777215	FFFFFF

FIGURE 5.5 The VB color constant map.

```
    'Binary Value
    Cells(vbColor + 1, 3).Value = vbColorValue(vbColor)
    'Hexadecimal Value
    Cells(vbColor + 1, 4).Value = hex(vbColorValue(vbColor))
Next vbColor
End Sub
```

A great deal of effort was expended creating the VB constant color map in terms of developing the algorithm for its construction (Figure 5.5). The level of effort in developing and explaining the algorithm was expended for two reasons. To effectively utilize the VB color constants, it is necessary to be able to produce on demand an array or arrays that contain both the name of the color constants, and the value of that color constant. The foregoing `DisplayVBColorCodes` subroutine can very easily be modified to place this information into an array where it can be utilized quite effectively.

The second reason behind showing the rationale for the development of the algorithm to compute the VB color constant values is that finding such relationships is a critical part of any automated data analysis process. When trying to formulate a relationship that describes a certain set of parameters, it is always helpful to approach the problem by taking the following steps: (1) Lay all of the parameters associated with the behavior to be modeled out as was done in Table 5.4. (2) Look for obvious relationships between the data items that could be utilized to form a model. (3) Look for discrepancies in relationships found. (4) Try to determine a way to adjust for any discrepancies found, such as adding an additional parameter. (In the color constant example we just saw, the *oddmult* parameter was created and utilized in the model.) (5) Write the equations for the model and test their validity. (6) Adjust the equations as necessary.

The `DisplayVBColorCodes` subroutine can be modified to produce an array that will hold both the VB color names and their color constant values. In the next example, a two-dimensional array is passed by reference to the subroutine, and both the color names and their decimal constant values are populated into the array.

```
Sub CreateColorArray(ByRef ColorMatrix())
'Create a Two Dimensional Array Which Contains Color Name and
Decimal Value
'ColorMatrix (X, 1) = Text: Color Name
'ColorMatrix (X, 2) = Number: Color Constant
Dim vbColor As Integer, ColorNames As String
Dim Colors() As String, vbColorValue(8) As Long
Dim oddmult As Integer
'The leading comma starts Array @ 1 vs 0 without use of Option
Base 1 Statement
```

```
ColorNames = ",
vbBlack,vbRed,vbGreen,vbYellow,vbBlue,vbMagenta,vbCyan,vbWhite"
Colors() = Split(ColorNames, ",")
ReDim ColorMatrix(8, 2)
For vbColor = 1 To 8
  Select Case vbColor
    Case 1
      vbColorValue(vbColor) = 0
    Case 2
      vbColorValue(vbColor) = 255
    Case Else
      If IsOdd(vbColor) = True Then
          'Odd Reliant Upon Index
          vbColorValue(vbColor) = vbColorValue(vbColor - 2 ^
          oddmult) * (255 + ((2 ^ oddmult) - oddmult))
          oddmult = oddmult + 1
      Else
          'Even Always Same Formula
          vbColorValue(vbColor) = vbColorValue(vbColor - 1) + 255
      End If
  End Select
  'Set Color Name
  ColorMatrix(vbColor, 1) = Colors(vbColor)
  'Set Color Value
  ColorMatrix(vbColor, 2) = vbColorValue(vbColor)
Next vbColor
End Sub
```

The second dimension of the passed array is what determines which parameter is stored. If the second dimension is a "1," then a text item storing a color name is held in that position. If the second dimension is a "2," then a number that is the color constant is held in that position. The foregoing subroutine can be tested rather easily by running the following subroutine:

```
Sub TestCreateColorArray()
Dim ColorMatrix()
Call CreateColorArray(ColorMatrix())
For ii = 1 To UBound(ColorMatrix(), 1)
  Debug.Print ii, ColorMatrix(ii, 1), ColorMatrix(ii, 2)
Next ii
End Sub
```

This will produce the following output in the immediate window:

```
1 vbBlack             0
2 vbRed               255
3 vbGreen             65280
4 vbYellow            65535
```

5	vbBluen	16711680
6	vbMagenta	16711935
7	vbCyan	16776960
8	vbWhite	16777215

Notice in the testing subroutine `TestCreateColorArray` the use of the `UBound` function. Because the array of interest `ColorMatrix` has more than one dimension, the `UBound` command allows the user to specify the dimension for which the upper index should be returned. In this instance, the second dimension is limited to two cases, but the first dimension holds the number of colors. Thus, the `UBound` function is utilized to return the total number of indexes of the first dimension.

Also take notice that, in the `CreateColorArray` subroutine, a leading comma is present in the `ColorNames` string. This is done so that when the `Split` function is utilized to parse out the Color Names string into the `Colors()` array, the index at which the colors will start will be 1 (the array will still have a variable for `Colors(0)`, but it will be an empty string). While it is tempting to declare the `Colors()` array like this: {Dim Colors(1 to 8)} to try to accomplish the same thing, it will not work. If an array is predimensioned with a specified lower limit, the `Split` function will fail.

5.3 FORM REUSE IN VBA PROJECTS

In the examples that follow, a number of the forms are very similar in nature. Rather than duplicate effort and recreate the forms that will be utilized to perform a similar task as forms that preexist in another project, it would be most effective to be able to reuse forms that currently already exist in the VBA environment. It is possible to export an existing form, make a copy of it, and then reimport the form into a VBA project. However, there is no documentation on how to accomplish this task. Because many of the example projects in this chapter did in fact reutilize forms in earlier examples, it is appropriate to discuss how to accomplish this task prior to explaining the sample applications.

The first step in copying a preexisting form for use in another VBA project is to export the form. This can be done by going to the Project Explorer, left-clicking on the form to be reused, and selecting Export File. It is often advantageous to export the form to its own folder: "C:\temp\VBAfoldername." Once this has been done, two files will be exported (Figure 5.6). Both will have the form name with the extensions .frm and .frx, respectively.

Once this has been done, the user should navigate to the directory where the form was exported to. The form file with the extension .frm should then be opened in a text editor such as Notepad or WordPad. This can be accomplished by right-clicking on the file with the extension .frm and selecting Open With and choosing WordPad or Notepad as editors. (A word of caution is in order here. The extension *.frm is associated with WordPerfect. If WordPerfect is installed on the machine being utilized and one double-clicks on the *.frm file, it will open in WordPerfect. DO NOT use WordPerfect to edit the *.frm file. WordPerfect will put some proprietary formatting on the file, and it will not be importable back into the Excel VBA environment.)

Once the .frm file has been opened, a new name for the form should be chosen. Every reference to the old form name must be replaced with the new form name in the *.frm file. The easiest way to accomplish this is to look for this line that is at the top of every *.frm file. (The numbers and letters following Begin may be different.)

Begin {C62A69F0-16DC-11CE-9E98-00AA00574A4F} frmShowRefEditErrs

At the end of this line the original exported form name is always found. Simply do a Find and Replace on the old form name with a selected new form name on the entire *.frm document. Now resave the *.frm document *as a text document only* using the new form name as the file name and retaining the *.frm extension (Example: NewFormName.frm). Once that has been accomplished,

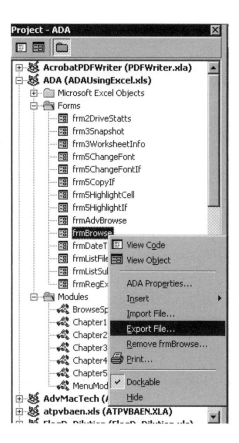

FIGURE 5.6 How to export a form in VBA.

change the name of the file with the *.frx extension to the new form (and file) name (Example: NewFormName.frx).

Once this process has been completed, essentially a carbon copy of the old form was created that has the new form name. This form can now be imported into any project by selecting the project in the Project Explorer (click on the area where the form is to be imported/located) where the form is to be imported to. Now select from the Menu File->Import File and choose either VB Files or Form Files from the drop-down menu. If the form does not import correctly, chances are that one of the following mistakes was made:

1. The *.frm file was not saved as type text from a generic text editor such as Notepad or WordPad. Do not use Word or WordPerfect to edit this file.
2. The *.frx file did not have its name changed to the new form name.
3. Both files (.frm, *.frx) were not located in the same directory when imported.

5.4   THE REFEDIT CONTROL AND ITS ASSOCIATED PROBLEMS

The samples given in this chapter make extensive use of the RefEdit control. The RefEdit control is shown in Figure 5.7. The RefEdit control is utilized to allow the user of an application to select a cell or a range of cells. It is the only control available within Excel to accomplish this task. Unfortunately, some very nasty bugs exist within the RefEdit control that Microsoft has known about for years that they have chosen not to fix. During the time frame this text was written, these bugs were not documented in Microsoft's online knowledgebase but were present in Microsoft's bug database. An application has been constructed that illustrates many of the bugs present within the RefEdit

FIGURE 5.7 The RefEdit control on the component palette.

control in the Workbook RefEditErrors.xls. This Workbook was created and sent to Microsoft by the author with the hope that they would fix the associated long-standing problems (bugs) within this control.

1. The most nasty bug is the fact that the RefEdit control will not work with a nonmodal form. If a form's modal property is set to False, and the form has a RefEdit control in it, Excel will hang up and then crash. If a form's .ShowModal property is set to FALSE at design time, the RefEdit control will not let go and return control of the macro to the form. This is true even after a selection has been made on the form and/or the Enter button has been pressed on the keyboard. This error can be demonstrated with the Workbook "RefEditErrors.xls" included on the accompanying CD-ROM (\Chapter 5 \Samples\RefEditErrors.xls) by choosing the option button "Hang and Crash Excel with RefEdit Control Use." If the "RefEdit Bugs Demo" button is pressed on the sheet with foregoing option set, the only way to regain control of Excel is to shut Excel down using Ctrl-Alt-Delete! This obviously greatly limits the utility of the control.

2. The Events `AfterUpdate` and `BeforeUpdate` never fire when utilized by a RefEdit Control. This is a debilitating bug for the following reasons. Immediately after a RefEdit Control has its contents updated, it would be prudent to run a check on the contents to ensure that the entered values point to where they should, or encompass a valid range for which they are supposed to reference. It might also be advantageous to color the current range pointed to in the RefEdit Control on the Worksheet by changing the color of the font or background color in the cells referenced by the control. Because the `AfterUpdate` Event does not fire for the RefEdit control, creating such a validation check is impossible. Conversely, immediately before updating the current contents of the RefEdit Control, it might be necessary to make some changes to the last region it pointed to prior to changing the contents of the RefEdit control. Perhaps the font in that Range should have its color returned to normal. Because the `BeforeUpdate` Event does not fire for the RefEdit control, creating such routines is not possible.

3. Setting the .Font property utilizing either the `Enter` or `Exit` events with the RefEdit control not only fails to work but causes the program to abnormally terminate as soon as focus is set to any other object on the form other than the RefEdit control. Simply uncomment out the marked VB code items in the Workbook "RefEditErrors.xls" to observe this effect. Note that the code in the following subroutines works fine when called from the Change Font Color button's click event!

```
Private Sub RefEdit_Enter()
    'MsgBox "RefEdit Enter Event Occurred", vbOKOnly +
    vbInformation, "Event Triggered"
```

FIGURE 5.8 A RefEdit control in a form.

```
'If RefEdit.Text <> "" Then Range(RefEdit.Text).Font.Color
= vbBlue

End Sub

Private Sub RefEdit_Exit(ByVal Cancel As MSForms.ReturnBoolean)
'MsgBox "RefEdit Exit Event Occurred", vbOKOnly +
vbInformation, "Event Triggered"
'If RefEdit.Text <> "" Then Range(RefEdit.Text).Font.Color
= vbRed
End Sub
```

With all of these associated problems with the RefEdit control, the reader may wonder why it is utilized in so many example applications or why it is even utilized at all. The answer, quite simply, is that it is the only game in town. A process of any degree of sophistication will require the user to select ranges within Worksheets, and the RefEdit control is the only mechanism available to accomplish this task. Therefore, building an application that is even remotely complex will require the developer to utilize the RefEdit Control and work around its associated problems. When a RefEdit control is placed in a form, it looks like the image shown in Figure 5.8.

When the RefEdit control in a form becomes active, the form will disappear, and a small rectangular box will appear that displays the region selected on the sheet. In addition, the region currently selected by the RefEdit control will have dashed lines surrounding it on the active sheet, indicating the current selection the user is making. A small box will also appear to the bottom right of the range selected, indicating the number of rows and columns currently selected in the RefEdit control. Figure 5.9 shows the RefEdit control when used in the Workbook RefEditErrors.xls.

Once a range has been selected in a RefEdit control, the logical question becomes how can the Range selected within the control be utilized within Excel VBA Code? There are a number of ways to utilize the Range returned by a RefEdit control, and those that lend themselves to repeated use will be discussed.

A sample application has been created to show the reader how to utilize the RefEdit control. It can be run by selecting ADA->Chapter 5->Ref Edit Control Use from the menu, which will cause the GUI shown in Figure 5.10 to appear. This application utilizes the range selected by the user with the RefEdit control in the GUI and then returns relevant information about the selected range when the "Range Info" button is pressed. The mechanisms that enable this example to work will now be discussed.

The first parameter returned tells the user if the range selected in the RefEdit control encompasses a single cell or a (group) range of cells. This information is determined with the `Single-CellInRange` function.

```
Function SingleCellInRange(RefEditTxt) As Boolean
'If a ":" is in the Range then More than one cell is in Range!
If InStr(RefEditTxt, ":") <> 0 Then
    SingleCellInRange = False
    Else
    SingleCellInRange = True
End If
End Function
```

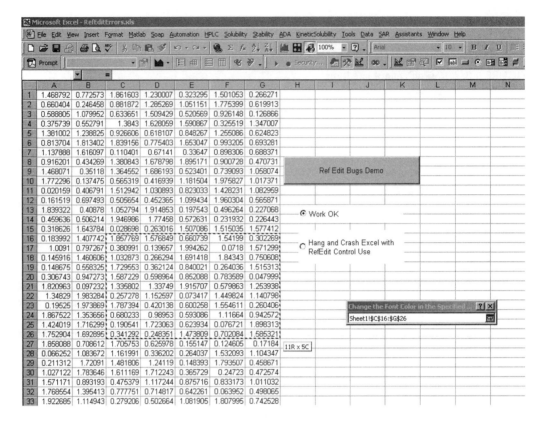

FIGURE 5.9 The RefEdit control in action.

This mechanism by which the function works is quite simple. The ranges returned by the RefEdit control will have a ":" in them if multiple cells are selected.

Examples: Sheet1!R11C5:R20C8   Sheet1!$E$11: $H$20

However, if only a single cell is selected, no ":" will be present in the range returned by the RefEdit control.

Examples: Sheet1!R13C13   Sheet1!$M$13

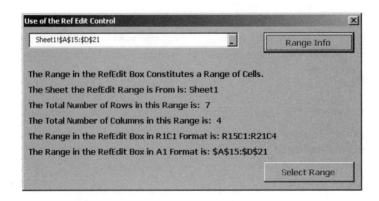

FIGURE 5.10 A sample application for utilization of the RefEdit control.

The `InStr` function will search for a character in a string and return the first instance of that character (positionwise from left to right) if that character is present in the specified string. If the character is not found within the string, the `InStr` function will return a 0. If the range returned by the RefEdit control does not contain a ":", then the range only encompasses a single cell.

It is easy to determine which sheet the RefEdit control returned the range from because that information is always given by the control. The sheet name is the first piece of information and is separated from the actual range by a "!". The function `SheetInRange` determines the sheet name the range references by utilizing the InStr function to determine the end of the sheet name and the `Mid` function to return the portion of the RefEdit range that contains the sheet name.

```
Function SheetInRange(RefEditTxt) As String
'Returns the SheetName in the RefEdit Controls Range
SheetInRange = Mid(RefEditTxt, 1, InStr(RefEditTxt, "!") - 1)
End Function
```

Ranges can, of course, be referenced in two ways: by the A1 method or the R1C1 method. Depending upon how the Range is to be utilized, it would be nice to have the capability to return the range in either format upon demand. This can be done rather easily by utilizing the following methodology. First, dimension three variables, one as a range and two as strings.

```
Dim CtrlRange As Range
Dim R1C1Address As String
Dim A1Address As String
```

Now the range within the RefEdit Control must be extracted. This can be done utilizing either the .Text or .Value properties.
Example:

```
CtrlText = frm5RefEditUse.RefEdit1.Text
CtrlValue = frm5RefEditUse.RefEdit1.Value
```

(Assuming `CtrlText` and `CtrlValue` are both String variables.)
Now the declared Range variable must be set to the range specified in the RefEdit control. This is done with the Set command.

```
Set CtrlRange = Range(CtrlText)
```

Once a range object variable has been set to the range contained within the RefEdit control, it is trivial to return the address in either the A1 or R1C1 format.

```
R1C1Address = CtrlRange.Address(, , xlR1C1)
A1Address = CtrlRange.Address(, , xlA1)
```

One word of caution is in order with regard to setting range object variables. The RefEdit control always returns ranges in the A1 format as an absolute address (recall that absolute addresses have $'s in them and will always point to the same cell). A range object variable can be set to a range depicted by a string variable representing a range in A1 style format, but it cannot be set to a range depicted by a string variable representing a range in R1C1 format. This is a bug in Excel.

Example:

```
'This Command Will Work
Set CtrlRange = Range("$A$12:$D$15")
'And this Command Will Not Work
Set CtrlRange = Range("R8C1:R10C3")
```

Taking this into account, the last parameters to be extracted are the number of rows and columns within the selected range. Two functions have been constructed to accomplish this. One returns the number of rows in the range, and the other returns the number of columns in the range.

```
Function RowsinRange(ByVal RangeAddress) As Integer
'Returns the Total Number of Rows in a Range
Dim topleft$, bottright$
Dim startrow As Integer, endrow As Integer
Dim CtrlRange As Range, ii As Integer
If InStr(RangeAddress, "$") <> 0 Then
    'Convert Range Address to R1C1 Format
    Set CtrlRange = Range(RangeAddress)
    RangeAddress = CtrlRange.Address(, , xlR1C1)
End If
If InStr(RangeAddress, ":") = 0 Then
    RowsinRange = 1
    Exit Function
    Else
    topleft$ = Mid(RangeAddress, 1, InStr(RangeAddress, ":") - 1)
    bottright$ = Mid(RangeAddress, InStr(RangeAddress, ":") +
    1, Len(RangeAddress))
End If
'Determine Starting Row
For ii = 1 To Len(topleft$)
    If Mid(topleft$, ii, 1) = "C" Then
        startrow = Val(Mid(topleft$, 2, ii - 1))
        Exit For
    End If
Next ii
'Determine Ending Row
For ii = 1 To Len(bottright$)
  If Mid(bottright$, ii, 1) = "C" Then
     endrow = Val(Mid(bottright$, 2, ii - 1))
     Exit For
  End If
Next ii
RowsinRange = (endrow - startrow) + 1
End Function

Function ColsinRange(ByVal RangeAddress) As Integer
'Returns the Total Number of Columns in a Range
Dim topleft$, bottright$
Dim startcol As Integer, endcol As Integer
Dim CtrlRange As Range, ii As Integer
If InStr(RangeAddress, "$") <> 0 Then
```

```
      'Set CtrlRange Range object to the range address passed
      Set CtrlRange = Range(RangeAddress)
      'Convert Range Address to R1C1 Format
      RangeAddress = CtrlRange.Address(, , xlR1C1)
   End If
   If InStr(RangeAddress, ":") = 0 Then
      ColsinRange = 1
      Exit Function
      Else
      topleft$ = Mid(RangeAddress, 1, InStr(RangeAddress, ":") - 1)
      bottright$ = Mid(RangeAddress, InStr(RangeAddress, ":") +
      1, Len(RangeAddress))
   End If
   'Determine Starting Column
   For ii = 1 To Len(topleft$)
   If Mid(topleft$, ii, 1) = "C" Then
      startcol = Val(Mid(topleft$, ii + 1, Len(topleft$)))
      Exit For
   End If
   Next ii
   'Determine Ending Column
   For ii = 1 To Len(bottright$)
     If Mid(bottright$, ii, 1) = "C" Then
        endcol = Val(Mid(bottright$, ii + 1, Len(bottright$)))
        Exit For
   End If
   Next ii
   ColsinRange = (endcol - startcol) + 1
   End Function
```

First, these functions look at the range passed to them and determine if it is an R1C1 type of range or an A1 type of range. It would create extra work to build in the ability to function using both types of ranges, so the mechanism was constructed to convert an A1 type of range to an R1C1 type of range. If an absolute type (A1) of range is passed, it will be converted to an R1C1 type of range. (Note, the mechanism built in looks for the "$" in the absolute type of range, so it will not function to convert a relative style A1 range. This suffices in this case because the RefEdit control always returns an absolute type of A1 range.)

A check is then done to determine if the range is limited to a single cell. If it is, then the number of rows or columns is set to one. If it is not, then the range is broken into two parts: the top left and bottom right corners of the range. Once that has been accomplished, the rows and columns can be extracted from the two parts of the range. A simple subtraction then determines the number of rows or columns.

A second button is present in this sample application titled "Select Range". If this button is pushed, the range present in the RefEdit box will be selected (or made active) on the active Worksheet. This is accomplished with one simple line of code:

```
Range(RefEdit1.Text).Select
```

5.5 HIGHLIGHTING AND COLORING CELL
FONTS AND BACKGROUNDS

One of the most useful ways to call attention to information of interest on a Worksheet is to apply color to either the text that contains the information or the background of the cell that holds the information. This is not a difficult task to accomplish utilizing VBA. Two sample applications have been constructed to show the reader how to accomplish these tasks. The first shows how to change the color of the font in a particular cell or group of cells.

Figure 5.11 shows the GUI for the Color Cell Font sample application, which can be run from the menu ADA->Chapter 5->Color Cell Font. The user must choose a selection on the Worksheet utilizing the RefEdit control. A color must then be selected from the drop-down list. Notice that this application makes use of the standard VB colors utilizing the VB color constants discussed earlier in this chapter. Once that has been done, the user also has the option of setting the font to bold by means of a checkbox. Finally, the user may change the color of the font in the range selected by the RefEdit Control by simply pressing the "Change Font Color" button. The mechanisms that allow this sample to work are now discussed.

Upon startup the `CreateColorArray` subroutine previously developed in this chapter is utilized to populate the combo box with the VB Color Constants.

```
Private Sub UserForm_Activate()
Dim ii As Integer
Call CreateColorArray(Colors())
For ii = 1 To UBound(Colors(), 1)
  frm5ChangeFont.ColorComboBox.AddItem Colors(ii, 1)
Next ii
End Sub
```

When the "Change Font Color" button is pressed, the program checks to determine if a valid range is present in the RefEdit control and if a valid color is present in the combo box. If both selections are valid, then the color of the font in the specified region is changed. The following snippet of code accomplishes this.

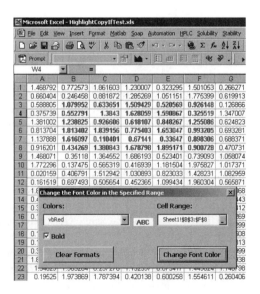

FIGURE 5.11 The Color Cell Font sample application.

```
Private Sub Button_ChangeColor_Click()
If RefEdit.Text = "" Then
   MsgBox "You Must Choose a Range!", vbOKOnly + vbExclamation,
   "No Range Selected!"
   Exit Sub
End If
If ColorComboBox.Text = "" Then
   MsgBox "You Must Choose a Highlighting Color", vbOKOnly +
   vbExclamation, "Highlighting Color Not Specified!"
     Exit Sub
End If
Range(RefEdit.Text).Font.Color =
Colors(ColorComboBox.ListIndex + 1, 2)
If CheckBox_Bold.Value = True Then
Range(RefEdit.Text).Font.Bold = True
End Sub
```

The "Clear Formats" button allows the user to erase all the formatting on the active Worksheet. This can very easily be accomplished with a single line of code.

```
Cells.ClearFormats
```

Recall that the RefEdit control cannot be utilized with nonmodal forms. This project is simple in nature in that it assumes that any formatting applied utilizing a RefEdit control was done to the current active Worksheet because a modal form will not allow the user to activate a new Workbook. Although a modal form will allow a user to jump around to different Worksheets, often when this is done while a RefEdit control is active, Excel will crash, and the user will see a message box like that shown in Figure 5.12. It is best to confine use of the RefEdit control to select ranges on the current active sheet.

Notice that a sample of the color the font is to be changed to is displayed right next to the combo box that selects the color of the font. This is accomplished quite easily with a label control and a single line of code that is executed whenever the contents of the Color Combo Box is changed.

```
Private Sub ColorComboBox_Change()
Color_Label.ForeColor = Colors(ColorComboBox.ListIndex + 1, 2)
End Sub
```

The second sample application demonstrates an equally important method of calling attention to Worksheet data, and that is highlighting the background of a cell or a particular group of cells. A really good color for this purpose is yellow because it gives the impression that someone took a highlighter and just colored in the background of the cells in the range of interest. This application was constructed utilizing the form reuse methods described earlier in this chapter. Because this application has nearly the same functionality as the previous application, there was no sense in creating a form from scratch.

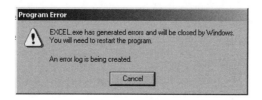

FIGURE 5.12 Possible result of changing Worksheets with an active RefEdit control.

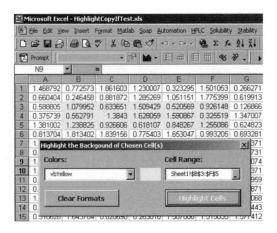

FIGURE 5.13 The Color Cell background sample application.

The Color Cell Background can be run from the menu by selecting <u>A</u>DA->Chapter <u>5</u>->Color Cell <u>B</u>ackground, which will activate the GUI shown in Figure 5.13. The only real difference between this application and the Color Cell Font sample application is the properties that have to be set in order to color the background interior of a Worksheets cell. When the "Highlight Cells" button is pressed, the following code is executed.

```
Private Sub Button_Highlight_Click()
If RefEdit.Text = "" Then
   MsgBox "You Must Choose a Range!", vbOKOnly + vbExclamation,
   "No Range Selected!"
   Exit Sub
End If
If ColorComboBox.Text = "" Then
   MsgBox "You Must Choose a Highlighting Color", vbOKOnly +
   vbExclamation, "Highlighting Color Not Specified!"
     Exit Sub
End If
Range(RefEdit.Text).Interior.Color =
Colors(ColorComboBox.ListIndex + 1, 2)
Range(RefEdit.Text).Interior.Pattern = xlSolid
End Sub
```

As in the previous instance, a valid range and color must be selected, and checking mechanisms are built into the code to ensure that they are present. To highlight the background of a cell, the cell's interior color must be set along with the interior pattern set to solid. The two highlighted areas of code just presented show how to accomplish this. It is worth mentioning that the same effort could be achieved by means of the "with–end with" construct as follows.

```
Range(RefEdit.Text).Select
With Selection.Interior
  .Color = Colors(ColorComboBox.ListIndex + 1, 2)
  .Pattern = xlSolid
End With
```

The only disadvantage to utilizing such a procedure is that the range must be *selected* prior to performing the action on the cells within the range, whereas in the previous instance, the range is merely *specified,* and the actions are performed upon the *specified* range without the need for *selecting* it.

These two sample applications provide the building blocks required to develop more sophisticated data mining procedures, which will be discussed in the subsequent sections of this chapter.

5.6 CREATING A HIGHLIGHT IF TOOL

The next logical step is to create some applications that will highlight entries in a Worksheet should they meet a specified criteria. Some examples have been created that perform a variety of useful functions. The next sample application will highlight the background of a cell to any color the user specifies, provided certain criteria are met. The thought process that went on before a single line of code was written to produce this application will now be discussed.

First, the user may not wish to perform the conditional highlighting over the entire Worksheet. Therefore, a mechanism should be in place to let the user decide whether to format the entire sheet or a portion of the sheet. Because most relational databases export data with each column containing a single parameter, it would be wise to allow the user the option of limiting the conditional highlighting to just one column.

Next comes what criteria should be checked. Probably, the often-sought comparisons would be those that are numerical in nature. Therefore, it would make sense to compare the contents of a cell to some value and highlight the cells' background if the cells' value were equal to, greater than, less than, or some combination thereof of the selected value. Thus, the operators =,>,>=,<,<=,<> should be utilized.

This brings about the question: how should the value be selected for comparison to these operators? Two possibilities come to mind for selection of a comparison value. The value for comparison could exist within a Worksheet and thus be pointed to by means of a range. The value could also be a discrete value that the user could type into a GUI. This application makes provisions for both instances.

With the foregoing considerations in mind, it is now possible to construct a GUI that will allow the user to set the above parameters. Looking at Figure 5.14, the user is prompted as to whether the formatting should be applied to the entire sheet or just a particular column. This is done by means of the option buttons contained in the "Look In" frame. If the user elects to just apply the formatting to a single column, the column to be acted on is specified by means of a spin button that increments or decrements the column to be acted on in a textbox immediately adjacent to the spin button.

The color to be utilized for highlighting the cells' background is specified in exactly the same manner as in the previous highlight cell background sample application.

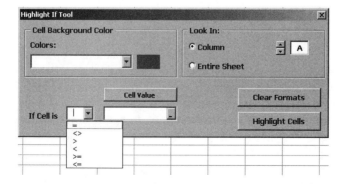

FIGURE 5.14 The Highlight If GUI.

The conditional statement that determines if a cells' value is to be highlighted required some thought for implementation in this GUI. Because the conditional statement should be able to make comparisons on both a static value specified by the user or a value that resides in a Worksheet, a mechanism had to be implemented in the GUI that allowed both possibilities to occur. In this instance, this was accomplished by means of having both a RefEdit Control (used to specify a single Worksheet cell), and a textbox (used to specify a discrete value). The RefEdit control and the textbox are placed on top of each other (in the same physical space) of the GUI. Depending upon which value is to be utilized, either the RefEdit control or the textbox is enabled and made visible. The control not to be utilized is then disabled and hidden. This is a very elegant way of accomplishing this task and allows real estate on the GUI to be conserved. To switch between the RefEdit control and the textbox, a button is utilized that changes state with each button press. The caption on the button identifies what state the GUI is in: "Cell Value" or "Textbox Value." The code attached to the button that performs this action is

```
Private Sub Button_PickValue_Click()
'Change (Toggle) State Each Time Button is Pressed
If ButtonState = False Then
   ButtonState = True
   Else
   ButtonState = False
End If

Select Case ButtonState
  Case True
     frm5HighlightIf.Button_PickValue.Caption = "Cell Value"
     frm5HighlightIf.RefEdit_Cell.Enabled = True
     frm5HighlightIf.RefEdit_Cell.Visible = True
     frm5HighlightIf.TextBox_StatVal.Enabled = False
     frm5HighlightIf.TextBox_StatVal.Visible = False
  Case False
     frm5HighlightIf.Button_PickValue.Caption = "Textbox Value"
     frm5HighlightIf.RefEdit_Cell.Enabled = False
     frm5HighlightIf.RefEdit_Cell.Visible = False
     frm5HighlightIf.TextBox_StatVal.Enabled = True
     frm5HighlightIf.TextBox_StatVal.Visible = True
  End Select
  End Sub
```

This provides a simple way to select the value that the conditional operators will use to compare the current cell value to. (Note that ButtonState is a Public or Global variable declared at the module level, although a variable declared with the `Static` parameter could have been utilized at the procedural level to provide the same functionality.)

Figure 5.15 illustrates the use of this application to highlight the background of all cells in Column 1 (Column A) whose cell value is >= the value contained in Column 1 row 19 ($A$19), which is 0.922237. Cells whose value is equal to or greater than the value of 0.922237 have had their background changed from white to yellow.

Figure 5.16 provides a similar illustration as was shown in Figure 5.15. Figure 5.16 illustrates the use of this application to highlight the background of all cells in Column 2 (Column B) whose

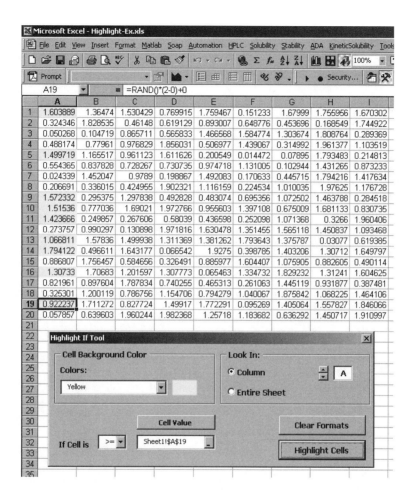

FIGURE 5.15 Example No. 1: the Highlight If application.

cell value is > the value contained in the textbox on the GUI, which in this case is 1.01. Cells whose values are greater than the value of 1.01 have had their background changed from white to red.

Figure 5.17 illustrates the use of this application to highlight the background of all cells in the entire Worksheet whose cell value is > the value contained in Column A row 1 ($A$1), which in this case is 1.603889. Cells whose value is equal to or greater than the value of 1.603889 have had their background changed from white to yellow.

The "Highlight Cells" button's click event contains the code that actually causes the cells to be highlighted, provided the value within a cell is in agreement with the condition specified in the GUI.

```
Private Sub Button_Highlight_Click()
Dim row As Long, colstpt As Integer
Dim colmax As Integer, rowmax As Long
'Test for Proper Selections
Select Case ButtonState
  Case True 'Cell Value
    If RefEdit_Cell.Text = "" Or InStr(RefEdit_Cell.Text, ":")
    <> 0 Then
```

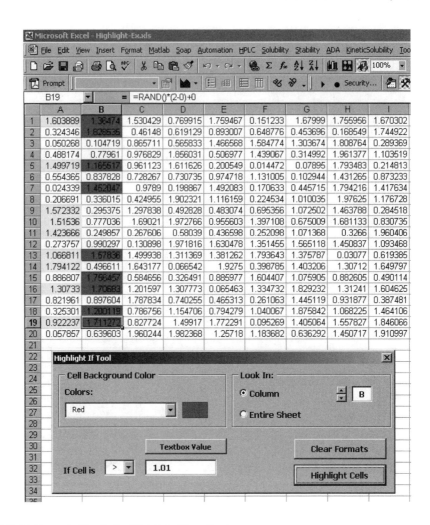

FIGURE 5.16 Example No. 2: the Highlight If application.

```
      MsgBox "You Must Choose a Range Comprising a SINGLE
      Cell!", vbOKOnly + vbExclamation, _

            "Improper Range Selected!"
        Exit Sub
      End If
      Compare = Range(RefEdit_Cell.Text).Value
  Case False 'TextBox Value
    If TextBox_StatVal.Text = "" Or
    IsNumeric(TextBox_StatVal.Text) = False Then
        MsgBox "You Must Specify a Comparison Value in the
        Textbox!", vbOKOnly + vbExclamation, "Improper Compare
        Value!"

        Exit Sub
    End If
    Compare = Val(TextBox_StatVal.Text)
End Select
```

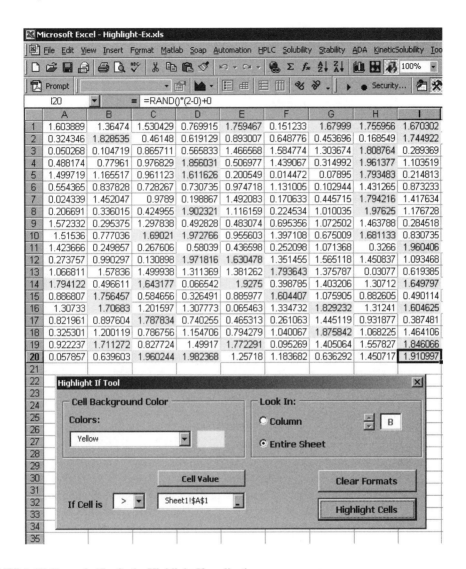

FIGURE 5.17 Example No. 3: the Highlight If application.

```
If ComboBox_Cond1.Text = "" Then
    MsgBox "You Must Choose a Comparison Operator!", vbOKOnly
    + vbExclamation, "Comparison Operator Missing!"
    Exit Sub
End If

If ColorComboBox.Text = "" Then
    MsgBox "You Must Choose a Highlighting Color", vbOKOnly +
    vbExclamation, "Highlighting Color Not Specified!"
    Exit Sub
End If

'Execute Highlighting
Select Case frm5HighlightIf.OptionButton_All
```

```
Case True
  'Search Entire Worksheet
  colstpt = 1: colmax = LastCol(ActiveWorkbook.Name,
  ActiveSheet.Name)
  rowmax = LastRow(ActiveWorkbook.Name, ActiveSheet.Name)
Case False
  'Just Search Specified colstptumn
  colstpt = Asc(TextBox_Col.Text) - 64: colmax =
  Asc(TextBox_Col.Text) - 64

  rowmax = LastRow(ActiveWorkbook.Name, ActiveSheet.Name)
End Select

For row = 1 To rowmax
  For col = colstpt To colmax
    If CheckCondition (Cells(row, col), Compare) = True Then
        Cells(row, col).Select
        With Selection.Interior
          .Color = Colors(ColorComboBox.ListIndex + 1, 2)
          .Pattern = xlSolid
        End With
    End If
  Next col
Next row
End Sub
```

Prior to executing, this subroutine makes some preliminary checks to determine if the user has made selections for all the necessary parameters prior to executing the highlighting algorithm. Although it requires some extra effort to incorporate such error-detection mechanisms, it is well worth the effort because it prevents both errors and unexpected results from occurring. Specifically, the subroutine checks for the following items prior to execution:

1. If the RefEdit control was to be utilized to pick a comparison value from the Worksheet, was only a single cell selected?
2. If a Comparison Value needs to be entered on the GUI, was this done?
3. Was a valid Comparison Operator chosen?
4. Was a color chosen for highlighting the background of those cells that met the specified criteria?

If all of these were chosen, then the macro will highlight the cells that meet the specified criteria. This is done by means of a looping structure. Prior to beginning the loop, a determination is made if the comparison operation should be made throughout the entire Worksheet or just a single column, and the looping parameters are set accordingly. Notice that the comparison operator is chosen by means of a Combo box. This makes writing a comparison statement problematic, as many different statements must be written depending upon the number of comparison operators. For clarity's sake, a function has been created to return a Boolean value of true or false depending upon if the condition tested for proves true. This greatly improves the readability of the Button_Highlight_Click subroutine.

```
Function CheckCondition(CellValue, CompareValue) As Boolean
  Select Case ComboBox_Cond1.Text
```

```
 Case "="
   If CellValue = CompareValue Then CheckCondition = True
 Case "<>"
   If CellValue <> CompareValue Then CheckCondition = True
 Case ">"
   If CellValue > CompareValue Then CheckCondition = True
 Case "<"
   If CellValue < CompareValue Then CheckCondition = True
 Case ">="
   If CellValue >= CompareValue Then CheckCondition = True
 Case "<="
   If CellValue <= CompareValue Then CheckCondition = True
End Select
End Function
```

5.7   CREATING A COLOR FONT IF TOOL

It is just as likely that the user would want to be able to set the font color of a cell to a certain color, provided the value in the cell met a certain condition or conditions. Because such an application is similar to the "Highlight Cell If" sample application created in the previous section, it only makes sense to reuse the form from the "Highlight Cell If" sample application. As shown in Figure 5.18, there are some minor differences in the GUI created from the earlier application, specifically that a checkbox has been added in the event that the user wants to turn the font bold, and that the color preview pane shows a colored font as opposed to a colored background.

Figure 5.19 illustrates an example of where the font in Column A (Column 1) has been changed to bold red when the value in a cell exceeds (>) the value in cell $A$1 (which is 1.603889). In this instance, only one cell matches this criteria, cell $A$14, and its font has been changed to bold red, whereas all the other cells' fonts remain unchanged.

Figure 5.20 illustrates an example of where the criteria is checked throughout the entire sheet. In this example, a cell's font is changed to bold green if its value is less than the specified value in the textbox in the GUI (1.01).

The only difference between this sample application and the previous sample application are the properties that are set. Here the cell's font properties of color and boldface are set as opposed to the cell's interior color and pattern properties. Thus, the lines in looping structure for the "Change Font in Cells" button are changed to

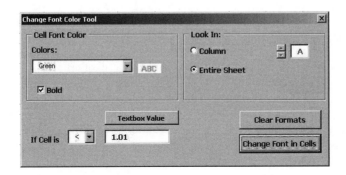

FIGURE 5.18 The Color Font If GUI.

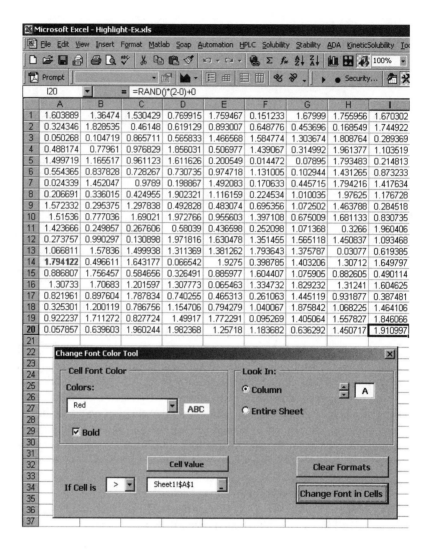

FIGURE 5.19 Example 1: the Color Font If sample application.

```
Cells(row, col).Font.Color = Colors(ColorComboBox.ListIndex + 1, 2)
If CheckBox_Bold.Value = True Then Cells(row, col).Font.Bold = True
```

5.8 CREATING A COPY IF/MOVE IF TOOL

The ability to highlight or draw attention to data based on a set of criteria is useful but sometimes not useful enough. This methodology can be taken a step further by allowing a user to either copy or move data out of one Worksheet and into another Worksheet based on a similar set of criteria (Figure 5.21). As was the case with the previous projects, some thought was required before designing this tool. First, it would be desirable to have the ability to choose between moving the data or copying the data with the same tool. This can be accomplished by means of a button that toggles between the move and copy states with each button press. It will also be necessary to pick which Workbook and Worksheet the information will be copied or moved to. This can be accomplished by means of a Combo box utilizing object variables as was done

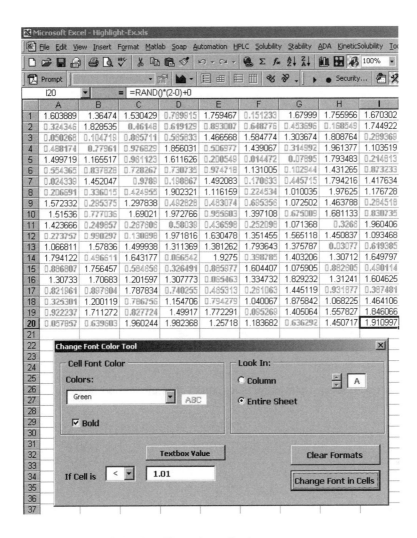

FIGURE 5.20 Example 2 - the Color Font If sample application.

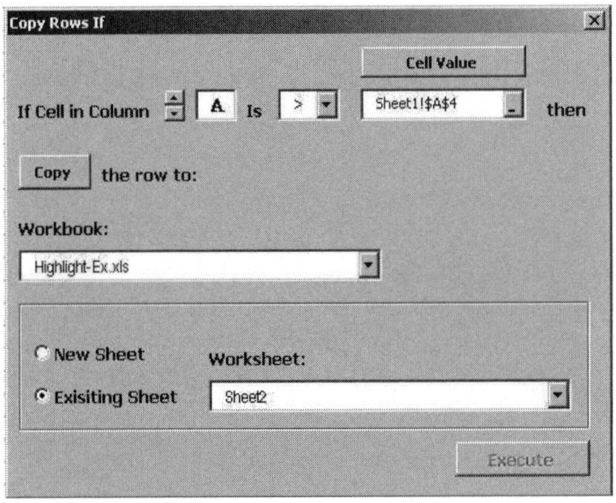

FIGURE 5.21 The GUI for the Copy/Move If sample application.

earlier in the chapter. It would also be worthwhile to have the ability to copy or move the data to a new sheet within an existing Workbook, or into a new Workbook altogether. As in the previous instances, the user should have the ability to pick the comparison value from either a Worksheet cell or specify a static value by means of a textbox. A button that toggles state is again utilized to determine how the comparison operator is specified.

Notice that when the user clicks the New Sheet option button, the existing Worksheets Combo box is grayed out and disabled. This is done on purpose, as it will not be necessary for the user to pick a Worksheet to copy or move the information to if it is to be copied or moved to a new sheet. Conversely, when the user clicks on the Existing Sheet option button, the existing Worksheets Combo box is enabled and repopulated with the Worksheets that exist within the chosen Workbook. The code that accomplishes this is placed within the click events of the option buttons.

```
Private Sub OptionButton_ES_Click()
frm5CopyIf.Label_Wksheets.Enabled = True
frm5CopyIf.ComboBox_Worksheets.Enabled = True
  Call PopulateWorksheets
End Sub

Private Sub OptionButton_NS_Click()
frm5CopyIf.Label_Wksheets.Enabled = False
frm5CopyIf.ComboBox_Worksheets.Enabled = False
End Sub
```

It is the PopulateWorksheets subroutine that actually fills the combo box control with the names of the Worksheets.

```
Sub PopulateWorksheets()
Dim VisibleWorksheets() As String

'Clear any Preexisting Entries
Call ClearWorksheetComboBox

'If No Workbook is Specified in the Chosen Combobox then Exit!
If ComboBox_Workbooks.Text = "" Then Exit Sub

If ComboBox_Workbooks.Text = "New Workbook" Then
    Call OptionButton_NS_Click
    frm5CopyIf.OptionButton_ES.Enabled = False
    frm5CopyIf.OptionButton_NS.Enabled = False
    Exit Sub
    Else
    'Call OptionButton_ES_Click
    frm5CopyIf.OptionButton_ES.Enabled = True
    frm5CopyIf.OptionButton_NS.Enabled = True
    frm5CopyIf.Label_Wksheets.Enabled = True
    frm5CopyIf.ComboBox_Worksheets.Enabled = True
End If
'Populate Visible Worksheet Combobox
TotalVisible = TotalVisibleWorksheets(VisibleWorksheets(),
ComboBox_Workbooks.Text)
If TotalVisible > 0 Then
```

```
    For ii = LBound(VisibleWorksheets()) To TotalVisible
      frm5CopyIf.ComboBox_Worksheets.AddItem
      VisibleWorksheets(ii)

    Next ii
  End If
End Sub
```

Notice that this subroutine contains a safety mechanism that ensures that a Workbook has been selected by the user. If the user chooses to populate a new Workbook with information to be copied or moved, he or she are not allowed to specify which sheet the information is to be copied to. Because every Workbook created in Excel must contain at least one new Worksheet (the user may specify how many Worksheets are created in every new Workbook from Tools->Options [General Tab]), the application will always copy or move the information to the first Worksheet in a newly created Workbook.

The PopulateWorksheets subroutine populates the Combo boxes on the form with the names of the existing visible Worksheets. The PopulateWorksheets subroutine is called when one of the following conditions occurs. The Workbook from which information is to be copied or moved to is changed, or the Worksheet the information is to be copied to is changed from a new Worksheet to an existing Worksheet.

It is the code in the "execute" button's click event that actually sets in motion the mining of the information and copies or moves it to the specified sheet.

```
Private Sub Button_Execute_Click()
Dim row As Long, rowmax As Long
Dim ColRef As Integer, trowptr As Long, frowptr As Integer

'Test for Proper Selctions
Select Case ButtonState
  Case True 'Cell Value
    If RefEdit_Cell.Text = "" Or InStr(RefEdit_Cell.Text, ":")
    <> 0 Then

      MsgBox "You Must Choose a Range Comprising a SINGLE
      Cell!", vbOKOnly + vbExclamation, "Improper Range
      Selected!"

      Exit Sub
    End If
    'Assign @ Time of Change!
    'Compare = Range(RefEdit_Cell.Text).Value
  Case False 'TextBox Value
    If TextBox_StatVal.Text = "" Or
    IsNumeric(TextBox_StatVal.Text) = False Then
      MsgBox "You Must Specify a Comparison Value in the
      Textbox!", vbOKOnly + vbExclamation, "Improper Compare
      Value!"
      Exit Sub
    End If
    Compare = Val(TextBox_StatVal.Text)
End Select
```

```
If ComboBox_Cond1.Text = "" Then
    MsgBox "You Must Choose a Comparison Operator!", vbOKOnly
    + vbExclamation, "Comparison Operator Missing!"
    Exit Sub
End If

If ComboBox_Workbooks.Text = "" Then
    MsgBox "You Must Choose a Workbook to Copy To!", vbOKOnly
    + vbExclamation, "Destination Workbook Missing!"
     Exit Sub
End If
If ComboBox_Worksheets.Text = "" And OptionButton_NS.Value <>
True And ComboBox_Workbooks.Text <> "New Workbook" Then
    MsgBox "You Must Choose a Worksheet to Copy To!", vbOKOnly
    + vbExclamation, "Destination Worksheet Missing!"
     Exit Sub
End If

If ComboBox_Workbooks.Text = "New Workbook" Then
    Workbooks.Add
    CopyToBook = ActiveWorkbook.Name: CopyToSheet =
    ActiveSheet.Name

    Else
    CopyToBook = ComboBox_Workbooks.Text
    If OptionButton_NS.Value = False Then
       CopyToSheet = ComboBox_Worksheets.Text
    End If
End If

'Is a new Sheet Required?
If OptionButton_NS.Value = True Then
    Call AddNewSheet
End If

Workbooks(CopyFromBook).Worksheets(CopyFromSheet).Activate
rowmax = LastRow(ActiveWorkbook.Name, ActiveSheet.Name)
ColRef = Asc(TextBox_Col.Text) - 64
For row = 1 To rowmax
  'Debug.Print
  Workbooks(CopyFromBook).Worksheets(CopyFromSheet).Cells(row,
  ColRef), ComboBox_Cond1.Text, Compare, _
  CheckCondition(Workbooks(CopyFromBook).Worksheets(CopyFromSh
  eet).Cells(row, ColRef), Compare)
    frowptr = frowptr + 1 'Always Increment this pointer
    If
    CheckCondition(Workbooks(CopyFromBook).Worksheets(CopyFrom
    Sheet).Cells(frowptr, ColRef), Compare) = True Then
```

```
          trowptr = trowptr + 1 'Only Increment this pointer with
          Copy or Move Event
          Select Case CopyMoveButton.Caption
            Case "Move"
              Debug.Print "Move to Sheet: "; CopyToSheet
              Call MoveRow (frowptr, CopyFromSheet, CopyFromBook,
              trowptr, CopyToSheet, CopyToBook, True, True)
                  frowptr = frowptr - 1 'Decrement this pointer as
                  row was deleted after move!
            Case "Copy"
               Debug.Print "Copy to Sheet: "; CopyToSheet
               Call CopyRow (row, CopyFromSheet, CopyFromBook,
               trowptr, CopyToSheet, CopyToBook, True)
              End Select
        End If
    Next row
End
End Sub
```

This subroutine has the usual error-checking mechanisms to make sure that the user selected the items necessary for the subroutine to function. It is the looping mechanism that accomplishes the copying and moving of the data; this is of particular interest. Notice that two pointers are utilized to designate the following: the row that the information is to be copied from, and the row that the information is to be moved to.

The frowptr (from row pointer) variable is utilized to point to the row *from* which the data will be copied or moved. The trowptr (to row pointer) variable is utilized *to* point to the row where the data will be placed after being copied or moved.

Two pointers are required for the following reasons. Information will not be copied or moved unless it meets the specified criteria. Therefore, the row in which information is copied from on one sheet will not necessarily be the row it is pasted to on the destination sheet. Notice that the trowptr is only incremented immediately prior to a copy or move action on the part of the subroutine. The trowptr (to row pointer) will attain a maximum value equal to the number of rows copied from the source Worksheet.

A further complication exists when trying to move the data. Because the information is first copied on the source sheet and then deleted on it (shifting the rows up), the row from which information should be copied or moved from will not necessarily be the same as the loop index. Notice that the frowptr (from row pointer) is always incremented at the start of each loop. In the case of simply copying the data, if it meets a certain condition, the frowptr value will match that of the loop index. In the case of moving the data, however, each time a row of data is deleted and removed from the source sheet, the frowptr value must be decremented by one. The reason for this is that, when a row of data is deleted, the next row of data *shifts upward* to fill the void left when the row was deleted. As soon as this occurs, the next row to be analyzed or possibly moved is now one less than the loop counter. After this occurs twice, the next row to be analyzed or possibly moved is now two less than the loop counter, and so on. The frowptr (from row pointer) is in a constant state of self adjustment, adjusting its value *downward* to compensate for data *shifting upward* when a row is moved from the source Worksheet to the destination Worksheet.

Two subroutines are actually responsible for moving or copying a row of data. The CopyRow subroutine will copy a row of information from any Worksheet in any Workbook to any Worksheet in any Workbook. Notice the Boolean passed parameter PreserveFormatting, which allows the user to decide if special attributes such as font color, background, etc., should be retained when the row is pasted into a new Worksheet.

```
Sub CopyRow(fromRow, fromSheet, fromBook, toRow, toSheet,
toBook, PreserveFormatting As Boolean)
'Copies a Row from one sheet (fromSheet) to another sheet (toSheet)
Workbooks(fromBook).Worksheets(fromSheet).Rows(fromRow).Copy
Select Case PreserveFormatting
  Case True
    Workbooks(toBook).Worksheets(toSheet).Rows(toRow).PasteSp
    ecial Paste:=xlAll

  Case False
    Workbooks(toBook).Worksheets(toSheet).Rows(toRow).PasteSpe
    cial Paste:=xlValues
End Select
End Sub
```

The MoveRow subroutine is a slightly more sophisticated cousin to the CopyRow subroutine. The key and critical difference here is that, when data are moved, the user has a choice of one or two options. The data that are moved can be (1) deleted from the original Worksheet (and data under it will shift up to fill the void) or (2) a "hole" can be left in the Worksheet where the data were moved from. The Boolean passed parameter named ShiftUp designates if the data should be shifted up after a move operation, or if a hole should be left in the Worksheet after the move operation.

```
Sub MoveRow(fromRow, fromSheet, fromBook, toRow, toSheet,
toBook, PreserveFormatting As Boolean, ShiftUp As Boolean)
'Moves a Row from one sheet (fromSheet) to another sheet (toSheet)
'ShiftUp = True, Delete Empty Row after Cut: ShiftUp = False,
Delete Empty Row after Cut
Workbooks(fromBook).Worksheets(fromSheet).Rows(fromRow).Copy
Select Case PreserveFormatting
  Case True
    Workbooks(toBook).Worksheets(toSheet).Rows(toRow).PasteSp
    ecial Paste:=xlAll
  Case False
    Workbooks(toBook).Worksheets(toSheet).Rows(toRow).PasteSp
    ecial Paste:=xlValues
End Select

Select Case ShiftUp
  Case True
    Workbooks(fromBook).Worksheets(fromSheet).Rows(fromRow).D
    elete Shift:=xlUp
  Case False
    Workbooks(fromBook).Worksheets(fromSheet).Rows(fromRow).Clear
End Select
End Sub
```

5.9   CREATING A WINDOWING TOOL

Thus far the examples have centered around making a determination if a single condition is met, and then taking actions based on the outcome of a single condition. More often than not, however, a result is desired that falls between *two* conditions, creating a "window" in which the desirable results will reside. Suppose, for example, that the ideal value one is looking for is 100, but an acceptable range of values could be ±10% of the ideal value. In such an instance, a desired value of (x) will fall between 90 and 110 ($90 \leq x \leq 110$) as illustrated in Figure 5.22. It is worth pointing out that the window in Figure 5.22 could be looked at in another way. The shaded region could also be viewed as a region of *exclusion*, in which case a desired value of (x) would be less than 90 or greater than 110 ($x < 90; x > 110$).

Although such analyses are useful, it is their rigidity that limits their usefulness. Creating a rigid boundary leads to all sorts of problems when automatically analyzing data. The difficulties begin when the value sought is not found within the specified window, yet a value exists just outside the boundary of the windows range. In the foregoing example, this would occur if a value existed at 89 or 111. Because these values fall outside the windows range, they would not be chosen. Invariably, what will happen when this occurs is that a user will complain that even though the value is outside the defined specification, they would have "let it slip by" because it was "close enough." It is easy to see how such analyses can lead to problems when scientists feel that the tool they are utilizing to analyze their data is excluding potential values of importance to them.

The way around this problem is to not utilize a single window but layered multiple windows, much like a Venn diagram that displays overlapping choices. Usually, this will consist of three windows. The first window is inclusive and contains a range around the optimal value that would be considered valid without question. The second window is also inclusive but encompasses a larger range around the optimal value. If the value sought is not found within the first window, the window can then be expanded out to the values of the second window. The second window contains values that could be termed acceptable but not optimal. The third window is exclusive and contains all values outside the second window. The third window encompasses the space of unreported values, meaning, it contains results that are absurd.

Looking at Figure 5.23, it is readily apparent that this method of windowing segregates the possible regions for a result to exist into three regions of space that can be described as ($n_l \leq x \leq n_u$) {Normal Range}, ($a_l \leq x \leq a_u$) {Acceptable Range}, and ($x > a_u$; $x < a_l$) {The Exclusion Region}; where

x = result sought, n_l = normal lower range limit, n_u = normal upper range limit, a_l = acceptable lower range limit, a_u = acceptable upper range limit.

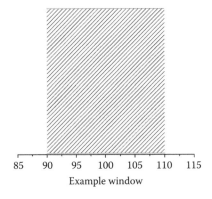

85   90   95   100   105   110   115

Example window

FIGURE 5.22 Sample window interval.

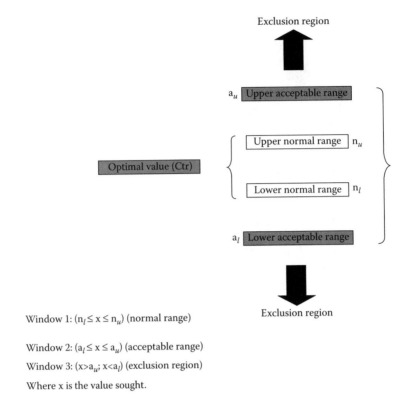

Window 1: $(n_l \le x \le n_u)$ (normal range)

Window 2: $(a_l \le x \le a_u)$ (acceptable range)

Window 3: $(x > a_u; \; x < a_l)$ (exclusion region)

Where x is the value sought.

FIGURE 5.23 A methodology for multiple windowing.

Clearly, any result falling within the acceptable range interval $(a_l \le x \le a_u)$ must be reported. The question now becomes: how does a result within the normal (or optimal) range be distinguished within a report from a result that falls within the acceptable range? Looking at the structure depicted in Figure 5.23, it is clear that results in the "acceptable" range closest to the optimal value (Ctr) are more desirable than results near the outer edge of the acceptable range (a_l or a_u). In other words, in the two intervals $(n_u - a_u)$ and $(n_l - a_l)$, the values closer to normal range limits are more desirable than the values closer to the acceptable range limits. One way of readily distinguishing the desirability of a result obtained in the acceptable range (or window) is to color the result based on its proximity to the normal or acceptable limits. Although any color scheme could be utilized, a green–yellow–red coloring scheme is readily understood (much like a stop light). Results closest to the normal limit would be colored green, whereas those in the middle of the (n_u-a_u) and (n_l-a_l) intervals would be colored yellow (for caution), and those closest to the edge of the acceptable range would be colored red (indicating a potential problem). Those within the normal range would have no color at all within a report.

The most desirable coloring scheme would be to have the colors change *gradually* from green to yellow to red over the entire region, depending upon where in the interval the result to be extracted was located. For example, a result extracted 3/4 of the way from the normal limit and 1/4 of the way from the acceptable limit would be colored orange (using a green to red coloring gradient; orange is 3/4 the distance from green to red). An example of such a coloring gradient is shown in Figure 5.24.

Creating such a gradient is not difficult, but the mechanism requires some thought and knowledge about how colors are created within the Excel VBA paradigm. One approach that seems intuitive but will not work is to take the starting decimal VB color code constant and adjust it to the value of the ending VB color code constant. Looking at Figure 5.25, it is clear that such an approach will not work for two reasons. First, notice that the decimal values between colors are not linear, which makes creating a consistent gradient all but impossible between different colors.

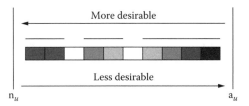

FIGURE 5.24 Interval coloring scheme.

More importantly, notice that the colors must be in the order shown in Figure 5.25 for any transition to occur at all. Thus, a green–yellow–red transition as described earlier would be impossible to implement as the decimal constants go from 65280 (green) to 65535 (yellow) to 255 (red).

Every color the eye can discern is made up of the combination or absence of the colors red, green, and blue. It is possible to construct a custom color in Excel VBA by directly manipulating the red, green, and blue color components (often referred to as the RGB color components). Table 5.5 shows the values of each RGB color component for the VB color constants. Notice that each color component can have a value from 0–255, and therefore the gradient effect of each color component on the overall composition color is in fact linear. When moving from one color to another color, there is the potential to have up to three degrees of freedom in terms of manipulating the RGB components. (Where a "degree of freedom" represents the total number of RGB color components which must be varied to transition from one color to another color.)

Obviously, the greater the number of degrees of freedom, the more complex an operation is to morph from one color to another. Accordingly, it would be wise to arrange the color shifts in such a manner that the transitions from one color to another color occur with only one degree of freedom differentiating one color to the next. This requires that the colors be arranged in a manner corresponding to the order in which their wavelength occurs in the color spectrum, as opposed to the order given in Table 5.5, which is from the Excel VB help documentation.

In Figure 5.26, it can readily be seen that following the order from one color to the next results in a change to only one of the three RGB components. For example, going from red to yellow is accomplished by changing the green value from 0 to 255. Using such a color shifting scheme makes it very easy to gradually transition from one color to the next. What is needed now is a subroutine that takes this a step further by creating an array that not only contains the RGB information but the parameters required for a successful gradual transition from one color to the next. Table 5.6 shows one method of representing the necessary information to implement such a transition.

Table 5.6 shows one method of representing this information. Column 1 shows the first array dimension index. Column 2 contains the color name. Columns 3 to 5 contain the RGB color component value for the current color. Column 6 contains an index to the RGB Component that must be altered to transition to the next color in the table (where R = 1, G = 2, B = 3). Column 7 contains the value the Red, Green, or Blue component must be changed *to* in order to transition from the current color in the table to the next color in the table.

Backgnd	Color	Dec	Hex
	[vbBlack]	0	0
vbRed	[vbRed]	255	FF
vbGreen	[vbGreen]	65280	FF00
vbYellow	[vbYellow]	65535	FFFF
	[vbBlue]	16711680	FF0000
vbMagenta	[vbMagenta]	16711935	FF00FF
vbCyan	[vbCyan]	16776960	FFFF00
vbWhite	[vbWhite]	16777215	FFFFFF

FIGURE 5.25 Visual Basic color code constants.

TABLE 5.5
RGB Color Components

Color	Red Value	Green Value	Blue Value
Black	0	0	0
Blue	0	0	255
Green	0	255	0
Cyan	0	255	255
Red	255	0	0
Magenta	255	0	255
Yellow	255	255	0
White	255	255	255

The subroutine `CreateRGBColorArray` duplicates the entries given in Table 5.6 and places them within an array where they may be referenced.

```
Sub CreateRGBColorArray(ByRef RGBColorMatrix())
'Creates an Array of Colors in Spectral Order with RGB Color Codes
'RGBColorMatrix(X,0) = Text: Color Name
'RGBColorMatrix(X,1) = Number: Red Component
'RGBColorMatrix(X,2) = Number: Green Component
'RGBColorMatrix(X,3) = Number: Blue Component
'RGBColorMatrix(X,4) = Number: Component to Change {1,2,3}
'RGBColorMatrix(X,5) = Number: Value to Change to {0,255}

Dim ColorNames As String, RGBCodes As String, Morph As String
Dim Colors() As String, Codes() As String, Change() As String
Dim Color As Integer, Component As Integer, RGBptr As Integer
Dim MorphPtr As Integer

'The leading comma starts Array @ 1 vs 0 without use of Option
Base 1 Statement
ColorNames = ",Black,Red,Yellow,Green,Cyan,Blue,Magenta,White"
Colors() = Split(ColorNames, ",")
RGBCodes =
",0,0,0,255,0,0,255,255,0,0,255,0,0,255,255,0,0,255,255,0,255
,255,255,255"
```

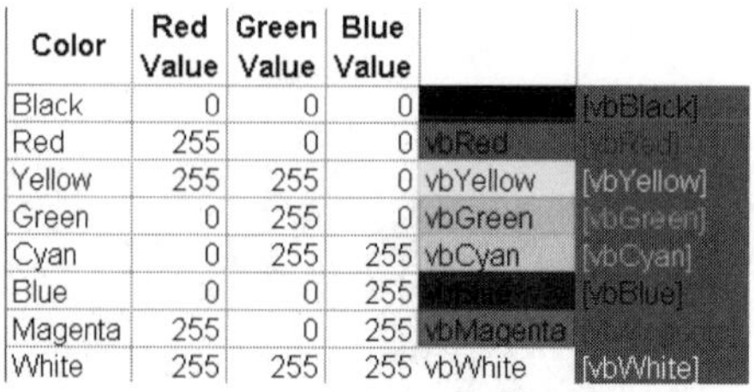

Color	Red Value	Green Value	Blue Value		
Black	0	0	0		[vbBlack]
Red	255	0	0	vbRed	[vbRed]
Yellow	255	255	0	vbYellow	[vbYellow]
Green	0	255	0	vbGreen	[vbGreen]
Cyan	0	255	255	vbCyan	[vbCyan]
Blue	0	0	255		[vbBlue]
Magenta	255	0	255	vbMagenta	
White	255	255	255	vbWhite	[vbWhite]

FIGURE 5.26 Single Degree of Freedom Color Shift Scheme.

TABLE 5.6
RGB Color Code Transition Information

Index	Color	R	G	B	Change	To
1	Black	0	0	0	1	255
2	Red	255	0	0	2	255
3	Yellow	255	255	0	1	0
4	Green	0	255	0	3	255
5	Cyan	0	255	255	2	0
6	Blue	0	0	255	1	255
7	Magenta	255	0	255	2	255
8	White	255	255	255	" "	" "

```
Codes() = Split(RGBCodes, ",")
Morph = ",1,255,2,255,1,0,3,255,2,0,1,255,2,255"
Change() = Split(Morph, ",")

ReDim RGBColorMatrix(8, 5)

'Populate RGBColorMatrix
For Color = 1 To 8
  RGBColorMatrix(Color, 0) = Colors(Color)
  For Component = 1 To 3
    RGBptr = RGBptr + 1
    RGBColorMatrix(Color, Component) = Codes(RGBptr)
Next Component
If Color <> 8 Then
    For Component = 4 To 5
      MorphPtr = MorphPtr + 1
        RGBColorMatrix(Color, Component) = Change(MorphPtr)
    Next Component
End If
Next Color
End Sub
```

The `CreateRGBColorArray` subroutine can easily be tested utilizing the following subroutine that will print out the array values to the immediate (debug) window.

```
Sub TestCreateRGBColorArray()
Dim ColorMatrix()
Call CreateRGBColorArray(ColorMatrix())
For ii = 1 To UBound(ColorMatrix(), 1)
  Debug.Print ii, ColorMatrix(ii, 0), ColorMatrix(ii, 1),
  ColorMatrix(ii, 2), ColorMatrix(ii, 3), _
          ColorMatrix(ii, 4), ColorMatrix(ii, 5)
Next ii
End Sub
```

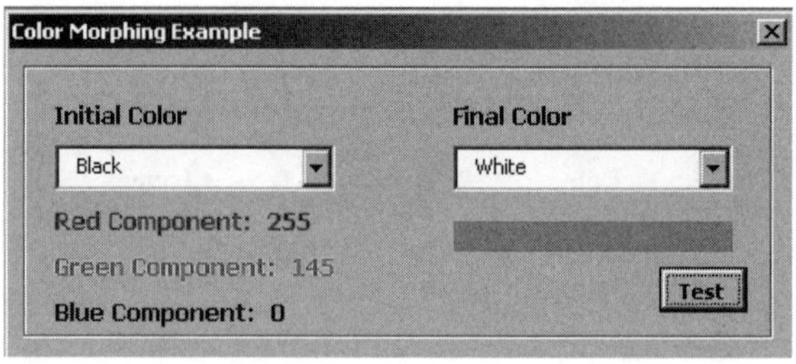

FIGURE 5.27 A color morphing sample application.

A sample application has been set up to demonstrate how the `CreateRGBColorArray` subroutine can be utilized to morph colors from one value to another value. It can be run by selecting the following from the menu: <u>A</u>DA->Chapter <u>5</u>-><u>M</u>orph. The sample application presents the GUI shown in Figure 5.27, in which the user specifies an initial starting color and a final ending color. Notice that the colors are presented in spectral order, and when the user selects an initial color, the final colors are limited to those following the chosen initial color. (The user could, however, create an application that would allow backward as well as forward morphing.)

When the test button is pushed, the program displays the current color in the box above the test button and begins morphing this color from the initial color to the final color. As the color is being transitioned, the RGB components are displayed in real time during the morphing process. The following subroutine is utilized to accomplish the morphing task.

```
Private Sub Button_Test_Click()
Dim ct As Integer 'Color Transitions
Dim sl As Integer 'Shading Level
Dim slvalue As Integer
Dim RGBCode As Long

If ComboBoxIColor.Text = "" Or ComboBoxFColor.Text = "" Then
   MsgBox "You Must Choose and Initial and Final Color!",
   vbOKOnly + vbInformation, _
      "Required Test Parameters Not Found!"
   Exit Sub
End If
For ct = frm5Morph.ComboBoxIColor.ListIndex + 1 To
(frm5Morph.ComboBoxFColor.ListIndex _
      + frm5Morph.ComboBoxIColor.ListIndex + 1)
For sl = 0 To 255
  If Colors(ct, 5) = 0 Then
      'Shading Value Decreasing
      slvalue = 255 - sl
      Else
      'Shading Value Increasing
      slvalue = sl
End If
Select Case Colors(ct, 4)
```

```
  Case 1
    RGBCode = RGB(slvalue, Colors(ct, 2), Colors(ct, 3))
    LabelRed = "Red Component: " & str(slvalue)
    LabelGreen = "Green Component: " & str(Colors(ct, 2))
    LabelBlue = "Blue Component: " & str(Colors(ct, 3))
  Case 2
    RGBCode = RGB(Colors(ct, 1), slvalue, Colors(ct, 3))
    LabelRed = "Red Component: " & str(Colors(ct, 1))
    LabelGreen = "Green Component: " & str(slvalue)
    LabelBlue = "Blue Component: " & str(Colors(ct, 3))
  Case 3
    RGBCode = RGB(Colors(ct, 1), Colors(ct, 2), slvalue)
    LabelRed = "Red Component: " & str(Colors(ct, 1))
    LabelGreen = "Green Component: " & str(Colors(ct, 2))
    LabelBlue = "Blue Component: " & str(slvalue)
  End Select
  'Display Color Using Label
  Label_ColorTest.backcolor = RGBCode
  frm5Morph.Repaint
  Call Idle_Time (0.01)
  Next sl
 Next ct
 End Sub
```

This subroutine is fairly straightforward. First, the number of color transitions (the variable ct) is determined by reading the indexes of the initial and final color selection Combo boxes. The shading level of the appropriate RGB component must then be adjusted to create the next color in the spectrum until the final color desired is reached. This is done in the second loop (the variable sl), and the shading level is denoted by the variable sl value that is adjusted up or down depending upon what is required for the current transition. The RGB Code is then created utilizing a Select Case statement that picks which RGB component is to be adjusted up or down. With each pass of the sl loop, the RGB Code is updated, and the RGB parameters are updated in the form in real time.

Notice that each time the label is updated that displays the current color transition, the form is repainted. This is necessary to ensure that the form does indeed display the current color assigned to the label. Also notice that a delay is built into the looping structure. This ensures that the transitions will not occur so quickly that the human eye cannot process them. The Idle_Time subroutine pauses the macro for the number of seconds specified in the passed parameter nSecond. It is important to note that, although the macro is pausing, background processes (such as repainting forms) are allowed to continue. This is accomplished by means of the DoEvents function.

```
Public Sub Idle_Time(ByVal nSecond As Single)
Dim t0 As Single, tn As Single
Dim dummy As Integer, temp$
t0 = Timer
Do While Timer - t0 < nSecond
    dummy = DoEvents()
 'If we cross Midnight, back up one day
```

```
If Timer < t0 Then
    t0 = t0 - CLng(24) * CLng(60) * CLng(60)
End If
Loop
tn = Timer
End Sub
```

5.10 LINEAR AND NONLINEAR MAPPING

Although the color morphing example given in the previous section of this chapter serves as an interesting academic example, its usefulness in real-world applications is somewhat limited. What is useful, however, is the ability to take a coloring scheme and map it to a range or region of interest. Using the previous example, if the ideal value one is looking for is 100, but an acceptable range of values could be ±10% of the ideal value, the desired value of (x) will therefore fall between 90 and 110 ($90 \leq x \leq 110$). In such an instance, the user might wish to map the color spectrum from green to red onto the region of interest ($90 \leq x \leq 110$). Here, the ideal color (green) would be mapped to the value of 100, and as results approach the limits of the acceptable region (90 or 110), the colors would turn from green to yellow to orange to red. Such a mapping can be done in one of two ways. The first and simplest is a linear map, whereby the colors change incrementally from the starting color to the ending color in a manner in which each successive step from the starting color to the ending color is identical. A more complicated means of mapping is a nonlinear map. Nonlinear maps are usually utilized with exponential functions that allow the initial (or ideal) color to dominate over the range for a longer interval prior to having the colors change more rapidly toward the ending color. Linear and potential nonlinear mapping schemes are shown in Figure 5.28.

Such a mapping scheme immediately raises several issues to the forefront, which arise from the fact that two differing intervals must be utilized to apply such a color-mapping scheme. The first interval is that which encompasses the range of expected values herein referred to as the interval range (IR). The second interval is that which encompasses all of the potential RGB color values over the spectrum to be applied, herein referred to as the color step range (CSR). This, of course, brings to light one of three possible cases with respect to the range of the two intervals.

1. IR > CSR
2. IR = CSR
3. IR < CSR

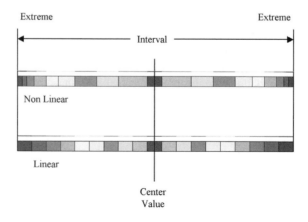

FIGURE 5.28 Linear vs. nonlinear mapping.

In case 1, the interval range is greater than the color step range. As a practical matter, there are more steps in the data range than the color range, which means some colors must be reused. In case 2, which is ideal, the number of color steps exactly equals the span of the interval range. Therefore, if the colors are mapped linearly, there will be exactly one color for each step in the range, and no colors will have to be either reused or discarded. (As a practical matter in life, this has about zero probability of occurring.) The last case, case 3, results when the interval range has fewer members than the number of color steps. In such an instance, the number of color steps utilized must be decimated. These cases apply when the mapping done is by means of a linear method. The nature of nonlinear mapping assures that colors will be both reused and decimated because the color range will be stretched out in one area and compressed in another.

A sample windowing application utilizing linear and nonlinear mapping functions has been created. This sample applies such a mapping down a column specified by the user. The range in which the mapping occurs can be specified by the user in one of three ways:

1. The range can span a percentage of a specified number.
2. The range can be ± a specified number.
3. The range can span the entire range of numbers in the column.

In case 1 and case 2, the range is centered around a specified number. The sample application allows the user to specify that number in one of two ways. The user may point to a number within the Worksheet via a RefEdit control, or the user may specify a particular number via a textbox located in the GUI. The user may select which means to pick the specific number by pushing a button that toggles between the possible states of a textbox or a RefEdit control. In the case of option 3, the range is specified by finding the minimum and maximum values within the specified column to be acted upon. In such an instance, the textbox and RefEdit control are grayed out as they would not be utilized. The minimum and maximum in any column can easily be determined by means of the following functions.

```
Function MaxValinColX(Workbook, Worksheet, columnX As Integer)
MaxValinColX =
Application.Max(Workbooks(Workbook).Worksheets(Worksheet).Ran
ge(Cells(1, columnX), Cells(Rows.Count, columnX)))
End Function

Function MinValinColX(Workbook, Worksheet, columnX As Integer)
MinValinColX =
Application.Min(Workbooks(Workbook).Worksheets(Worksheet).Ran
ge(Cells(1, columnX), Cells(Rows.Count, columnX)))
End Function
```

In case 1 and case 2, it would be advantageous to have the ability to dynamically change the center value. For instance, it would be lucrative to have the ability to change the center value by taking values down a column specified by the RefEdit Control. In the static instance, a single center value is used specified by the RefEdit Control. When the dynamic option is chosen, the center value will change down the column specified, starting with the row position identified in the RefEdit control.

The final consideration is the color scheme utilized for the mapping function. The Combo boxes show the colors in the order that they appear in the color spectrum. When the initial color is picked, the final colors are limited to only those that follow after the initial color. However, this may or may not be the order in which they should appear. For instance, the color order is as follows: Black, Red, Yellow, Green, Cyan, Blue, Magenta, White. If the user were to pick Red as the initial color, their selection would be limited to the following colors as the final color in the gradient: Yellow,

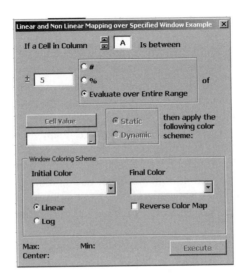

FIGURE 5.29 Sample application using linear and nonlinear mapping.

Green, Cyan, Blue, Magenta, White. A very popular color scheme is to mimic that of a stoplight. Optimal values would be coded as green, and as values progressed toward less desirable values, they would be shaded toward red (undesirable). Using this coloring order, it would be impossible to implement such a scheme because red precedes green in the manner that the color spectrum is presented in the Combo boxes. Therefore, it would be desirable to have the ability to reverse the color map, which, in this case, is accomplished by means of a checkbox.

The final consideration before coding this project is how to apply the color gradient by means of a mapping function. The simplest mapping scheme would obviously be linear. In such an instance, a uniform step size would be utilized across the entire range of values, from one extreme to the other. A more complicated means of mapping would involve some sort of logarithmic scaling function whereby the colors would change slowly from the center value outward but then more rapidly as they approached an extreme as illustrated in Figure 5.28. In the sample application, both methodologies are demonstrated. The user may select whether to utilize linear or nonlinear mapping functions by means of an option button on the GUI.

The sample application can be selected by choosing ADA->Chapter 5->Windowing from the menu, which will bring up the GUI shown in Figure 5.29.

Looking at Figure 5.29, the user must select a column of numerical data to act on. The user must then decide on how to pick the range (or window) that the data lies in. If the user selects a percentage or numerical amount above or below a specified value, then he or she must specify the center value and the variance around that value. If the user selects to apply a mapping gradient across the entire range, then the macro will determine the maximum and minimum values in the specified column of interest. It will then automatically calculate the center point of the range of data within the column. For specified center values, the user can choose to dynamically change the value, that is, as each numerical entity in the column of interest is evaluated, the center value is changed to the next value in the next row from the column in which the original center value was specified. The final specifications relate to the application of the window mapping function. Here the user chooses a starting or ending color, and has the option to reverse the order of the chosen coloring scheme. The user must also specify whether to apply a linear or logarithmic mapping function. Once all of these parameters have been chosen, the mapping function will be applied to the specified column by pressing the Execute button. The code that accomplishes these tasks will now be broken down and discussed.

Upon pressing the Execute button, the first task to be accomplished is to determine how many rows are present in the column in which the mapping function is to be applied. This can be accomplished quite simply with this line of code as discussed in earlier chapters.

```
rowmax = LastRowinColX(ActiveWorkbook.Name, ActiveSheet.Name,
Asc(TextBox_Col) - 64)
```

The next task is to create the Color Step Range (CSR) discussed earlier in this section. The following snippets of code accomplish this:

```
CSRange = CreateColorRange(ColorRange())

Function CreateColorRange(ByRef ColorRange()) As Long
Dim ct As Integer 'Color Transitions
Dim sl As Integer 'Shading Level
Dim slvalue As Integer
Dim RGBCode As Long

Dim ColorIndex As Integer
Dim RC As Integer, GC As Integer, BC As Integer

If ComboBoxIColor.Text = "" Or ComboBoxFColor.Text = "" Then
    MsgBox "You Must Choose and Initial and Final Color!",
    vbOKOnly + vbInformation, "Required Test Parameters Not Found!"
    Exit Function
End If

For ct = frm5Window.ComboBoxIColor.ListIndex + 1 To _
        (frm5Window.ComboBoxFColor.ListIndex +
        frm5Window.ComboBoxIColor.ListIndex + 1)

            For sl = 0 To 255
              CreateColorRange = CreateColorRange + 1
              If Colors(ct, 5) = 0 Then
                  'Shading Value Decreasing
                  slvalue = 255 - sl
              Else
                  'Shading Value Increasing
                  slvalue = sl
              End If
            Select Case Colors(ct, 4)
              Case 1
                RGBCode = RGB(slvalue, Colors(ct, 2),
                Colors(ct, 3))
              Case 2
                RGBCode = RGB(Colors(ct, 1), slvalue,
                Colors(ct, 3))
              Case 3
                RGBCode = RGB(Colors(ct, 1), Colors(ct, 2),
                slvalue)
```

```
                    End Select
                    ReDim Preserve ColorRange(1 To CreateColorRange)
                    ColorRange(CreateColorRange) = RGBCode
              Next sl
        Next ct
     End Function
```

The function `CreateColorRange` accomplishes two tasks. First, it creates a series of incremental RGB color code steps from the initial color to the final color chosen for the transition range. These steps are stored in the array `ColorRange()` that is passed by reference from the calling function. The function returns a long integer value that represents the total number of elements in the array `ColorRange()`, which is, in fact CSR. Notice, again, the use of the `Preserve` keyword when using the `ReDim` function. Its use is essential in that if the `Preserve` keyword is omitted, all previously assigned values in the array will be erased!

A subroutine has been created to calculate the parameters within IR necessary to implement the mapping function. These parameters are passed by reference in the subroutine and are defined as follows:

IRMax — the maximum value of the interval range
IRMin — the minimum value of the interval range
IRange–IRMax–IRMin — the span of the interval range
CtrPt — the center point of the interval range

To calculate the above parameters, the subroutine requires the following values be passed to it by value:

CVal — the Center Value the user specifies in the GUI
Rval — the Range Value the user specifies in the GUI

The CalcWindow subroutine that follows performs the actual calculations.

```
Sub CalcWindow(ByVal CVal, ByVal RVal, ByRef IRMax, ByRef
IRMin, ByRef IRange, ByRef CtrPt)
If frm5Window.OBNumb.Value = True Then
    'Window within Numerical Range of Target Number
    IRMax = CVal + RVal
    IRMin = CVal - RVal
    ElseIf frm5Window.OBPct.Value = True Then
    'Window within Percentage Range of Target Number
    IRMax = CVal + CVal * (RVal / 100)
    IRMin = CVal - CVal * (RVal / 100)
    ElseIf frm5Window.OB_EOER.Value = True Then
    'Window adjustment over entire data range
    IRMax = MaxValinColX (ActiveWorkbook.Name,
    ActiveSheet.Name, Asc(TextBox_Col) - 64)
    IRMin = MinValinColX (ActiveWorkbook.Name,
    ActiveSheet.Name, Asc(TextBox_Col) - 64)
    End If
    frm5Window.LabelMax = "Max: " & str(IRMax)
```

```
frm5Window.LabelMin = "Min: " & str(IRMin)
IRange = IRMax - IRMin
'Determine Center Point
CtrPt = ((IRMax - IRMin) / 2) + IRMin
frm5Window.LabelCtr = "Center: " & str(CtrPt)
End Sub
```

Notice that, if the range is to span the entire data range of the selected column, it is necessary to have a means to find both the minimum and maximum values present within that column. This can easily be done utilizing the following two functions.

```
Function MaxValinColX(Workbook, Worksheet, columnX As Integer)
MaxValinColX =
Application.Max(Workbooks(Workbook).Worksheets(Worksheet).Ran
ge(Cells(1, columnX), Cells(Rows.Count, columnX)))
End Function
Function MinValinColX(Workbook, Worksheet, columnX As Integer)
MinValinColX =
Application.Min(Workbooks(Workbook).Worksheets(Worksheet).Ran
ge(Cells(1, columnX), Cells(Rows.Count, columnX)))
End Function
```

The CalcWindow() subroutine will only be called once if the window is to be centered around a static data point. If the window is to be centered around a dynamic data point that will change with every row in the column, a means must be put in place to change the center point for each successive value in the column. For dynamic data pointing, the row and column the starting point pointed to by the RefEdit control must be determined. For both the static and dynamic cases, the following snippet of code is utilized:

```
'Center Value (CVal) will Change!
If OBDynamic.Value = True Then
    RefEdCol = ColNoInRange (RefEdit_Cell.Text)
    RefEdRow = RowNoInRange (RefEdit_Cell.Text)
    Else
    'Static, Calculate Parameters Only Once
    Call CalcWindow(CVal, RVal, IRMax, IRMin, IRange, CtrPt)
End If
```

Notice that, for the static case, the `CalcWindow` subroutine is called once to establish the parameters `IRMax`, `IRMin`, `IRange`, and `CtrPt`. In the dynamic instance, the `CalcWindow` subroutine must be called multiple times for each successive change in the center value. In order to accomplish this, the macro needs to determine the starting row and column of the initial data point. The following two functions are able to return this information.

```
Function ColNoInRange(RefEditTxt) As Single
'Pass the Text within the RefEdit Control. Will Return the
Column Number
'Set CtrlRange Range object to the range specified in the
RefEdit control.
Set CtrlRange = Range(RefEditTxt)
ColNoInRange = CtrlRange.Column
```

```
End Function

Function RowNoInRange(RefEditTxt) As Single
'Pass the Text within the RefEdit Control. Will Return the
Column Number
'Set CtrlRange Range object to the range specified in the
RefEdit control.

Set CtrlRange = Range(RefEditTxt)
RowNoInRange = CtrlRange.row
End Function
```

The last and most important phase of this example is the methodology required to actually implement the color mapping scheme. The data that will be mapped will encompass a window, which will have a minimum and a maximum value. The window will also obviously have a center point, which, in this macro, is stored in the variable `CtrPt`. The closer a discrete value is to this center point, the closer its shade will match the ideal (or initial) color. The closer a discrete value is to either extreme (the maximum or minimum value of the window), the closer its shade will match the final color. Recall that all possible colors in the color step range are stored in the `ColorRange()` array. What is required now is a formula that calculates the deviation from the ideal value (or center point `CtrPt`), which can be applied in a meaningful way to the color step range to determine what color should be assigned to a particular discrete point. Such a formula has been derived in Equation 5.3.

$$deviation\ from\ ideal = \frac{Abs(CtrPt - current\ value)}{(IRange/2)} \qquad (5.3)$$

To demonstrate how to apply Formula 5.3, an example case will be shown. Suppose the Interval Range (`IR`) is from 80 to 120, in which case the center point (`CtrPt`) would be 100 and the Interval Range (`IRange`) would be 40 (the width of the interval or 120–80).

Table 5.7 can very easily be constructed by simply constructing a Worksheet with the numbers 80–120 in column A and then applying the following formula to the first row of column B and dragging it down.

=ABS(((MEDIAN($A$2:$A$42)-A2)/((MAX($A$2:$A$42)-MIN($A$2:$A$42))/2)))

Upon careful inspection, note that this formula is, in reality, identical to that of (5.3). (Remember that the $ notation indicates that cell references will be fixed when dragged. In this instance, the only reference that will vary is one that references the current value.) Also notice that, in Table 5.7, the numbers closest to the center point (100) are green, and as values progress toward either extreme, they change toward red (i.e., green->yellow->orange->red). Further notice in Table 5.7 that the formula returns a value of 0 at the ideal value (or center of the window) and that this value incrementally increases and reaches a limit of 1 at either extreme of the window spectrum. A value of 0 therefore denotes a 0% deviation from the ideal value, whereas a value of 1 indicates a 100% deviation (within the allowable window) from the ideal value.

Although Formula 5.3 works fine for a linear stepwise progression of values over the designated window, a more complicated formula must be devised for logarithmic treatment of data. Often, what is desired is to maintain the ideal colors over a wider interval around the center point and then proceed toward the nonideal colors quite rapidly at the two extreme edges of the window (as shown in the top part of Figure 5.28).

Looking at Figure 5.30, notice that taking the log of numbers 1–10 produces a scale from 0–1 with the spacing desired. What is required for this application is an algorithm to convert the linear

---

TABLE 5.7
Deviation from Ideal Calculations

Current Value	Deviation from Ideal
80	1.000
81	0.950
82	0.900
83	0.850
84	0.800
85	0.750
86	0.700
87	0.650
88	0.600
89	0.550
90	0.500
91	0.450
92	0.400
93	0.350
94	0.300
95	0.250
96	0.200
97	0.150
98	0.100
99	0.050
100	0.000
101	0.050
102	0.100
103	0.150
104	0.200
105	0.250
106	0.300
107	0.350
108	0.400
109	0.450
110	0.500
111	0.550
112	0.600
113	0.650
114	0.700
115	0.750
116	0.800
117	0.850
118	0.900
119	0.950
120	1.000

---

scale of 0–1 to a logarithmic scale such as that shown in Figure 5.30. To do this, the log scale from 1–10 must be utilized. A formula must now be established to convert a linear number with a range of (0–1) to a number with a range of (1–10), of which the log will be taken.

$$(\{0 - 1\}*9) + 1 \tag{5.4}$$

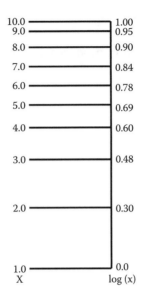

FIGURE 5.30 Linear-to-log transformation scale (1–10).

The converted number must have a minimum value of 1 and a maximum value of 10. Equation 5.4 provides a means of doing this. Here, a linear step value in a range from 0 to 1 is multiplied by 9, the result of which is added to 1.

Although the conversion in Equation 5.4 works, it does not accomplish a logarithmic scaling factor suitable for the tools purpose when a base 10 logarithm is applied to the original number converted using the equation. Recall that when the original deviation from the ideal factor is 0, the ideal color is utilized and, as the deviation from the ideal factor approaches 1, the color shifts toward the final (or nonideal) color. What is desired is to take the original deviation from the ideal value, elongate the transition values in the earlier part of the deviation from the ideal value (0–0.7), and compress the transition values in the latter portion of the deviation from the ideal value (0.7–1.0). This can be accomplished by multiplying the original deviation from the ideal value by $\log_{10}$ of the number converted using Formula 5.4. This, in essence, produces a logarithmic deviation from the ideal value shown in the final column of Table 5.8. Notice that if the log deviation from the ideal value is utilized as the multiplier against the CSR, a much greater deviation from the ideal

TABLE 5.8
Linear-to-Log Conversion Table

Deviation from Ideal (devidl)	Log Scale x	Log(x)	Log(x) * Devidl = Log Deviation Ideal
1.000	10	1.000	1.000
0.889	9	0.954	0.848
0.778	8	0.903	0.702
0.667	7	0.845	0.563
0.556	6	0.778	0.432
0.444	5	0.699	0.311
0.333	4	0.602	0.201
0.222	3	0.477	0.106
0.111	2	0.301	0.033
0.000	1	0.000	0.000

value will be required to shift away from the ideal color and that the majority of color transitions toward the nonideal colors will be compressed into the regions closest to the edges of the window or the nonideal regions.

The two-color mapping schemes are illustrated in Table 5.9. If the sample application is run, notice that it takes much longer for the logarithmic mapping scheme to deviate from the ideal color (green), whereas the linear mapping scheme shifts equidistant increments of color over the entire range.

TABLE 5.9
Linear vs. Log Color Map Schemes

Linear	Log	Deviation from Ideal
80	80	1.000
81	81	0.950
82	82	0.900
83	83	0.850
84	84	0.800
85	85	0.750
86	86	0.700
87	87	0.650
88	88	0.600
89	89	0.550
90	90	0.500
91	91	0.450
92	92	0.400
93	93	0.350
94	94	0.300
95	95	0.250
96	96	0.200
97	97	0.150
98	98	0.100
99	99	0.050
100	100	0.000
101	101	0.050
102	102	0.100
103	103	0.150
104	104	0.200
105	105	0.250
106	106	0.300
107	107	0.350
108	108	0.400
109	109	0.450
110	110	0.500
111	111	0.550
112	112	0.600
113	113	0.650
114	114	0.700
115	115	0.750
116	116	0.800
117	117	0.850
118	118	0.900
119	119	0.950
120	120	1.000

The final snippet of code that executes when the "Execute" button is pushed will now be discussed. It is in this segment of code that the mapping function parameters are calculated.

```
For row = 1 To rowmax
   'If Dynamic Values Must be recalculated with each loop pass!
   If OBDynamic.Value = True Then
       'Dynamic, Recalculate Parameters for each loop pass
       CVal =
       Workbooks(ActiveWorkbook.Name).Worksheets(ActiveSheet.Name).
       Cells(RefEdRow + row - 1, RefEdCol).Value
       Call CalcWindow(CVal, RVal, IRMax, IRMin, IRange, CtrPt)
   End If
       curval =
       Workbooks(ActiveWorkbook.Name).Worksheets(ActiveSheet.Name).
       Cells(row, Asc(TextBox_Col) - 64).Value
       devidl = Abs((CtrPt - curval) / (IRange / 2))
         Select Case CheckBox_RCM.Value
           Case True
           If OBLog.Value = True Then
               'Log Scale
               devidl = ConvertToLog (devidl)
               ColorIndex = Int((devidl) * CSRange)
               Else
               'Linear Scale
               ColorIndex = Int((devidl) * CSRange)
           End If
           Case False
           If OBLog.Value = True Then
               'Log Scale
               devidl = ConvertToLog (devidl)
               ColorIndex = Int(devidl * CSRange)
               Else
               'Linear Scale
               ColorIndex = Int(devidl * CSRange)
           End If
         End Select
       If ColorIndex < 1 Then ColorIndex = 1
       If curval <= IRMax And curval >= IRMin Then
       Workbooks(ActiveWorkbook.Name).Worksheets(ActiveSheet.Name).
       Cells(row, Asc(TextBox_Col) - 64).Font.Color =
       ColorRange(ColorIndex)
       'Debug.Print row, curval, ColorIndex, CSRMax,
       ColorRange(ColorIndex)
       Debug.Print row, OBLog.Value, ColorIndex
       End If
   Next row
```

The foregoing code is somewhat cryptic in nature, so some explanation is in order. The variable curval stores the value of the current cell to be mapped. The variable devidl stores the calculated value of the deviation from the ideal (or center point: CtrPt). This value is calculated using the previously discussed Equation 5.3. The ColorIndex is always calculated by multiplying the deviation from the ideal (a value from 0 to 1) by the number of color steps over the CSR. In the event that the color map should be reversed (for example, ideal color = green, final color = red), the variable devidl must be inverted by setting its value equal to 1 itself. If a logarithmic mapping function is desired, a function has been created to perform the conversion between the linear mapping scheme and the logarithmic mapping scheme. The ConvertToLog function accomplishes this using the formula discussed in (5.4) and Table 5.8.

```
Function ConvertToLog(index) As Single
Dim adj As Single
'Convert 0-1 to 1-10
adj = (index * 9) + 1
'Gives Natural Log (ln)
adj = Log(adj)
'Convert to Base 10 Logarithm
adj = adj / Log(10)
ConvertToLog = adj * index
Debug.Print index, ConvertToLog
End Function
```

Finally, a provision needs to be made for the case when the variable devidl is equal to 0. Because a devidl value of 0 will produce a ColorIndex value of 0, and the lowest ColorIndex value can be 1, a provision has been made to change the ColorIndex value to 1 if its current value is less than 1.

A test sheet has been created named "WindowTest.xls." Here, a series of columns has been created with values between 80 and 120. This was easily accomplished utilizing the following function:

$$=100+RANDBETWEEN(-20,20)$$

A button has also been placed on this sheet that will clear the sheet of any formatting changes that have been made during testing. The code attached to this button to accomplish this is

```
Columns("A:H").Select
Selection.Font.ColorIndex = 0
```

Figure 5.31 depicts the use of the example program to perform dynamic nonlinear mapping on column A utilizing the values in column B as reference (center) points. Looking at Figure 5.31, the optimal color for a value closest to a center point is red. Numbers within column A that are within ± 5.0 of the value in column B will be color coded. Those numbers in column A that are nearly identical to those in column B will be colored red. Any number that is within ± 5.0 of the value in column B will be colored in some manner from red to green. Those in the middle will be orange, then yellow, and finally shades of green as the edges of the specified window (± 5.0 of the value in column B) are approached. Those values that lie outside of the specified window receive no coloring at all.

Notice that, when running the sample program, it conveniently remembers which colors were selected from the Combo boxes. This can be done by utilizing the ListIndex property of the Combo box. Each time the sample application is run, the values present in the color Combo boxes are stored in the Windows Registry by means of these statements:

```
SaveSetting "ColorMap", "Color", "IColor",
frm5Window.ComboBoxIColor.ListIndex
SaveSetting "ColorMap", "Color", "FColor",
frm5Window.ComboBoxFColor.ListIndex
```

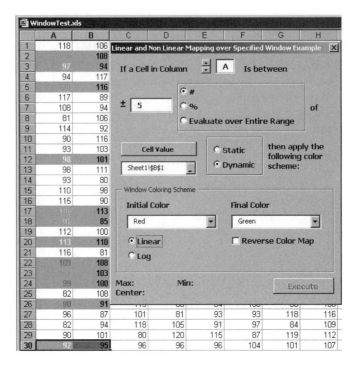

FIGURE 5.31 Dynamic example of linear mapping — columns A and B.

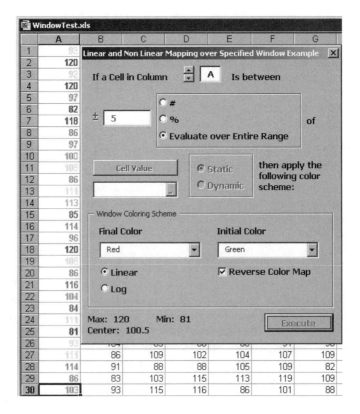

FIGURE 5.32 Linear color mapping over entire value range.

When the sample program is first started, the saved values (if they exist) are loaded by means of these statements:

```
frm5Window.ComboBoxIColor.ListIndex =
Val(GetSetting("ColorMap", "Color", "IColor"))
frm5Window.ComboBoxFColor.ListIndex =
Val(GetSetting("ColorMap", "Color", "FColor"))
```

Figure 5.32 shows an example of linear color mapping over the entire range of values in column A. Notice that, in this example, the color map was reversed (by checking the checkbox). Values closest to the center (100.5 — shown on GUI) are green, and as values deviate from the center value, they change from green to yellow to orange and finally to red at the extremes of the window. Notice that the maximum and minimum values found in the column to be mapped are determined and shown in the GUI (Max = 120, Min = 81, Center = 100.5).

Although this example uses linear mapping to color the elements in a column, keep in mind that such an algorithm could be altered to perform many other tasks. For example, instead of coloring the element in a column, a row or column could be copied, moved, or written to a new file depending upon where it was located in the mapping scheme. The mapping scheme algorithm can also be altered to fit whatever parameters the user sees fit to model.

5.11 AUTOMATICALLY LOADING AND EXTRACTING DATA FROM COMPLEX DIRECTORY STRUCTURES

Chances are that an application with any degree of sophistication will require data from many sources. In such an instance, it is necessary for an application to have the ability to automatically load data files from specified locations and extract the required data from the imported files to complete a specified task the macro was developed to accomplish. Instrumentation will often dump data with certain parameters to differently named folders. Going back to the example cited earlier for an HPLC with four methods, each of which has a different percentage of methanol in solution (40, 50, 60, and 70%), the following directory structure might be created.

Top folder: XYZ
Sub folders under XYZ: method40, method50, method60, method70

Now suppose that, after the instrument is run, each folder will contain a data file with the data acquired for each percentage of methanol. The challenge here is to have the ability to "point" the macro toward a file structure and have the program make a rational decision as to what files should be imported and what data should be extracted. In many instances, files are automatically named by machines with a date–time stamp format. For illustrative purposes, suppose three compounds are run on an HPLC, and the directory structure set up for storage is that shown in Figure 5.33. Within this directory structure, suppose that each data file is saved with a date–time stamp for a filename like "YYYYMMDD-HHMMSS.xls." Thus, the filename 20040219-150100.xls represents a filename created on February 19, 2004, at 3:01:00 PM (in military 24-h time 15:01:00). In this example, this file is stored on the accompanying CD-ROM in the directory:

\Chapter 5\Samples\AutoLoadDirs\Compound1\method40\

In this example, one data file is created each minute, and the files are stored by methods under the compound name, after which they then increment to the next compound name and are again stored by method under the current compound name. The data files in this example follow this format:

\Chapter 5\Samples\AutoLoadDirs\Compound1\method40\20040219-150100.xls
\Chapter 5\Samples\AutoLoadDirs\Compound1\method50\20040219-150200.xls
\Chapter 5\Samples\AutoLoadDirs\Compound1\method60\20040219-150300.xls
\Chapter 5\Samples\AutoLoadDirs\Compound1\method70\20040219-150400.xls
\Chapter 5\Samples\AutoLoadDirs\Compound2\method40\20040219-150500.xls

For this example, the data files created by the HPLC contain the information shown in Figure 5.34.

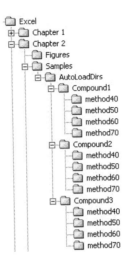

FIGURE 5.33 Directory structure for Auto Load files example.

Each data file contains four columns that store the following information: Compound, Area, Retention Time, and Height. For the purpose of this example, suppose that it is desired to create a macro that will automatically open every data file for each compound (at all the various methanol percentages 40, 50, 60, and 70) and create a report that extracts for each method the retention time corresponding to the largest area with the largest peak height as a secondary selection factor. (In other words, the first selection criteria is the area; should two areas be equal, choose the retention time corresponding to the area that has the greatest height.) The final goal here would be to produce a report such as that depicted in Figure 5.35.

The development of a macro capable of creating the report shown in Figure 5.35 requires the writer to develop code capable of completing the following processes. The macro must first be pointed to a file structure, and from there it must automatically be able to navigate through the underlying file structure and open all the data files required to construct the report depicted in Figure 5.35. Each data file must be sorted first by area and then by height. Once this has been accomplished, the retention time corresponding to the peak with the greatest area can be extracted from each data file. Once this has been done, the final report should be saved and possibly printed.

Figure 5.36 shows the sample GUI utilized to execute this macro. It is simplistic in design in that the user merely points to a file structure and clicks the "Extract from Files" button to run the sample macro. The first major step in the process of this macro is to determine the total number of files that must be accessed and the full path to each file's location. The function `Create_File_List` constructed in Chapter 2 will accomplish this task quite nicely by placing all of this information into an array.

```
TotalFiles = Create_File_List(frm5AutoLoad.TextBox_Path.Text,
"xls", FileNames(), True, True, True)
```

Upon execution of this line of code, the total number of Excel (*.xls) files (in this case 12) will be stored in the variable `TotalFiles`. The name and location of each of these files (or the

	A	B	C	D
1	Compound	Area	Retention Time	Height
2	Compound1	0.022	0.931	3.713
3	Compound1	33.757	0.402	1.123
4	Compound1	53.408	0.175	9.521
5	Compound1	43.538	0.497	6.653
6	Compound1	8.113	0.176	5.394
7				

FIGURE 5.34 Example data file for Auto Load files example.

E20	▼	=		
	A	B	C	D	E
1	Compound	Retention Times			
2		Method 40	Method 50	Method 60	Method 70
3	Compound1	2.317	0.756	1.434	2.358
4	Compound2	2.069	0.990	1.650	2.402
5	Compound3	0.213	0.662	1.058	1.615

FIGURE 5.35 Example of desired report.

full path to these files) will be stored in the array `FileNames()`. In this example, the following elements would be stored in FileNames():

```
1 C:\Author\Books\Excel\Chapter
5\Samples\AutoLoadDirs\Compound1\method40\20040219-150100.xls

2 C:\Author\Books\Excel\Chapter
5\Samples\AutoLoadDirs\Compound1\method50\20040219-150200.xls

3 C:\Author\Books\Excel\Chapter
5\Samples\AutoLoadDirs\Compound1\method60\20040219-150300.xls

4 C:\Author\Books\Excel\Chapter
5\Samples\AutoLoadDirs\Compound1\method70\20040219-150400.xls

5 C:\Author\Books\Excel\Chapter
5\Samples\AutoLoadDirs\Compound2\method40\20040219-150500.xls

6 C:\Author\Books\Excel\Chapter
5\Samples\AutoLoadDirs\Compound2\method50\20040219-150600.xls

7 C:\Author\Books\Excel\Chapter
5\Samples\AutoLoadDirs\Compound2\method60\20040219-150700.xls

8 C:\Author\Books\Excel\Chapter
5\Samples\AutoLoadDirs\Compound2\method70\20040219-150800.xls

9 C:\Author\Books\Excel\Chapter
5\Samples\AutoLoadDirs\Compound3\method40\20040219-150900.xls

10C:\Author\Books\Excel\Chapter
5\Samples\AutoLoadDirs\Compound3\method50\20040219-151000.xls

11C:\Author\Books\Excel\Chapter
5\Samples\AutoLoadDirs\Compound3\method60\20040219-151100.xls

12C:\Author\Books\Excel\Chapter
5\Samples\AutoLoadDirs\Compound3\method70\20040219-151200.xls
```

While on the subject of creating file lists, it would also be lucrative to have a function that would create a list of directories below a file structure that is pointed to. The function `CreateDirectory List` whose source code follows will accomplish this task. Notice that, like its big brother, the

FIGURE 5.36 Autoload and Extract sample GUI.

`Create_File_List` function created in Chapter 2, this function also relies upon recursion (a function or subroutine calling itself) to process all the subdirectories that lie below the file structure pointed to.

```
Function CreateDirectoryList(ByVal DirPath$, SubDirs() As
String) As Integer
'Function Extracts all subdirectories below the specified
DirPath$ into
'array SubDirs() and returns the total number of subdirectories
'Search and Extract from SubDirs set BDoSubs = True

Static subDirCount As Integer
Dim RecCount As Integer, RecSubDirs() As String
Dim SearchString As String
Dim fname
Dim i As Long
'Always Process Sub Directories
Const bDoSubs As Boolean = True

'make certain the directory path ends in a \
If Right(DirPath$, 1) <> "\" Then
    DirPath$ = DirPath$ & "\"
End If

SearchString = DirPath$ & Dir(DirPath$ & "*.*", vbDirectory)

Do While SearchString <> DirPath$
 If Not (Right(SearchString, 2) = "\." Or Right(SearchString,
 3) = "\..") Then
    If GetAttr(SearchString) = vbDirectory Then
        'do this if a subdirectory
        If bDoSubs Then
        ' add to array of directories
        subDirCount = subDirCount + 1: RecCount = RecCount + 1
        ReDim Preserve SubDirs(1 To subDirCount): ReDim Preserve
        RecSubDirs(1 To RecCount)
        SubDirs(subDirCount) = SearchString:
        RecSubDirs(RecCount) = SearchString

        Debug.Print subDirCount, SubDirs(subDirCount) ',
        RecSubDirs(RecCount)

        End If
    End If
  End If
fname = Dir()
SearchString = DirPath$ & fname
Loop
'call recursively to process SubDirs
If subDirCount > 0 And bDoSubs Then
    For i = 1 To RecCount
```

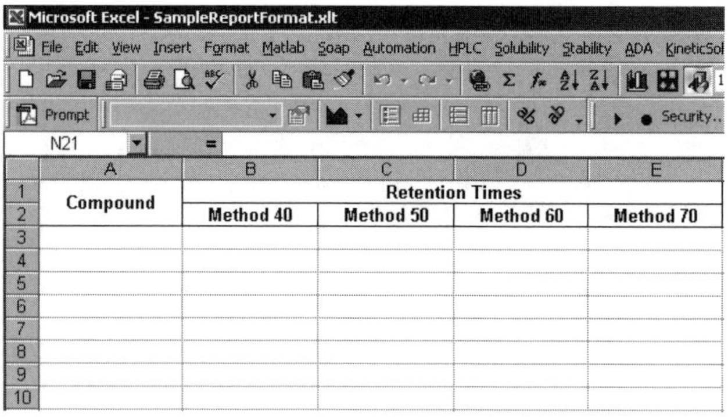

FIGURE 5.37 Template file utilized in Autoload and Extract example.

```
        CreateDirectoryList RecSubDirs(i), SubDirs()
    Next i
End If
CreateDirectoryList = subDirCount
End Function
```

The next step is to load in a template that contains the report format desired, which is shown in Figure 5.37. The code to accomplish this is

```
'Load Template with Sample Report
ReportBook = Open_Template ("C:\Author\Books\Excel\Chapter
5\Samples\SampleReportFormat.xlt", "c:\temp\tempname.xls")

Function Open_Template (fulltemplatepath$, fullintpath$) As String
'Opens an Excel Template File and Then Saves it to a Temporary
Filename (fullintpath$)

'Returns the Name of the Template Workbook for storage to
Public Variable

'Open Template for Data Storage
If FileExists(fulltemplatepath$) <> True Then
    MsgBox "You must install the file: " & fulltemplatepath$,
    vbCritical + vbOKOnly, "File: " & fulltemplatepath$ & " Not Found"

    End
End If

If DirectoryExists(Extract_Path(fullintpath$)) = False Then
    Call CreateDataDir(Extract_Path(fullintpath$))
End If

'Open the Template
Workbooks.Open Filename:=fulltemplatepath$
'Save the Template as a new Workbook with an Intermediate name
'(This prevents corruption of the Original Template)
ActiveWorkbook.SaveAs Filename:=fullintpath$,
FileFormat:=xlNormal _
```

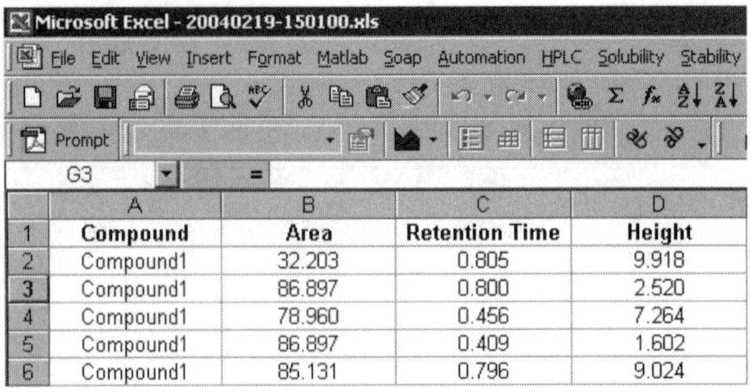

	A	B	C	D
1	Compound	Area	Retention Time	Height
2	Compound1	32.203	0.805	9.918
3	Compound1	86.897	0.800	2.520
4	Compound1	78.960	0.456	7.264
5	Compound1	86.897	0.409	1.602
6	Compound1	85.131	0.796	9.024

FIGURE 5.38 Sample data file.

```
, Password:="", WriteResPassword:="",
ReadOnlyRecommended:=False, _
CreateBackup:=False
'Return Workbook/Worksheet Name for storage to Global (Public)
Variable
Open_Template = ActiveWorkbook.Name
End Function
```

Notice that the function checks to see if the specified template exists, and terminates the macro if it does not. If the specified template does exist, it is loaded into Excel, and the name of the Workbook is returned by the function for storage into a public variable.

The last phase to complete this sample project is to open, sort, and extract the information required from all of the sample data files (Figure 5.37). The snippet of code which accomplishes this will be discussed at length.

```
'Loop through all reports and extract pertinent information
rowctr = 3: colctr = 1 'Start Populating Template on Row 3, col 2
For rptnbr = 1 To TotalFiles
  colctr = colctr + 1
  'Load the first report
  DataBook = AutoLoadFile(FileNames(rptnbr))
  Call SortWorksheet (DataBook,
  Workbooks(DataBook).Sheets(1).Name, 1, 4, 1, 2, False, 4, False)
  'Write in RT Value
  Workbooks(ReportBook).Worksheets(1).Cells(rowctr, colctr) = _
  Format(Workbooks(DataBook).Worksheets(1).Cells(2, 3), "#0.000")
```

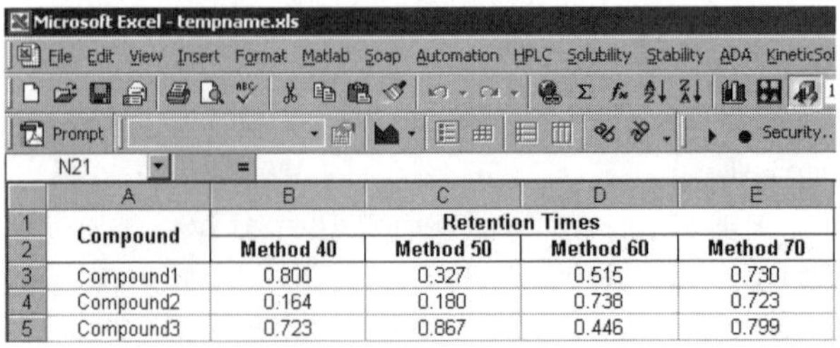

	A	B	C	D	E
1	Compound	Retention Times			
2		Method 40	Method 50	Method 60	Method 70
3	Compound1	0.800	0.327	0.515	0.730
4	Compound2	0.164	0.180	0.738	0.723
5	Compound3	0.723	0.867	0.446	0.799

FIGURE 5.39 Final data file created by Autoload and Extract sample.

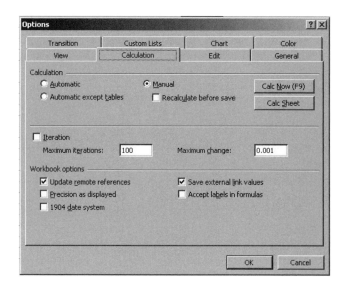

FIGURE 5.40 Setting the Calculation option in Excel.

```
'Write in Compound Number
Workbooks(ReportBook).Worksheets(1).Cells(rowctr, 1) =
Workbooks(DataBook).Worksheets(1).Cells(2, 1)
   'Increment row counter - reset column counter
   If rptnbr Mod 4 = 0 Then rowctr = rowctr + 1: colctr = 1
   Workbooks(DataBook).Close (False)
Next rptnbr
```

Looking at Figure 5.37, it is clear that data should be populated into the template starting at row 3 and column 2. The compound numbers should be added to the template starting at row 3, column 1. To accomplish this, two variables are utilized as row and column counters (`rowctr` and `colctr`). Notice that when the remainder of the loop divided by 4 is equal to zero, the row is incremented and the column number is reset back at the beginning. With each loop iteration, a new data file is opened, its contents sorted utilizing the `SortWorksheet` subroutine developed in Chapter 2, and the proper retention time is extracted from that data file and placed in the report. The data file is then closed, and the loop increments and repeats the process until all files have been processed.

Looking at Figure 5.38, notice that the largest area (86.897) is repeated twice. The Worksheet is sorted first by area and then by peak height. The area with the peak height of 2.52 is selected, and its corresponding retention time of (0.800) is selected and placed into the final report as shown in row 3, column 2 (B) of Figure 5.39.

One final point that must be mentioned before the user attempts to run this example is the Calculation setting option in Excel. If the Calculation setting in Excel is set to automatic, it will appear that this sample does not function correctly. This is because, after processes such as sorting occur in a Worksheet, Excel will automatically recalculate all the values derived from formulas within the Worksheet. The calculation setting can be changed in Excel by selecting Tools->Options and choosing the Calculations tab as shown in Figure 5.40.

Obviously the auto loading and extraction example shown previously are built relying heavily upon material covered in the previous chapters. It is also somewhat rudimentary in that it has no built-in checks. A "real" application of this type would check the reasonableness of the values extracted, the existence of the proper files, and so forth. It is left to the reader to modify the above example to suit their own needs.

6 Creating Custom Report Worksheets

6.1 INTRODUCTION

Thus far the discussion has centered around obtaining, classifying, and analyzing data. These tasks are useless unless the data that have been so painstakingly analyzed can be stored and presented to the final users in a useful format. That is the purpose of this section. It teaches the reader not only how to make a report Worksheet but how to create a truly useful report Worksheet that is able to show or hide data pertinent to a particular situation at any given time.

In this section, a sample report will be created for an instructor of an MBA class who wishes to have a means of creating a nice report of student grades. For this example, the class in question will have two tests and a final exam. The class will also have six quizzes, of which the top three quiz scores will be averaged to form a final quiz grade. For the sake of discussion, suppose the instructor is a really nice guy and wants to utilize one of three grading schemes and will give the student the top score earned under any of the schemes.

Table 6.1 illustrates the three possible grading schemes for the class. Each method gives the student different weighting options. Method 1 is more or less the traditional grading scheme seen in many colleges today. Each component of the class carries an equal weight, with the exception of the final exam, which is double the weight of any of the other components. This professor realizes, however, that anyone can have one bad test, so method 2 drops the lowest test score and places a little heavier weight on the remaining elements in the course. The professor came up with method 3 for his professional students who may not have had the time to balance the responsibilities of work and home enough to prepare every class for a potential quiz. This method eliminates the quiz grades from the grading component entirely. Sadly, those students who do not do well on final exams are left out in the cold, and are only able to reduce the weight of that grading component by 10% under any method.

Looking at the problem from the instructor's standpoint, it would be nice to have a report with a limited amount of information that would give the instructor at a glance a good idea of what each student's relative standing in the class is. That would greatly aid his or her ability to decide who to call upon in class. At the same time, the instructor must have access to all of the grading information for the report to be useful. It will be shown how to construct a report that allows both paradigms to be possible.

6.2 USE OF TEMPLATES WHEN CREATING CUSTOM REPORTS

Using a template for creating each new report can save the developer a lot of time under certain circumstances. If the report is unisectional, a template can be created to hold the information for the entire report at once. This will save the user from having to write code to add text for column headings and help him or her write code to format the headings with things such as bold font and underlining. Figure 6.1 is an example of just such a report.

TABLE 6.1
Possible Grading Schemes

	Percentage Component of Total Grade				
Grading Method	Test 1	Test 2	Quiz	Final	Total
1	20	20	20	40	100
2	25		25	50	100
3	30	30		40	100

Often, however, things are not quite as neat as shown in Figure 6.1. A multiple-section report is often necessary. Multiple-section reports become problematic because, unless the number of elements in them will be fixed every time, only the top portion of the report can be prefabricated in the template. Subsequent sections of the report must have their headers created using code or copied from another Worksheet, as will be shown later in this chapter. A bisectional report, illustrated in Figure 6.2, in which one section is located below another section, will only allow prefabrication of the top portion of the report.

Figure 6.2 shows not only the salaries of the employees but also the results of the most recent drug test. Notice that this report shows little correlation between salary and drug use, as both a top level earner (the CEO) and an average salaried worker (the secretary) were the only two to test positive for drug use. It was subsequently learned that they were having an affair with each other, at which point they were summarily terminated.

Table 6.2 lists the parameters that must be present in the sample report to be created for the MBA class sample application. First to be listed will be the student's name, along with the test and final scores, followed by the scores of all the quizzes and the average of the three highest quiz scores. Next will follow the results of grading the students under each of the three possible schemes, along with the student's final grade (the highest score under the three possible grading methods), with a final column at the end for standardized instructor comments. The comments can be added by means of a drop-down box, where the contents of the drop-down box will be added programmatically. At the top of the report will be a header section that will contain information about the course and the instructor.

Now comes the tricky part. The amount of information in Table 6.2 is a little overwhelming to look at in a single glance. The instructor really does not need to see each individual quiz score to get a good idea of a student's standing at a quick glance. After all, only the top three quiz scores are utilized in the average, and under one grading method, the quiz scores do not even contribute toward the final grade. None the less, the quiz scores are a component of the class, and they must be available in the final report.

	A	B	C	D
1	Last Name	First Name	Title	Salary
2	Bunker	Archie	Foreman	42,000
3	Fuller	Jeffrey	CEO	230,000
4	Pelletier	Jeffrey	Director	80,000
5	Reebeck	Elise	Secretary	43,000
6	Reed	Micheal	Assembler	40,000
7	Smith	Don	Machinist	59,000
8				

FIGURE 6.1 A single-section report example.

	A	B	C	D
1	Last Name	First Name	Title	Salary
2	Bunker	Archie	Foreman	42,000
3	Fuller	Jeffrey	CEO	230,000
4	Pelletier	Jeffrey	Director	80,000
5	Reebeck	Elise	Secretary	43,000
6	Reed	Micheal	Assembler	40,000
7	Smith	Don	Machinist	59,000
8				
9	Last Name	First Name	Drug Test	
10	Bunker	Archie	Negative	
11	Fuller	Jeffrey	Positive	
12	Pelletier	Jeffrey	Negative	
13	Reebeck	Elise	Positive	
14	Reed	Micheal	Negative	
15	Smith	Don	Negative	

FIGURE 6.2 A bisectional report example.

6.3 PREPARATION OF DUAL-VIEW REPORTS

The solution to this problem is to create a report that is capable of being rendered in two views on a single Worksheet. This is a much simpler task than it sounds. For the example in this section, it would be desirable to have the capability to hide all of the quiz grades and just show the quiz average.

Figure 6.3 shows a sample report with all of the parameters that make up the class shown. Notice that this is a fairly large Worksheet. Figure 6.4 shows the exact same report, but in this report the quiz grades are not visible. This creates a report that is much more readable for someone just taking a quick glance.

Figure 6.4 is the exact same report as in Figure 6.3; it just has some of its columns *hidden*. However, creating such a dual-view report is not just as simple as hiding the columns. Notice that the header crosses the area of space on the Worksheet encompassed by the columns that are hidden.

TABLE 6.2
Sample Report Parameters

Parameter Name

Last name
First name
Test 1
Test 2
Final
Quiz 1
Quiz 2
Quiz 3
Quiz 4
Quiz 5
Quiz 6
Quiz avg
Grade 1
Grade 2
Grade 3
Class grade
Comments:

Advanced Financial Management
Instructor: Al Chapman
Course No. MGMTADVFIN002
Location: Springfield, MA
Time: TR 6:00 - 9:00 PM
TAF12776
Division 02

Quiz Grades Shown

Grade Analysis

Print Report

Last Name	First Name	Test 1	Test 2	Final	Quiz 1	Quiz 2	Quiz 3	Quiz 4	Quiz 5	Quiz 6	Quiz Avg	Grade 1	Grade 2	Grade 3	Class Grade	Comments:
Argento	Crocetta	85	100	96	73	48	91	95	100	51	95.3	94.5	96.8	93.9	96.8	
Bissett	Brian	90	95	100	52	63	89	88	98	65	91.7	95.3	96.7	95.5	96.7	
Buckley	Russell	100	97	88	61	89	82	61	53	75	82.0	91.0	89.5	94.3	94.3	
Campos	Gus	65	85	77	47	47	89	80	92	84	88.3	78.5	81.8	75.8	81.8	
Dees	Robert	77	82	75	85	49	90	57	74	78	84.3	78.7	79.1	77.7	79.1	
Dyjak	Pamela	70	80	75	89	87	80	88	70	55	88.0	77.6	79.5	75.0	79.5	
Fisher	Ellen	35	15	12	84	48	75	89	57	50	82.7	31.3	35.4	19.8	35.4	
Harrison	Paige	92	95	97	52	81	52	97	68	66	82.0	92.6	92.8	94.9	94.9	
Kuhn	Max	93	88	98	98	77	72	86	92	53	92.0	93.8	95.3	93.5	95.3	
Lambert	Charon	90	77	89	93	58	60	55	95	58	82.7	85.5	87.7	85.7	87.7	
Maclelland	Cynthia	100	77	88	81	80	78	73	49	97	86.0	87.8	90.5	88.3	90.5	
Onorato	Sandra	77	87	89	68	84	55	81	70	80	81.7	84.7	86.7	84.8	86.7	
Pelletier	Jeffrey	65	66	95	67	100	62	67	56	78	81.7	80.5	84.4	77.3	84.4	
Potter	David	100	100	100	83	63	96	65	93	97	95.3	99.1	98.8	100.0	100.0	
Racine	Ned	13	22	62	45	78	98	52	85	59	87.0	49.2	58.3	35.3	58.3	
Rafuse	Charles	88	72	94	85	93	54	50	46	79	85.7	86.7	90.4	85.6	90.4	
Reed	Micheal	97	92	63	76	73	59	65	62	57	71.3	77.3	73.6	81.9	81.9	
Rekos	Russell	100	100	100	76	51	80	52	63	70	75.3	95.1	93.8	100.0	100.0	
Tanski	Lucy	70	80	90	66	96	86	45	81	82	88.0	83.6	87.0	81.0	87.0	
Turner	Howard	80	70	92	95	63	45	69	66	53	76.7	82.1	85.2	81.8	85.2	
Vasilchik	Tiffany	90	70	95	100	95	80	88	67	97	97.3	89.5	94.3	86.0	94.3	

FIGURE 6.3 Sample report shown with all parameters shown.

When the columns are hidden, the header must be fixed to reflect the status of the Worksheet (which columns are visible and which are not).

When parts of a Workbook are hidden, the data is no longer visible but it is not deleted from the Workbook. If the Workbook is saved and closed, the hidden data remains hidden the next time the Workbook is opened. If the Workbook is printed, Excel will not print the hidden parts.

Both Workbooks and Worksheets can be hidden in Excel. This can be done to reduce the number of windows and sheets on the screen, and to prevent unwanted changes. An additional use would be to hide sheets that contain sensitive data, or to hide a Workbook containing macros so that the macros are available to run but no window appears for the macro Workbook. The hidden Workbook or Worksheet will remain accessible but not visible, and other Workbooks or Worksheets can utilize the information contained within them.

More useful to report creation is the ability to hide rows and columns within a worksheet. By hiding selected rows and columns, users do not have to view extraneous data within a report or

Advanced Financial Management
Instructor: Al Chapman
Course No. MGMTADVFIN002
Location: Springfield, MA
Time: TR 6:00 - 9:00 PM
TAF12776
Division 02

Quiz Grades Hidden

Grade Analysis

Print Report

Last Name	First Name	Test 1	Test 2	Final	Quiz Avg	Grade 1	Grade 2	Grade 3	Class Grade	Comments:
Argento	Crocetta	85	100	96	95.3	94.5	96.8	93.9	96.8	
Bissett	Brian	90	95	100	91.7	95.3	96.7	95.5	96.7	
Buckley	Russell	100	97	88	82.0	91.0	89.5	94.3	94.3	
Campos	Gus	65	85	77	88.3	78.5	81.8	75.8	81.8	
Dees	Robert	77	82	75	84.3	78.7	79.1	77.7	79.1	
Dyjak	Pamela	70	80	75	88.0	77.6	79.5	75.0	79.5	
Fisher	Ellen	35	15	12	82.7	31.3	35.4	19.8	35.4	
Harrison	Paige	92	95	97	82.0	92.6	92.8	94.9	94.9	
Kuhn	Max	93	88	98	92.0	93.8	95.3	93.5	95.3	
Lambert	Charon	90	77	89	82.7	85.5	87.7	85.7	87.7	
Maclelland	Cynthia	100	77	88	86.0	87.8	90.5	88.3	90.5	
Onorato	Sandra	77	87	89	81.7	84.7	86.7	84.8	86.7	
Pelletier	Jeffrey	65	66	95	81.7	80.5	84.4	77.3	84.4	
Potter	David	100	100	100	95.3	99.1	98.8	100.0	100.0	
Racine	Ned	13	22	62	87.0	49.2	58.3	35.3	58.3	
Rafuse	Charles	88	72	94	85.7	86.7	90.4	85.6	90.4	
Reed	Micheal	97	92	63	71.3	77.3	73.6	81.9	81.9	
Rekos	Russell	100	100	100	75.3	95.1	93.8	100.0	100.0	
Tanski	Lucy	70	80	90	88.0	83.6	87.0	81.0	87.0	
Turner	Howard	80	70	92	76.7	82.1	85.2	81.8	85.2	
Vasilchik	Tiffany	90	70	95	97.3	89.5	94.3	86.0	94.3	

FIGURE 6.4 Sample report with quiz grades hidden.

can be prevented from seeing data that others should not be allowed to see. Rows and Columns can be hidden at will by setting the Hidden property to True. The HideColRange subroutine illustrates how to hide a selected range of columns on the current active Worksheet.

```
Sub HideColRange(colrng$)
  Columns(colrng$).Select
  Selection.EntireColumn.Hidden = True
End Sub
```

The reader should be aware that Charts and Worksheets can have their Visible property set to a special value termed xlVeryHidden. The xlVeryHidden property hides the object so that the only way to make it visible again is by resetting this property in VBA code to xlVisible (or more specifically, xlSheetVisible). The user viewing the object is not capable of making the object visible. Such functionality can be useful to keep various users from viewing portions of information contained within a Workbook. Some sample code follows for invoking these properties using Excel VBA.

```
Sub HideWorksheet()
  Worksheets(1).Visible = Hide   'Use Hide or False
End Sub

Sub ReallyHideWorksheet()
  Worksheets(1).Visible = xlVeryHidden
End Sub

Sub UnHideWorksheet()
  Worksheets(1).Visible = True
End Sub

Sub Toggle_Hidden_Visible()
'Use this to toggle between hidden and visible
  Worksheets(1).Visible = Not Worksheets(1).Visible
End Sub

Sub UnhideAllWorksheet()
Dim sh As Worksheet
For Each sh In Worksheets
  sh.Visible = True
Next
End Sub

Sub ReallyHideAllSheets()
On Error Resume Next
Dim sh As Worksheet
For Each sh In Worksheets
  sh.Visible = xlVeryHidden
Next
End Sub
```

For the sample in this text, a number of things must happen to transition from Figure 6.3 to Figure 6.4.

1. The appropriate Columns ("E:L") must be hidden.
2. The Header must be reformatted.
3. The Quiz Grades must be calculated (they may have been modified when all the columns were shown!).
4. The button caption should be changed to: "Quiz Grades Hidden."
5. The position of the print button must be moved (to the left).

Similarly, when transitioning from Figure 6.4 to Figure 6.3, the following steps must be taken.

1. The appropriate Columns ("E:L") must be unhidden or made visible.
2. The Header must be reformatted.
3. The button caption should be changed to: "Quiz Grades Shown."
4. The position of the print button must be moved (to the right).

These steps are taken when the "Quiz Grades Shown/Quiz Grades Hidden" button is pressed. The caption may change, but the button executes the same VBA code no matter which caption is shown. Note that the button merely toggles state (and Worksheet view) with each push.

```vba
Private Sub Hide_Button_Click()
Select Case Hide_Button.Value
  Case True
    'Show All Columns (Columns Hidden @ Start)
    Columns("E:L").Select
    Selection.EntireColumn.Hidden = False
    'This Copies the whole Header over
    Call CopySection(ActiveWorkbook.Name, "HdrVisible", 1, 1,
    7, 20, ActiveWorkbook.Name, ActiveSheet.Name, 1, 1)
    Print_Button.Left = 641.25
    Hide_Button.Caption = "Quiz Grades Shown"
  Case False
    'Hide Columns with Extraneous Info (QUIZ GRADES)
    '(All Columns All Shown @ Start)
    'Unhide Columns
    Call ClearUnHiddenRange(ActiveSheet.Name, 7, 15)
          Call HideColRange("F:K")
          'This Copies the whole Header over
          Call CopySection(ActiveWorkbook.Name, "HdrHidden",
          1, 1, 7, 20, ActiveWorkbook.Name, ActiveSheet.Name, 1, 1)
          Print_Button.Left = 631.25
          Hide_Button.Caption = "Quiz Grades Hidden"
          Call CalcQuizAvg
End Select
End Sub
```

The code for the button just given actually resides within the Worksheet object itself — in this case the "ClassGrades" Worksheet. Up until now, all VBA code has either resided within a module or a form, but it can reside in objects such as Worksheets as well! Notice that the code just given makes calls to two subroutines. The *CopySection* subroutine does something rather tricky. It adds a new header by copying it from one of two hidden Worksheets named "HdrVisible" or "HdrHidden," depending upon if the Quiz grades are visible or hidden, respectively.

```
Sub CopySection(ByVal frmwkbook, ByVal frmwksht, ByVal r1,
ByVal c1, ByVal r2, ByVal c2, ByVal towkbook, _
          ByVal towksht, ByVal row, ByVal col)
'Copies the "Chunk" Bounded by r1,c1,r2,c2 in frmwkbook to
towkbook starting at row,col
'Select "from" Workbook and Worksheet
ActivateWorkbook (frmwkbook)
Sheets(frmwksht).Visible = True
Worksheets(frmwksht).Select
Range(Cells(r1, c1), Cells(r2, c2)).Select
Selection.Copy
'Select "to" Workbook and Worksheet
ActivateWorkbook (towkbook)
Worksheets(towksht).Select
Range(Cells(row, col), Cells(row, col)).Select
ActiveSheet.Paste
Range("A1").Select
Sheets(frmwksht).Visible = False
End Sub
```

The CopySection subroutine must perform one trick in order for it to work. When a Worksheet, chart, column, or row is hidden, it cannot be selected from VBA code! So, in order for the above subroutine to work, the Worksheets that hold the headers to be copied must be made visible (just for an instant) in order to select and copy the preformatted headers. Although this produces a quick flash on the screen, it can save a lot of coding and a great deal of headaches.

6.4 EXECUTING CALCULATIONS UPON CHANGING VIEWS

The CalcQuizAvg subroutine is what calculates the quiz grades. It also adds a group of standardized comments to the final report that the user may select by means of a drop-down box. Calculating the quiz scores is problematic here because only the top three quiz grades are to be utilized. If only the top quiz score were sought, the Max function could be utilized. This leaves the developer with two options. The first is to simply sort the quiz scores and select the highest three. The second would be to use the Max function on the data set, then eliminate the selected value from the data set, and take the Max function again two more times. The former shall be utilized because it is simpler.

```
Sub CalcQuizAvg()
Sheets("QuizSort").Visible = True
For ii = 12 To LastRow("SampleReport.xls", "ClassGrades")
  Workbooks("SampleReport.xls").Worksheets("ClassGrades").Cell
  s(ii, 12) = _
  CalcQuizScore ("SampleReport.xls", "ClassGrades", 6, 11, ii)
Next ii
Worksheets("ClassGrades").Activate
```

```
Sheets("QuizSort").Visible = False
Call CommentDropDn (Range(Cells(12, 17),
Cells(LastRow("SampleReport.xls", "ClassGrades"), 17)))
End Sub
```

Notice that the CalcQuizScore function is called once for every student in the report. The function returns a single precision number truncated to two decimal places that represents the quiz score for that student. In order to calculate the quiz score, however, the top three quiz scores must be known. To accomplish this, the quiz scores are first copied to a hidden Worksheet named "QuizSort." The HorizontalSort subroutine then sorts all of the known quiz scores on the "QuizSort" Worksheet (notice that this leaves the order of the Quizzes on the original report unchanged.) The top three quiz scores are then taken from the "QuizSort" Worksheet and can be used to calculate the quiz score for each student. Recall that, when using hidden Worksheets, they must be made visible to select ranges and perform operations upon them.

```
Function CalcQuizScore(ByVal wkbook, ByVal wksht, ByVal
startcol As Integer _
, ByVal endcol As Integer, ByVal row As Long) As Single

Call CopyBlockTo(ActiveWorkbook.Name, "ClassGrades", row,
startcol, row, endcol, _
    ActiveWorkbook.Name, "QuizSort", row, startcol)
Call HorizontalSort (wkbook, "QuizSort", startcol, endcol, row,
True)
'Debug.Print
Workbooks("SampleReport.xls").Worksheets("QuizSort").Cells(row,
endcol).Value
CalcQuizScore =
(Workbooks("SampleReport.xls").Worksheets("QuizSort").Cells(row,
endcol).Value + _
Workbooks("SampleReport.xls").Worksheets("QuizSort").Cells(row,
endcol - 1).Value + _
Workbooks("SampleReport.xls").Worksheets("QuizSort").Cells(row,
endcol - 2).Value) / 3
'Truncate to 2 digits
CalcQuizScore = Format(CalcQuizScore, "###.00")
End Function

Sub HorizontalSort(ByVal wkbook, ByVal wksht, ByVal startcol
As Integer _
, ByVal endcol As Integer, ByVal row As Long, SortAscending
As Boolean)

Dim Order
'Select Workbook and Worksheet
ActivateWorkbook (wkbook)
Worksheets(wksht).Select
Range(Cells(row, startcol), Cells(row, endcol)).Select
If SortAscending = True Then Order = xlAscending Else Order
= xlDescending
```

```
'Perform Horizontal Sort on row across specified columns
Selection.Sort Key1:=Range(Cells(row, startcol), Cells(row,
startcol)), Order1:=Order, header:=xlGuess, _

    OrderCustom:=1, MatchCase:=False,
Orientation:=xlLeftToRight

End Sub
```

The last objective upon changing views was to add the standardized comments to the last column in the report. Table 6.3 shows the comments that the instructor desires to have available within the report. What is needed is a mechanism to allow these comments to be added to a range of cells so that they can be selected by means of a drop-down box.

```
Sub CommentDropDn(rng As Range)
'Adds Drop Down List of Standard Comments
Dim Cmt(10) As String, ii As Integer, CmtList$
'Assign up to 10 specific Comments to Array Here
Cmt(1) = "Perfect Attendance"
Cmt(2) = "Poor Attitude"
Cmt(3) = "Poor Attendance"
Cmt(4) = "Active Participant"
Cmt(5) = "Presented for Group"

For ii = 1 To UBound(Cmt())
 If ii < UBound(Cmt()) Then
 CmtList$ = CmtList$ & Cmt(ii) & Chr(44)
 Else
 CmtList$ = CmtList$ & Cmt(ii)
 End If
Next ii

rng.Select
'Range(rng$).Select
Selection.HorizontalAlignment = xlLeft
  With Selection.Validation
   .Delete
```

TABLE 6.3
Desired Comments for Report

Index	Comment Text
1	"Perfect Attendance"
2	"Poor Attitude"
3	"Poor Attendance"
4	"Active Participant"
5	"Presented for Group"

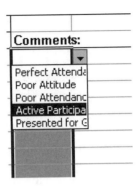

FIGURE 6.5 Creating a drop-down list of values in a cell.

```
.Add Type:=xlValidateList, AlertStyle:=xlValidAlertStop,
Operator:= _
xlBetween, Formula1:=CmtList$
.IgnoreBlank = True
.InCellDropdown = True
.ShowInput = True
.ShowError = True
End With
End Sub
```

In this particular instance, only five comments were assigned. The user could modify the array of the subroutine to allow as many comments as desired. Notice that all the comments to be made available in the drop-down box are stored within a string (CmtList$) and that they are separated from one and other by means of commas (Chr(44) = ","). The preceding subroutine creates an effect illustrated in Figure 6.5.

6.5 ANALYSIS WITHIN REPORT WORKSHEETS

The created report, as it currently stands, is useful for the application it was created for; however, some additional functionality can be added that enhances its value to the user. Recall that the instructor utilized the highest of three grading methods to calculate the final grade for each student. The instructor would probably like to know how often each grading method was utilized to determine which grading schemes work best for a particular class. Does it benefit the student more to drop all the quiz scores or the lowest exam grade? The answers to such questions could help the instructor make the class more conducive to learning.

A simple routine can be set up within this report to highlight the grading method that was utilized for each student. This will instantly give the instructor a visual indication as to the relative use of each grading scheme. If the user presses the "Grade Analysis" button on the ClassGrades Worksheet, the grading method (1, 2, or 3) will be highlighted on the ClassGrades Worksheet as illustrated in Figure 6.6.

```
Private Sub Button_Grade_Click()
'Determine Which Grading Scheme (1,2, or 3) was utilized
For ii = 12 To LastRow("SampleReport.xls", "ClassGrades")
  For jj = 0 To 2
```

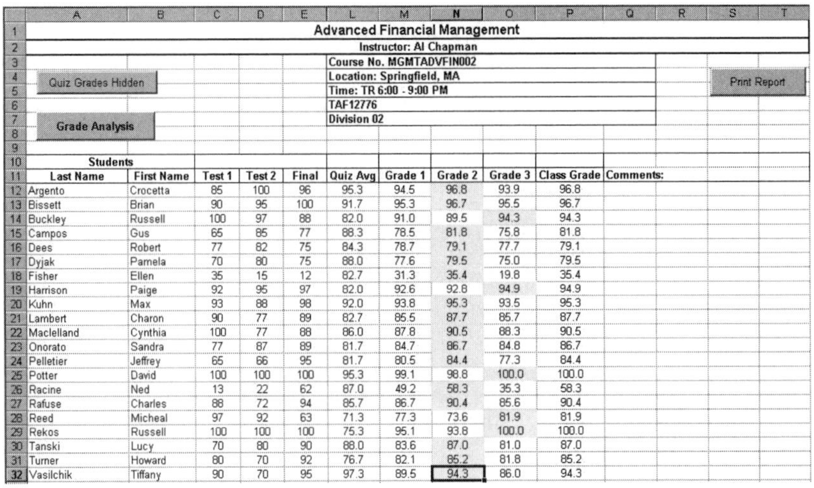

Last Name	First Name	Test 1	Test 2	Final	Quiz Avg	Grade 1	Grade 2	Grade 3	Class Grade	Comments:
Argento	Crocetta	85	100	96	95.3	94.5	96.8	93.9	96.8	
Bissett	Brian	90	95	100	91.7	95.3	96.7	95.5	96.7	
Buckley	Russell	100	97	88	82.0	91.0	89.5	94.3	94.3	
Campos	Gus	65	85	77	88.3	78.5	81.8	75.8	81.8	
Dees	Robert	77	82	75	84.3	78.7	79.1	77.7	79.1	
Dyjak	Pamela	70	80	75	88.0	77.6	79.5	75.0	79.5	
Fisher	Ellen	35	15	12	82.7	31.3	35.4	19.8	35.4	
Harrison	Paige	92	95	97	82.0	92.6	92.8	94.9	94.9	
Kuhn	Max	93	88	98	92.0	93.8	95.3	93.5	95.3	
Lambert	Charon	90	77	89	82.7	85.5	87.7	85.7	87.7	
Maclelland	Cynthia	100	77	88	86.0	87.8	90.5	88.3	90.5	
Onorato	Sandra	77	87	89	81.7	84.7	86.7	84.8	86.7	
Pelletier	Jeffrey	65	66	95	81.7	80.5	84.4	77.3	84.4	
Potter	David	100	100	100	95.3	99.1	98.8	100.0	100.0	
Racine	Ned	13	22	62	87.0	49.2	58.3	35.3	58.3	
Rafuse	Charles	88	72	94	85.7	86.7	90.4	85.6	90.4	
Reed	Micheal	97	92	63	71.3	77.3	73.6	81.9	81.9	
Rekos	Russell	100	100	100	75.3	95.1	93.8	100.0	100.0	
Tanski	Lucy	70	80	90	88.0	83.6	87.0	81.0	87.0	
Turner	Howard	80	70	92	76.7	82.1	85.2	81.8	85.2	
Vasilchik	Tiffany	90	70	95	97.3	89.5	94.3	86.0	94.3	

Advanced Financial Management — Instructor: Al Chapman — Course No. MGMTADVFIN002 — Location: Springfield, MA — Time: TR 6:00 - 9:00 PM — TAF12776 — Division 02 — Quiz Grades Hidden — Grade Analysis — Print Report — Students

FIGURE 6.6 Frequency of grading methods highlighted.

```
If
Workbooks("SampleReport.xls").Worksheets("ClassGrades").Cells
(ii, 16) = _
Workbooks("SampleReport.xls").Worksheets("ClassGrades").Cells
(ii, 13 + jj) Then
  'Match - Highlight Background
      Call HighlightCell("SampleReport.xls", "ClassGrades",
      ii, 13 + jj, 6)
  Exit For
  End If
 Next jj
Next ii
End Sub
```

All that the above subroutine does is compare the final grade value with each of the three grading methods to determine which one was utilized. The grading method that was utilized is then highlighted. Taking a quick look at the results, grading method 1, which utilizes all of the components in the class, was not utilized for a single student. Every student benefited from either dropping the lowest test grade (method 2), or dropping the Quiz average (method 3), and reweighing the other elements present within the course.

Notice that the final grade can be altered quite a bit by choosing one method over another. In the case of Ms. Fisher, her grade nearly doubled from a low of 19.8 (using method 3) as opposed to a high of 35.4 (using method 2). Sadly, even when given the best of circumstances, she was unable to attain a passing grade. Looking at the other extreme, there are instances when using one grading method over another makes little difference. Take the case of Mr. Potter, with final calculated grades of 99.1, 98.8, and 100 using each of the three methods, respectively. No matter what grading method is used in his case, an A will be his grade. A similar case exists for that of Mr. Dees, whose grades of 78.7, 79.1, and 77.7 put him in solid C+ territory.

Although the above analysis provides a good quick-and-dirty means of looking at the data, the instructor is probably going to want a more comprehensive analysis on the results of his classes. The beauty of standardized reports is that, when reports conform to a known format, it becomes very easy to write macros to perform analyses on them. This can be taken a step further to simplify

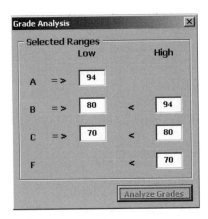

FIGURE 6.7 Grade analysis GUI tool.

the analysis process as well. If the analysis to be performed will be the same time and time again, it is easy to set up a template to accomplish the task and minimize the use of VBA coding.

In this instance, the professor would probably want to set the criteria for each grade level: what constitutes an A, B, C, or F. Once the criteria were chosen, it would probably be beneficial for the instructor to know what the frequency distribution of grades is in terms of how many A's, B's, C's, or F's are awarded for a chosen set of criteria. Are too many students receiving A's or F's? A bar chart could be utilized to provide a nice visual display of the distribution of grades. Similarly, what is the percentage breakdown of which grading methods give students the best grades? A pie chart would provide an appropriate picture of which methods are most frequently utilized. From this information the instructor could get some insight as to what tools are best to aid students in acquiring the knowledge taught in class. Are quizzes an ineffective learning tool?

A stand-alone macro has been set up to analyze the sample report provided in this chapter. It can be run by selecting ADA->Chapter 6->Grade Analysis from the Excel menu. Doing so will bring up the analysis GUI shown in Figure 6.7.

Figure 6.7 allows the user to set the range for each grade. In this instance, the following breakdown is utilized.

A hidden sheet is provided in the SampleReport.xls named "Analysis." This hidden sheet greatly simplifies doing all the tasks mentioned in the preceding text. Here, the bar and pie charts are already preformatted and set up. All the user must do is

1. Unhide the Sheet.
2. Write the Number of Grade Occurrences A, B, C, and F in cells B11:B14.
3. Write the Number of Times Each Grading Method is utilized in cells B2:B4.

Once this has been done the pie and bar charts will automatically update themselves. After running the macro with the default set of grading parameters (Table 6.4), the user will be provided the following analysis (Figure 6.8):

An analysis of the grade frequency is shown in Figure 6.9. Notice that the number of A's and B's are nearly equal at 8 and 9, respectively. Perhaps the course needs to be made a little harder next time around. The number of failures or near failures (F's and C's) is equal at 2 each for a total of 4 for the entire class.

Looking at the methods utilized to calculate the final grades (Figure 6.10), notice that method 1 was never utilized for a single student. Every student in this class benefited more by having his or her grade calculated by either dropping one exam score or dropping the Quiz grades. Not a single person was able to achieve a higher score by averaging all the components of the class across

TABLE 6.4
Numerical Grade Range

Grade	Range	
	Low	High
A	94	100
B	80	93
C	70	79
F	0	69

	A	B	C
1		Total	Percentage
2	Method 1	0	0
3	Method 2	16	76.190476
4	Method 3	5	23.809524
5	Total	21	
6			
7			
8			
9			
10	Grade	Total	
11	A	8	
12	B	9	
13	C	2	
14	F	2	

FIGURE 6.8 Analysis Worksheet of grading breakdown.

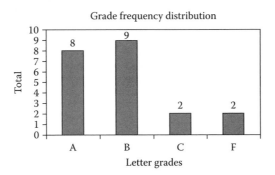

FIGURE 6.9 Grade frequency using default criteria.

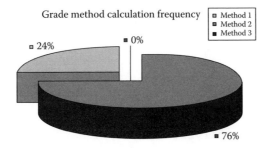

FIGURE 6.10 Frequency of grading methods.

the board. Perhaps this grading method is redundant and should be eliminated? Further notice that most students benefited by using method 2 for calculating the final grade by a margin of nearly 3 to 1 (76% vs. 24%). This tells the professor that the quizzes in class were an effective learning tool because only 5 students achieved a higher grade using method 3, which eliminates quiz scores, as opposed to the 16 who benefited more by dropping the lowest exam score using method 2.

Code to automate the preceding analysis using the provided template follows.

```
Sub GradeAnalysis()
'Analyze Frequency Distributions of Grades
Dim GradeA As Integer, GradeB As Integer
Dim GradeC As Integer, GradeF As Integer
Dim GradeMethod1 As Integer, GradeMethod2 As Integer,
GradeMethod3 As Integer
On Error GoTo Catch_Error
'Make Sure a Report is Open
If ActiveWorkbook.Name <> "SampleReport.xls" Then
    MsgBox "You must first open the file SampleReport.xls prior
    to macro Execution!", _
        vbOKCancel + vbCritical, "Macro Terminating"
    End
End If
'Unhide Analysis Sheet
ActiveWorkbook.Sheets("Analysis").Visible = True
'Count Number of Times Grading Scheme (1,2, or 3) was utilized
For ii = 12 To LastRow("SampleReport.xls", "ClassGrades")
  For jj = 0 To 2
    If
    Workbooks("SampleReport.xls").Worksheets("ClassGrades").Cells
    (ii, 16) = _
    Workbooks("SampleReport.xls").Worksheets("ClassGrades").Cells
    (ii, 13 + jj) Then
    Select Case jj
      Case 0
        GradeMethod1 = GradeMethod1 + 1
      Case 1
        GradeMethod2 = GradeMethod2 + 1
      Case 2
        GradeMethod3 = GradeMethod3 + 1
    End Select
    Exit For
  End If
  Next jj
Next ii
'Write Results to Analysis Worksheet
Workbooks("SampleReport.xls").Worksheets("Analysis").Cells(2, 2)
= GradeMethod1
```

```
Workbooks("SampleReport.xls").Worksheets("Analysis").Cells(3, 2)
= GradeMethod2

Workbooks("SampleReport.xls").Worksheets("Analysis").Cells(4,
2) = GradeMethod3
'Determine Grading Distribution
For ii = 12 To LastRow("SampleReport.xls", "ClassGrades")
    Select Case
    Workbooks("SampleReport.xls").Worksheets("ClassGrades").Cells
    (ii, 16)

      Case Val(frm6Grades.TextBox_AL) To 100
        GradeA = GradeA + 1
      Case Val(frm6Grades.TextBox_BL) To
      Val(frm6Grades.TextBox_BH)
        GradeB = GradeB + 1
      Case Val(frm6Grades.TextBox_CL) To
      Val(frm6Grades.TextBox_CH)
        GradeC = GradeC + 1
      Case 0 To Val(frm6Grades.TextBox_FH)
        GradeF = GradeF + 1
    End Select
Next ii
'Write Results to Analysis Worksheet
Workbooks("SampleReport.xls").Worksheets("Analysis").Cells(11, 2)
= GradeA
Workbooks("SampleReport.xls").Worksheets("Analysis").Cells(12, 2)
= GradeB
Workbooks("SampleReport.xls").Worksheets("Analysis").Cells(13, 2)
= GradeC
Workbooks("SampleReport.xls").Worksheets("Analysis").Cells(14, 2)
= GradeF
'Hide the form
frm6Grades.Hide
Exit Sub
Catch_Error:
  MsgBox "You must first open the file SampleReport.xls prior
  to macro Execution!", _
    vbOKCancel + vbCritical, "Macro Terminating"
  End
End Sub
```

6.6 BASIC FORMATTING TECHNIQUES

When creating reports, there are some tasks that present themselves time and time again. It will be beneficial to the reader to have a tool kit of VBA routines to accomplish these tasks for the construction of their own specialized reports.

Analyzed data is a valuable resource. However, it is only valuable as long as it is accurate. When a process is performed, some people perform the process better than others. When reports are constructed, it is often a good idea to make a notation as to just who worked on the report. (This can even be done without their knowledge to a very hidden Worksheet!) The best way to determine that is to use the username of the person logged into the machine on which it was prepared. Such a step eliminates having to ask the users who they are and prevents them from lying as well. The current user logged onto a machine is readily available using the Environ function. The following function will return a string indicating who was working on the report at the time the function was called.

```
Function PreparedBy() As String
  'Who worked on this report?
  PreparedBy = "Prepared By: " & Environ("UserName")
End Function
```

If a report will be worked on multiple times, it might be prudent to make multiple calls to a function such as PreparedBy to keep track of every user who modified or worked on the report. If something were to come to light, for example, a particular employee was analyzing data under the influence of alcohol or drugs, the reports prepared by that user over a certain date range could be easily purged out. The Environ function can take a number of parameters as an argument, all of which are listed in Table 6.5.

Another useful functionality is to have the ability to determine the last time a report was saved. The following function will return the date and time any file (including an Excel Workbook) was saved. It will also return a variant array named Elements that breaks up the elements returned.

```
Function LastSaved(fullpath As String, Elements) As String
  'Return Date & Time a File was saved
  LastSaved = FileDateTime(fullpath)
  Elements = Split(LastSaved, " ")
End Function
```

This function can easily be tested utilizing the following code. (The actual filepath will need to be changed to reflect the location of the SampleReport.xls file on the reader's machine.)

```
Sub TestLastSaved()
Dim filele As Variant
Debug.Print
LastSaved("C:\Author\Books\Excel\20051128\Manuscript\Chapter
6\Samples\SampleReport.xls", filele)
For ii = 0 To UBound(filele)
  Debug.Print ii, filele(ii)
Next ii
End Sub
```

When tested, the subroutine will return an output as follows that first lists all the information lumped together and then the array of elements.

```
11/28/2005 1:20:04 PM
011/28/2005
11:20:04
2PM
```

TABLE 6.5
Environ Function Arguments

Environ Arguments

ALLUSERSPROFILE
APPDATA
AVENGINE
CLIENTNAME
CommonProgramFiles
COMPUTERNAME
ComSpec
FP_NO_HOST_CHECK
HOMEDRIVE
HOMEPATH
INCLUDE
INOCULAN
LIB
LOGONSERVER
NUMBER_OF_PROCESSORS
OS
Path
PATHEXT
PROCESSOR_ARCHITECTURE
PROCESSOR_IDENTIFIER
PROCESSOR_LEVEL
PROCESSOR_REVISION
ProgramFiles
SESSIONNAME
SystemDrive
SystemRoot
TEMP
TMP
USERDOMAIN
USERNAME
USERPROFILE
VS71COMNTOOLS
WecVersionForRosebud.FF0
windir

These functions provide critical system information such as dates, times, and usernames, which are perfectly suitable for inclusion on headers or footers in Worksheets. The next subroutine will allow the developer to add a header or a footer *to every sheet* contained in the specified Workbook. The subroutine allows the developer to add text to the left, middle, and right sides of the header or footer. It can easily be modified to perform the operation on a single sheet should that type of functionality be required.

```
Sub AddHeaderOrFooter(Workbook$, ltxt$, mtxt$, rtxt$, header
As Boolean)
'Places a Universal Header or Footer on all sheets in a
specific Workbook$
Dim ws As Worksheet
```

```
'Activate the Workbook in which Footers are to be added
Call ActivateWorkbook(Workbook$)
'Loop through each sheet, activate, and set footers
For Each ws In Worksheets
  Sheets(ws.Name).Activate
  GoSub Format_Sheet
Next
'All done
Exit Sub
Format_Sheet:
'This section will add the Footer to each selected sheet!
Select Case header
  Case True
  With ActiveSheet.PageSetup
    .LeftHeader = ltxt$
    .CenterHeader = mtxt$
    .RightHeader = rtxt$
  End With
  Case False
  With ActiveSheet.PageSetup
    .LeftFooter = ltxt$
    .CenterFooter = mtxt$
    .RightFooter = rtxt$
  End With
End Select
Return
End Sub
```

When creating reports with VBA, a very common need is to add additional columns or rows before preexisting columns or rows. The next two subroutines handle the process easily.

```
Sub InsertColumns(wkbook, wksht, colno, number)
'Subroutine inserts "number" of blank columns in wksht before
column number (colno)
Dim ii As Integer
Columns(colno).Select
For ii = 1 To number
  Selection.Insert Shift:=xlToRight
Next ii
End Sub

Sub InsertRows(wkbook, wksht, rowptr, number)
'Subroutine inserts "number" of blank rows in wksht @ rowptr
'Pass Rowptr by Reference to retain ending row position
Dim RRange$
'Select Workbook and Worksheet
```

```
ActivateWorkbook (wkbook)
Worksheets(wksht).Select
'Determine Range
RRange$ = Trim$(str$(rowptr)) & ":" & Trim$(str$(rowptr +
number - 1))
Rows(RRange$).Select
Selection.Insert Shift:=xlDown
rowptr = rowptr + number + 1
End Sub
```

Often, many reports must be prepared, and they will all share much of the same information. In such an instance, it may be easier to copy a preexisting sheet, delete the items not needed, and add any items that are required. The following subroutine will create a copy of any sheet to a location in the Workbook specified by the user.

```
Sub CopyWorksheet(SourceBook, CopySheet, DestBook, PivotSheet,
beforepivot As Boolean, newsheetname)
'Make a Copy of (copysheet) contained within (sourcebook) to
destination Workbook (destbook)
'before sheet (beforesheet)
Select Case beforepivot
Case True
  Workbooks(SourceBook).Worksheets(CopySheet).Copy
  Before:=Workbooks(DestBook).Sheets(PivotSheet)
Case False
  Workbooks(SourceBook).Worksheets(CopySheet).Copy
  After:=Workbooks(DestBook).Sheets(PivotSheet)
End Select
ActiveSheet.Name = newsheetname
End Sub
```

Existing Worksheets can also be renamed at will utilizing the next subroutine.

```
Sub RenameSheet(wkbook, wksheet, newname)
'Renames the specified Worksheet in Workbook to (newname)
Workbooks(wkbook).Worksheets(wksheet).Name = newname
End Sub
```

When a user is working on a report, or when a macro is executing, it can be very useful to provide the user with messages or instructions that appear on Excel's status bar. The next subroutine will write any message desired to Excel's status bar.

```
Sub StatusBarMsg(ByVal Msg$)
'This will Display a Message to Excel's Status Bar
'Make Sure Status Bar Display is Turned on
If Application.DisplayStatusBar = False Then
Application.DisplayStatusBar = True
'Display String Passed in Msg$ Variable
Application.StatusBar = Msg$
End Sub
```

Columns are a special type of entity in Excel because often a column will only hold data of a specific type. When numerous rows of data are present in a Worksheet, it is good to have all the columns sized large enough such that none of the data in them is truncated (if practical). The next subroutine will autosize a Worksheet such that all columns will be able to hold the largest row of data without truncating it.

```
Sub AutoSizeSheet(wkbook, wksht)
'Select Workbook and Worksheet
ActivateWorkbook (wkbook)
Worksheets(wksht).Select
'Select All Cells
Cells.Select
'Autosize Columns
Cells.EntireColumn.AutoFit
'Unselect Cells
Range("A1").Select
End Sub
```

Sometimes, a column will contain formulas, and the user will wish that the column contained static values that will not change. The following subroutine will convert any column to hold values only. The value that the cell will hold will be the value the formula generated when the subroutine is invoked. This subroutine is especially useful when copying a Worksheet to create a report.

```
Sub ConvColToValues(wkbook, wksht, Col)
'Convert a Given Column (with formulas) to Values only
Workbooks(wkbook).Worksheets(wksht).Columns(Col).Select
Selection.Copy
Selection.PasteSpecial Paste:=xlValues
End Sub
```

The text data type is unique in that it retains data in exactly the same manner as it is extended. It does not try to convert dates or times (or what the application "thinks" are dates and times) to certain formats. Text-type columns are excellent because anything added to them is never converted to anything different. The following subroutine will convert an existing column to type text.

```
Sub SetColTypeAsText(ByVal wkbook, ByVal wksht, ByVal Col)
'Sets a column format to a specific type
Workbooks(wkbook).Worksheets(wksht).Columns(Col).Select
Selection.NumberFormat = "@"
End Sub
```

This subroutine can be taken a step further by converting an entire sheet to type text as follows.

```
Sub SetWkshtTypeAsText(ByVal wkbook, ByVal wksht)
'Sets a column format to a specific type
Workbooks(wkbook).Worksheets(wksht).Cells.Select
'Cells.Select
Selection.NumberFormat = "@"
End Sub
```

Dates are problematic because there are so many different ways to express them. Is a four-digit or two-digit year desired? The following subroutine allows a column to be set as the date type to a format specified by the developer.

```
Sub FormatDate(ByVal wkbook, ByVal wksht, ByVal Col, ByVal
ddmmyy As String)
'Macro Sets a column to Type Date with custom format
Workbooks(wkbook).Worksheets(wksht).Columns(Col).Select
Selection.NumberFormat = ddmmyy
Columns(Col).EntireColumn.AutoFit
End Sub
```

There are many different kinds of formats that can be applied to cells, rows, and columns in Excel. Once a type of format has been set up on one cell, it can be a real time-saver to copy the format to another cell or range of cells. The next subroutine allows the developer to do this.

```
Sub CopyCellFormat(fromstartRow, fromstartCol, fromendRow,
fromendCol, fromSheet, _
      fromBook, tostartRow, tostartCol, toendRow, toendCol,
      toSheet, toBook)
'Copy Formats Only from One Range to Another Range
Workbooks(fromBook).Worksheets(fromSheet).Select
Workbooks(fromBook).Worksheets(fromSheet).range(Cells(fromstar
tRow, fromstartCol), Cells(fromstartRow, fromstartCol)).Copy
Workbooks(toBook).Worksheets(toSheet).Select
Workbooks(toBook).Worksheets(toSheet).range(Cells(tostartRow,
tostartCol), Cells(toendRow, toendCol)).PasteSpecial
Paste:=xlFormats
End Sub
```

Whenever reporting values, it is a good idea to set the precision for the numbers being reported. If a value can only be accurately calculated to two decimal places, it is meaningless (and misleading) to display a number with eight decimal places. The following two subroutines allow the precision to be set for both a cell and a column.

```
Sub SetColPrecision(ByVal wkbook, ByVal wksht, ByVal ColNumber,
ByVal decplaces)

'Set the number of Decimal Places for a particular Column in
a Worksheet
Dim numforstr As String, ii As Integer
ActivateWorkbook (wkbook)
Sheets(wksht).Activate
Columns(ColNumber).Select
'This used to work?
'Selection.NumberFormat = decplaces
numforstr = "0."
For ii = 1 To decplaces
  numforstr = numforstr & "0"
Next ii
```

```
Selection.NumberFormat = numforstr
End Sub

Sub SetCellPrecision(ByVal wkbook, ByVal wksht, ByVal row,
ByVal Col, ByVal decplaces)
'Set the number of decimal places for a particular cell
Dim numforstr As String, ii As Integer
ActivateWorkbook (wkbook)
Sheets(wksht).Activate
Range(Cells(row, Col), Cells(row, Col)).Select
numforstr = "0."
For ii = 1 To decplaces
 numforstr = numforstr & "0"
Next ii
Selection.NumberFormat = numforstr
End Sub
```

The next subroutine serves two purposes. From a practical standpoint, it erases any ranges that may exist on any sheets within a Workbook. This is important because, if a range preexists when a function is run in VBA code, it may generate an error. It is good practice to eliminate all ranges in Workbooks prior to saving them as reports. From an aesthetic standpoint, Worksheets look nicer without preexisting shaded blue ranges on them.

```
Sub SetActiveCell()
'Choose Cell "A1" on Every Worksheet (don't leave set on
formula cells)
For Each ws In Worksheets
 Sheets(ws.Name).Activate
 range("A1").Select
Next
End Sub
```

In a report, appearance is everything, so it is important to be able to set both the characteristics of the text and its alignment within a cell or range of cells. The next subroutines allow the developer to change text to bold, italic, or underlined, as well as specify how text should be aligned in a cell or region of cells.

```
Sub SetFontBold(CellRange As range)
  CellRange.Select
  Selection.Font.Bold = True
End Sub

Sub SetFontItal(CellRange As range)
  CellRange.Select
  Selection.Font.Italic = True
End Sub

Sub UnderlineText(CellRange As range)
  CellRange.Select
```

```
      Selection.Font.Underline = xlUnderlineStyleSingle
End Sub

Sub FontAlignment(CellRange As range, Left As Boolean, _
   Center As Boolean, Right As Boolean)
CellRange.Select
If Left = True Then
   With Selection
     .HorizontalAlignment = xlLeft
   End With
End If
If Center = True Then
   With Selection
     .HorizontalAlignment = xlCenter
   End With
End If
If Right = True Then
   With Selection
     .HorizontalAlignment = xlRight
   End With
End If

End Sub
```

Notice that, in the above subroutine to "center across columns," the HorizontalAlignment property is set to xlCenter, but the Range must encompass a group of cells rather than a single cell.

Similarly, it is also helpful to be able to set the size and type of font over a range of cells. If the font name is not specified when the following subroutine is called, it is assumed to be Arial, Excel's default font.

```
Sub SetFontSpecs(CellRange As range, Size As Integer, Optional
FontName As String = "Arial")
CellRange.Select
  With Selection.Font
    .Name = FontName
    .Size = Size
  End With
End Sub
```

The following technique can be very beneficial if the reports a macro produces must conform to a governmental or quality control standard that ensures they will not be tampered with (Example: Drug Industry 21 CFR part 11). This method will only work if the sheet it is invoked on is protected. A sheet can easily be protected in code using the *.Protect* method. Example:

```
ActiveSheet.Protect
```

To prevent a user from changing the values in a range of cells a report has calculated, simply specify the range and lock it as shown in the following code snippet.

```
Range("A1").Locked = true
```

6.7 AUTOMATICALLY EMAILING REPORTS

The reader can have the greatest report in the world, but it is meaningless if the right people cannot obtain it. Reports can be sent by one of three ways utilizing Excel VBA. They are listed here in order of increasing versatility.

1. A Single Worksheet can be mailed from within Excel.
2. An Entire Workbook can be mailed from within Excel.
3. Microsoft Outlook® can be harnessed from within Excel to mail reports.

Sample Excel scripts will be provided to accomplish each of these tasks. Mailing a Worksheet is the way to go in terms of raw simplicity. There is no file attachment, and the data is sent in a space/tab delimited format. If a report consists of a single sheet within a Workbook, this is not a bad option. In order to do this, a range of cells must be specified that are to be sent. The simplest way to accomplish this is via the *.UsedRange* property. An active Worksheet within Excel can be mailed by selecting <u>A</u>DA->Chapter <u>6</u>->**M**ail Worksheet. (The address will need to be changed in VBA code from "name@domain.com" to the desired destination.)
(Note: This subroutine will not work with Excel versions prior to 2002.)

```
Sub Mail_Worksheet()
'This will mail the Active Sheet as Text
'Select the range of cells used on the active Worksheet.
ActiveSheet.UsedRange.Select
' Show the envelope on the ActiveWorkbook.
ActiveWorkbook.EnvelopeVisible = True
With ActiveSheet.MailEnvelope
 .Introduction = "This is a sample Worksheet."
 .item.To = "name@domain.com"
 .item.Subject = "Email Subject Line Text Here"
 .item.Send
End With
End Sub
```

If necessary, the entire Workbook can be mailed to anyone right from within Excel. The Workbook is sent as an attachment within the message. If a Workbook is open within Excel, it can be mailed by selecting <u>A</u>DA->Chapter <u>6</u>->Mail <u>Workbook</u>. (The address will need to be changed in VBA code from "name@domain.com" to the desired destination.)

```
Sub Mail_Workbook()
  'This will mail the Active Workbook as an Attachment
  ActiveWorkbook.SendMail "name@domain.com", "Email Subject
  Line Text Here"
End Sub
```

The ultimate in flexibility is to use Microsoft Outlook to mail a report. This method gives developers an additional degree of freedom. Not only can they mail an Excel Workbook but any file that resides on the user's system can be mailed as well by simply specifying its full path.

In order to harness Outlook from Excel, a Reference to the Outlook Object Library (msoutl9.olb for version 9.0) must exist within the current project as shown in Figure 6.11.

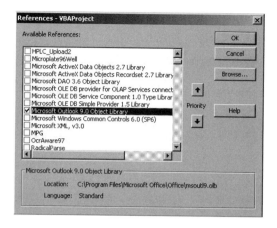

FIGURE 6.11 Referencing the Outlook Object Library.

This functionality can be tested by selecting ADA->Chapter 6->Outlook Mail. (The address will need to be changed in VBA code from "name@domain.com" to the desired destination, as well as the path to the file that is to be attached and mailed.)

```
Sub Send_Msg_Using_Outlook()
Dim objOL As New Outlook.Application
Dim objMail As MailItem
Dim CRLF As String
CRLF = Chr(13) + Chr(10)
Set objOL = New Outlook.Application
Set objMail = objOL.CreateItem(olMailItem)
With objMail
 .To = "name@domain.com"
 .Subject = "Automated Mail Response"
 .Body = "This is an automated message generated in Excel
 VBA." & CRLF & _
   "A Worksheet Cell Item Can be send by utilizing the commented
   code below: " '& _
   'Format(range("A1").Value, "$ #,###.#0") & "."
     .Attachments.Add "C:\temp\junk.xls", , 0
     '0 = Hidden Attachment
     '1 = Beginning of Email
     .Display
End With
'To Actually Send Message Uncomment Next Statement
objMail.Send
Set objMail = Nothing
Set objOL = Nothing
End Sub
```

6.8 SUMMARY

Reports are the means by which all of the analyses the reader learns to accomplish in the text will be conveyed to their end users. The reader is introduced to the most common and useful tasks required to create effective reports such as the use of templates, preparation of multiview reports, performing calculations within reports, basic formatting of reports, and distribution of reports by electronic means.

7 Introduction to Microsoft Access

7.1 INTRODUCTION

Thus far, the discussion has centered around utilizing data from files for automated data analysis. The reality is that although data are often *first generated* in a file type format, it will rarely be stored or archived as a file. The principal impediment to storing information as a series of files is that it is not readily searchable. Although it is true that there are applications that will seek out information contained in a series of files, the scale of efficiency is not nearly comparable to a modern database. Thus, information that started in a file type format is often loaded into a modern relational database, where it can be searched and compared against other like elements of the database with relative ease.

A relational database is used to manage a collection of data. The software is more correctly termed a relational database management system, or RDBMS. Relational databases are based on a relational model, meaning all data is represented as a mathematical relation, or more precisely, a subset of the Cartesian product of n sets. In such a mathematical model, reasoning (or queries) can return two possible states for each proposition, either true or false. Data is operated on by means of relational calculus and algebra.

Relational databases store data inside tables, and nothing more. All operations in a relational database on data are done on the tables themselves, or produce other tables as a result. The developer will never see anything except tables (and the values they contain).

Data, like everything else in life, are always changing. Not only do discrete values change but the underlying structure the data resides in may change as well. This is especially true when data are retained over an extended period of time.

Normalization is a method to avert problems in a relational database due to duplication of data values and when modification of structure and/or content has taken place. Databases are normalized to ensure data consistency and stability and to eliminate redundancy. Normalization ensures consistent updatability and maintainability of the data and avoids anomalies that result in ambiguous data or inconsistent results.

Microsoft Access is one type of relational database. Oracle is another type. Usually, Oracle is utilized for very large projects, whereas Microsoft Access is utilized for small- to intermediate-sized projects. Most large corporations utilize Oracle for their primary database, although often ancillary groups will utilize Microsoft Access or similar products for smaller projects. Access and Oracle are by no means the only relational databases that exist. They are, however, two of the most popular relational databases that exist. Others such as FileMaker Pro and many open source alternatives are popular, especially within niche markets. Because Access is part of the Microsoft family, it will be utilized for the majority of sample applications that show how to utilize Excel to access information within a database.

7.2 ELEMENTS OF A RELATIONAL DATABASE

Using Microsoft Access, a variety of information pertaining to a single project can be stored within a single database file. Each file within Access can contain separate storage containers called tables. A table is a collection of data about a specific topic, such as products or suppliers. Using a different table for each topic means that data are stored only once, which makes the database more efficient and reduces data-entry errors. Tables organize data into columns (also referred to as fields) and rows (also referred to as records).

A table is just a set of rows and columns. This is very important because a set does not have any *predefined* order for its elements. Each row is a set of columns, each of which can have only one possible value. All rows from the same table have the same set of columns, although some columns may have NULL values, i.e., some values in that row may not be initialized. Note that a NULL value for a string column is different from an empty string. You should think about a NULL value as an "unknown" value.

The next logical question is: if data can be stored within many different tables in the same database, how can all the different tables relate to each other to extract required information? The answer is: the common field. A common field is a column that is present within each table that must be related to. Using a common field, Microsoft Access can bring together the data from many different tables for viewing, editing, or printing.

To illustrate the concept of the table, a sample database will be created for a clock shop. In this instance, two tables will be created — one for products (named "Products") and one for suppliers (named "Suppliers"). The products table will contain information relating to the products available for sale at the clock shop. The fields or columns in the products table will be: MANUFACTURER, MODELNAME, MODELNUMBER, PRICERETAIL, PRICECON-SUMER, PRICECOST, PICTURE, TYPE, and DESCRIPTION. Notice that every field name is in capital letters. This is not required, but many database administrators choose to name columns in uppercase letters (it takes the guesswork out of what case a letter is in a database query, which will be covered in depth later in the chapter). Another common nomenclature is to separate descriptors in field names by means of an underscore character "_." Example: MODELNAME becomes MODEL_NAME.

The Supplier's table will contain information relating to each supplier the clock shop acquires stock from. The fields in this table will be: NAME, MANUFACTURER, ADDRESS, CITY, STATE, ZIP, PHONE, FAX, and WEBSITE. The MANUFACTURER is the common field between the products and suppliers tables. If the contact information is desired for a specific product, a query can be made between the two tables utilizing the MANUFACTURER field. In some cases the NAME of the supplier will be identical to the name of the MANUFACTURER.

A table can easily be created in Access by clicking on the "Create Table in Design View" icon as illustrated in Figure 7.1. Before creating a table, the user must decide what data type each field will be.

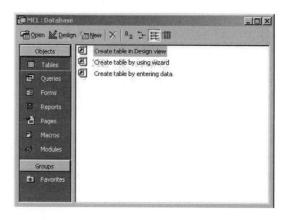

FIGURE 7. 1 Creating a Table in the Design View.

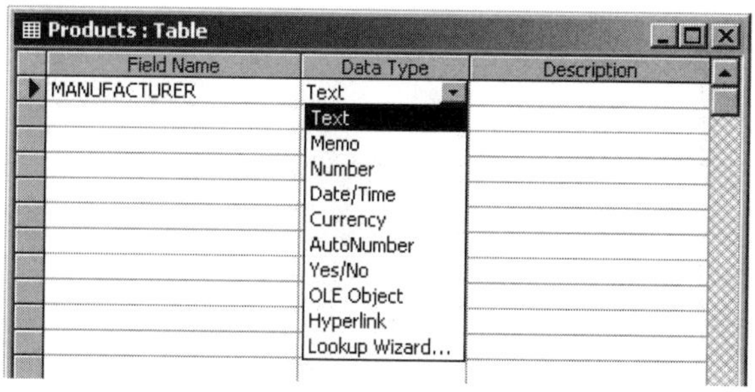

FIGURE 7.2 Data types available in Microsoft Access.

In this example, five data types will be utilized: number, text, currency, hyperlink, and memo (Figure 7.2). A number must be a number and not contain any other characters. Therefore, in the case of the field MODELNUMBER, which might conceivably contain letters, the text data type must be used. The only difference between text and memo is that memo allows more information to be stored in the field. A text field is limited to 50 characters, whereas a memo field is limited to about 50 paragraphs. The reader should be aware that many difficulties exist converting Access database tables to Oracle database tables. One tool the author has utilized to make such conversions easier is the Intelligent Converters' Access to Oracle tool that can be referenced at this website: http://www.convert-in.com/. A special difficulty exists in converting Access Memo data types to Oracle because Oracle really does not have an equivalent field. If the memo data type in Access is 4000 characters or less, it can be converted to the Oracle varchar2 data type. The programmers at Intelligent Converters modified the Access to Oracle tool to accomplish this for the author at no charge!

Separation of the products table from the suppliers table allows products from a manufacturer to be added or deleted at will without compromising or duplicating the contact information about the supplier. Thus, there is *one degree of separation* between the products and the suppliers.

Figure 7.3 illustrates the Products Table built in the Design View within Microsoft Access. Microsoft Access has two views for database elements: the Design View and the Datasheet View. The Design View is utilized when designing a new table. Design View shows the column names or fields *in row format* — somewhat counterintuitive. There are three columns in Design View that show:

1. The field (or column) name
2. The data type of the column
3. An optional description of the column

Notice that in Figure 7.3 the columns Field Name, Data Type, and Description are shown in order. In the Design View this will be the only information displayed. No information (or records)

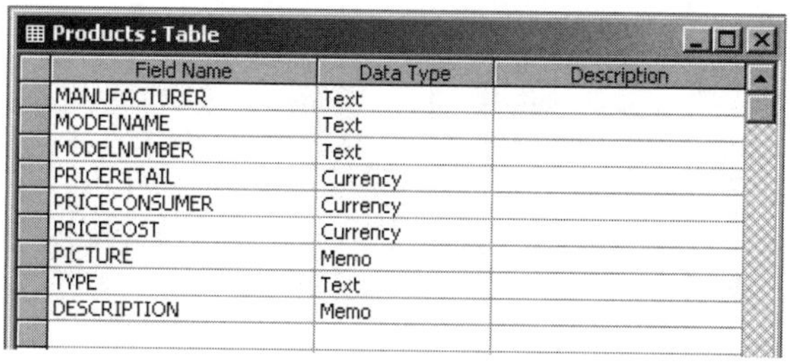

FIGURE 7.3 The Products Table in Design View.

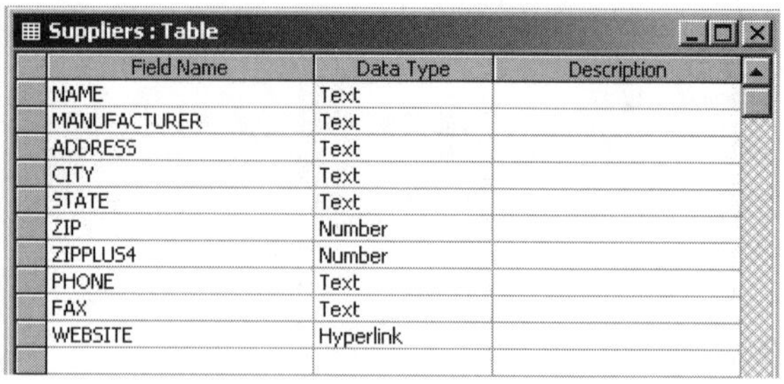

Field Name	Data Type	Description
NAME	Text	
MANUFACTURER	Text	
ADDRESS	Text	
CITY	Text	
STATE	Text	
ZIP	Number	
ZIPPLUS4	Number	
PHONE	Text	
FAX	Text	
WEBSITE	Hyperlink	

FIGURE 7.4 The Suppliers Table in Design View.

contained within the database are shown in the Design View. Similarly, Figure 7.4 shows the Suppliers Table laid out in Design View.

The final concept to be covered is that of the key. Simply stated, a key is a value (or group of values) that can be utilized to identify a particular record or group of records within a database. A primary key is a value that can be used to identify a single record in a table. Primary keys should be immutable, meaning that its value cannot be changed during the course of normal operations of the database. A primary key can be a normal attribute that is guaranteed to be unique (such as Social Security Number in a table with no more than one record per person), or it can be generated by the Database Manager as a globally unique identifier. A plurality of keys can be utilized in combination to create a primary key. A key consisting of attributes of many keys is referred to as a compound key. For the sake of simplicity, a compound key is *usually* not appropriate to utilize as a primary key.

In terms of the sample application, some thought has to be given as to how to set up the keys for each table. The Suppliers Table will only utilize one field as its primary key. The NAME column will be utilized as the primary key for the Suppliers Table. Although a supplier may carry more than one manufacturer, or two suppliers may be in the same city, each supplier will have a unique name. Setting a single primary key is easy; the user must view the table in Design View, choose a single field (shown as a row), and either right-click and select "Primary Key" from the menu or press the primary key button on the Access toolbar (Figure 7.5).

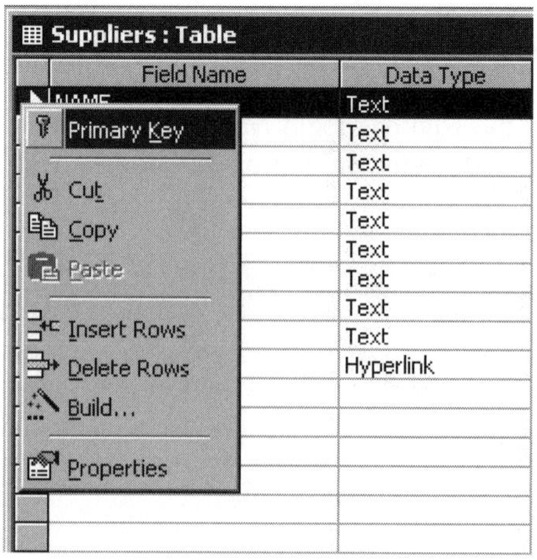

FIGURE 7.5 Selecting a single primary key for the Suppliers Table.

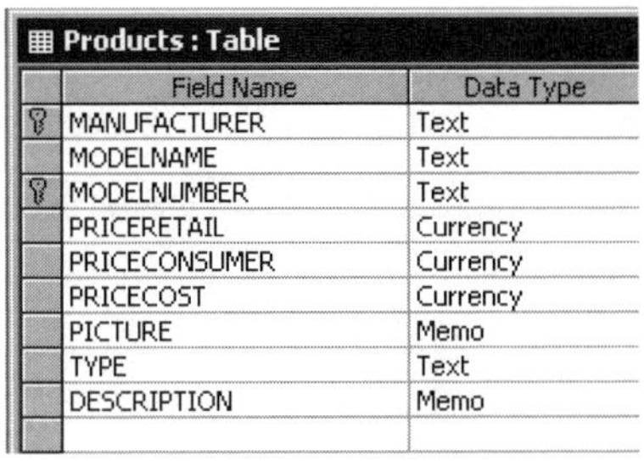

Products : Table

Field Name	Data Type
MANUFACTURER	Text
MODELNAME	Text
MODELNUMBER	Text
PRICERETAIL	Currency
PRICECONSUMER	Currency
PRICECOST	Currency
PICTURE	Memo
TYPE	Text
DESCRIPTION	Memo

FIGURE 7.6 Compound primary key designation for Products Table.

The Products Table is somewhat more problematic. Although assigning a single field for a primary key is desirable, it is not always possible. The Products Table represents just such a case. Here, many products may have the same manufacturer, and some manufacturers might even duplicate model names or model numbers. The only way to guarantee uniqueness is by utilizing a plurality of keys. In this instance, if both the MANUFACTURER and the MODELNUMBER are jointly referenced, they must produce a unique row entry. No manufacturer will have two products that utilize the same model number. Doing so would make it impossible for the manufacturer to then know what the consumer desired, defeating the whole purpose of having a model number to begin with.

Therefore, both MANUFACTURER and MODELNUMBER must be designated as primary keys. Setting multiple primary keys is also easy. The user must view the table in Design View and choose the fields (shown as rows) to be designated as primary keys, and click on them holding down the CTRL key. Holding down the CTRL key will allow the user to select multiple rows. Once multiple rows have been selected, the user simply either right-clicks and selects "Primary Key" from the menu or presses the primary key button on the Access toolbar. When compound or multiple primary keys are selected, a key will appear next to each field designated as a subset constituting the primary key as shown in Figure 7.6.

7.3 CONNECTING TO AN MS ACCESS DATABASE

Whether the user builds the database or wishes to access a database constructed by means of a third party, a mechanism must be put in place which allows Excel to access the database from the VBA environment. For Excel to access a database, it needs the following information about the database:

1. The database name (For MS Access a *.mdb file name)
2. The location of the database (the path to the *.mdb file)

Excel can obtain this information by utilizing a Data Source Name (DSN). A DSN provides connectivity to a database through an ODBC driver. ODBC stands for Open DataBase Connectivity and is an interface for ODBC drivers. Many applications utilize ODBC drivers to allow other programs to access their data. The DSN contains, at a minimum, the database name, directory, and database driver, and may contain other information such as a UserID and password. Once a DSN is created for a particular database, the DSN can be utilized by any application (that supports DSN/ODBC) to query information from the database.

To access information in a database, the application must utilize software components called drivers. Data Sources (ODBC) is instrumental in adding and configuring these drivers. To open Data Sources (ODBC) from the Start menu, choose Settings, then click Control Panel. Double-click Administrative Tools, and then double-click Data Sources (ODBC). Figure 7.7 shows the ODBC Data Source Administrator.

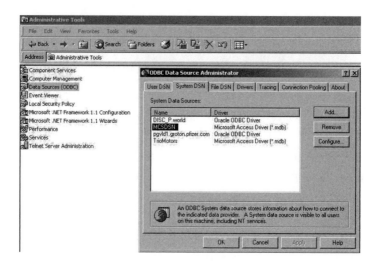

FIGURE 7.7 Opening the ODBC Data Source Administrator.

The sample application requires a DSN Named MCSDSN. To create this DSN, push the Add button on the System DSN tab of the ODBC Data Source Administrator shown in Figure 7.7. The GUI shown in Figure 7.8 will appear. Select Microsoft Access Database (*.mdb) from the listbox that appears and press the finish button. Fill in Data Source Name with "MCSDSN" and for Description type "Sample DSN Connection." Under database press the select button and navigate to the path of the (MCL.mdb) file, which must be copied from the CD-ROM to the computer the sample will be run on. Then click OK. The MCSDSN Data Source Name has just been created.

The problem with DSN connections is that they must be configured properly on every computer that will execute code seeking to make a DSN connection to the specified database. This can rapidly become a logistical nightmare if a very large user base is to utilize an application. Fortunately, there is a method to establish a connection to a database without using a DSN, commonly referred to as a "DSN-less connection." A DSN-less connection specifies the necessary information right in the application. With a DSN-less connection, the developer is free to use connection standards other than ODBC, such as OLE DB. DSN-less connections resolve the issue of maintaining data sources on several computers.

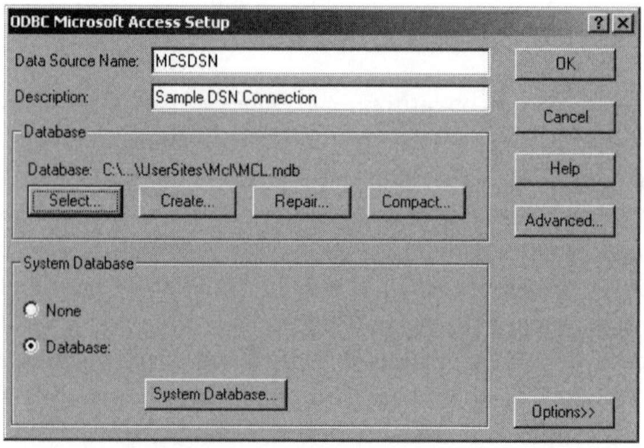

FIGURE 7.8 Creating the MCSDSN Data Source Name.

When using an ACCESS Database, it is best to try to avoid using a system DSN. They are much slower because they go through ODBC, which then uses the Jet Drivers to access the database. They also have to do a registry lookup that further adds to the performance gap. When DSN-less connections are utilized, the Jet Drivers are accessed directly, and the performance increase can be dramatic.

Because all the connection requirements are specified within the code, DSN-less connections require no setup when using the ODBC Data Source Administrator as shown earlier. All requirements are specified at run time in code. Here is an example DSN-less connection string for Microsoft access:

connstring = "ODBC;DRIVER={Microsoft Access Driver(*.mdb)};DBQ= C:\temp\ ADA\MCS. mdb"

where:
   ODBC — Specifies an interface to the ODBC drivers.
   DRIVER — Specifies which database application driver needs to be utilized; in this case, Microsoft Access. Other popular database applications such as Oracle and Filemaker Pro have their own specific drivers as well.
   DBQ — The path to the database file.

A DSN-less connection string to an Oracle database would look something like this:

"Driver={Microsoft ODBC for Oracle};Server=Clock_Store.world; Uid=myUserID; Pwd=myPassword;"

Notice that, in the above example, if the database is password protected, the developer may specify the user ID and password for access to the database without the user being prompted for them. Further notice that Oracle DSN-less connections utilize a server name (specified by SERVER =) as opposed to a filename. Although the center of focus here is MS Access, Oracle is so popular it is worth mentioning how to create a DSN-less connection string for Oracle databases.

Some important limitations are worth mentioning. The Excel 2000 implementation of OLE DB does not support background refresh, parameter queries, or page fields that retrieve data for each item separately. The most egregious offender in this category is the lack of support for background refreshing. In practical terms, what this means is that if a query is built from or triggered from a form, the form must be hidden and unloaded prior to executing the query. Otherwise an error will result.

Now some code to support use of the above concepts using VBA. A DSN connection using the previously defined MCSDSN DSN can be made utilizing the following code:

```
Sub Test_Get_Data()
    'Will NOT work when subroutine invoked from a Forms button press
    'OR when query is to be Executed as a Background Process
    DoEvents
    'Define SQL query string
    sqlstring = "SELECT NAME FROM Suppliers"
    'Define connection string and reference File DSN.
    connstring = "ODBC;DSN=MCSDSN"
    'Create a QueryTable in worksheet beginning with cell A1.
    With ActiveSheet.QueryTables.Add(Connection:=connstring, _
        Destination:=Range("A1"), Sql:=sqlstring)
```

```
      DoEvents
      .Refresh
   End With
   End Sub
```

If the preceding code snippet is run, every Name will be returned in the Suppliers Table starting in spreadsheet position A1. What is returned is determined by the SQL String. SQL Strings will be covered in the next section of this chapter. The foregoing snippet of code, however, will not function if it is triggered from an object on a form such as a button. Obviously, this is a limitation on the utility of database support available within the Excel paradigm. The code can work when executed from a form object by making the following change.

```
Sub Test_Get_Data2()
   'Unload Form - The Excel 2000 implementation of OLE DB does
   not support background refresh
   Unload frm7Access
   'Will work when called from Form button press or as a
   Background Process
   DoEvents
   'Define SQL query string
   sqlstring = "SELECT NAME FROM Suppliers"
   'Define connection string and reference File DSN.
   connstring = "ODBC;DSN=MCSDSN"
   'Create a QueryTable in worksheet beginning with cell A1.
   With ActiveSheet.QueryTables.Add(Connection:=connstring, _
   Destination:=Range("A1"), Sql:=sqlstring)
   DoEvents
   .Refresh
   End With
End Sub
```

Because the form is unloaded prior to execution of the Query, the Query is not run as a background process and thus will function correctly. Note that frm7Access should be replaced with the name of the form that has the object that triggers the code. The final case is using a DSN-less connection string, which is the most desirable way to connect to a database as all the connection parameters are specified within the application and does not require that a DSN be set up on the computer executing the code.

```
Sub Test_Get_Data_DSNless()
   'Unload Form - The Excel 2000 implementation of OLE DB does
   not support background refresh
   Unload frm7Access
   'Will work when called from Form button press
   DoEvents
   'Define SQL query string
   sqlstring = "SELECT NAME FROM Suppliers"
   'Define a DSN-less Connection String.
   connstring = "ODBC;DRIVER={Microsoft Access Driver
   (*.mdb)};DBQ=C:\temp\ADA\MCS.mdb"
```

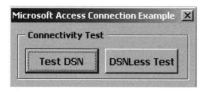

FIGURE 7.9 Test application to establish a sample connection.

```
'Create a QueryTable in worksheet beginning with cell A1.
 With
Workbooks(ActiveWorkbook.Name).Worksheets(ActiveSheet.Name).
QueryTables.Add(Connection:=connstring, _
    Destination:=Range("A1"), Sql:=sqlstring)
    DoEvents
    .Refresh
    End With
End Sub
```

The preceding sample connections can be tested by selecting ADA->Chapter 7->Access Connection. Doing so will bring up the GUI shown in Figure 7.9. Pressing either button will establish a connection to the sample Microsoft Access database from Excel. (For this sample application to work on the reader's machine, the sample Microsoft Access database "MCS.mdb" must be copied to the location "C:\temp\ADA\" from the CD-ROM.) Although pressing either button will establish a connection and query the sample database for the Supplier names, one button utilizes a DSN to create the connection, and the other creates a DSN-less connection. To create a connection using a DSN, remember that the DSN must be set up on the machine running the code as described earlier. DSN-less connection strings require no connectivity setup on the host machine and are thus less complicated and preferred methods of connection.

7.4 QUERIES: HOW TO RETRIEVE INFORMATION IN DATABASES USING SQL

Connecting to a database is meaningless unless the information desired can be extracted from it. The way information is extracted from a database is through a Query. Queries utilize their own language known as SQL or Structured Query Language. Entire texts have been written on writing and constructing SQL Queries. Fortunately, there are a few simple constructs that will allow most users to extract the information they require for automated data analysis and report preparation. SQL varies depending upon the database utilized. Thus, code that works with Microsoft Access may not function correctly (or at all) with Oracle.

Information is queried from databases utilizing some form of the Select From SQL construct. Recall that in the previous section, when setting up database connectivity, the following SQL command was utilized:

"SELECT NAME FROM Suppliers"

The SELECT command is followed by what fields (or columns) in the database the user wishes to access. In this instance, the user wants to view the field named "NAME." The FROM command is used to identify where the query is to pull the information from (usually a table). Here, the user wishes

TABLE 7.1
Data Types in Microsoft Access

Data Type	Use	Size
Text	Text or combinations of text and numbers, such as addresses. Also numbers that do not require calculations, such as phone numbers, part numbers, or postal codes.	Max = 255 characters
Memo	Lengthy text and numbers, such as notes or descriptions.	Max = 64,000 characters
Number	Numeric data to be used for mathematical calculations, except calculations involving money (use Currency type). Set the FieldSize property to define the specific Number type.	1, 2, 4, or 8 bytes. 16 bytes for Replication ID (GUID) only
Date/Time	Dates and times.	8 bytes
Currency	Currency values. Use the Currency data type to prevent rounding off during calculations. Accurate to 15 digits to the left of the decimal point and 4 digits to the right.	8 bytes
AutoNumber	Unique sequential (incrementing by 1) or random numbers automatically inserted when a record is added.	4 bytes. 16 bytes for Replication ID (GUID) only
Yes/No	Fields that will contain only one of two values, such as Yes/No, True/False, On/Off.	1 bit
OLE Object	Objects (such as Microsoft Word documents, Microsoft Excel spreadsheets, pictures, sounds, or other binary data) created in other programs using the OLE protocol that can be linked to or embedded in a Microsoft Access table. You must use a bound object frame in a form or report to display the OLE object.	Up to 1 GB (limited by disk space)
Hyperlink	Field that will store hyperlinks. A hyperlink can be a UNC path or a URL.	Up to 64,000 characters
Lookup Wizard	Creates a field that allows you to choose a value from another table or from a list of values using a combo box. Choosing this option in the data type list starts a wizard to define this for you.	

Courtesy Microsoft.

to query all the NAMES in the Suppliers Table. The Select and From SQL commands are case insensitive. The case of the fields and tables should be identical as to how they appear in the database.

Note that the foregoing command will return *all* of the NAMEs in the Suppliers Table. Another way of putting it is that the query will return all the records (or rows) for the name field (or column). Table 7.2 shows the entire contents of the Suppliers Table.

If the user wished to query the entire contents of the database, the SQL query could be modified as follows:

"SELECT * FROM Suppliers"

The "*" in the statement effectively tells the database to return all columns in the table. This is a very popular form of the select statement and is widely utilized to determine information about field names residing in tables.

TABLE 7.2
The Suppliers Table in the MCL.mdb Database

NAME	MANUFACTURER	ADDRESS	CITY	STATE	ZIP	ZIPPLUS4	PHONE	FAX	WEBSITE
Howard Miller	Howard Miller	860 East Main Avenue	Zeeland	MI	49464	1300	616-772-7277		#http://www.howardmiller.com/#
Chelsea Clock Company	Chelsea Clock Company	284 Everett Avenue	Chelsea	MA	02150	1598	800-284-1778		#http://www.chelseaclock.com/#
Bulova Corporation	Bulova Corporation	One Bulova Avenue	Woodside	NY	11377		800-228-5682		#http://www.bulova.com/brands/office/office.aspx#
Hentschel	Hentschel	564 Weber St. N. - Unit 2	Waterloo	Ontario, Canada	N2L 5C6		866-565-1972		#http://www.hentschelclock.com#
Hermle Black Forest Clocks	Hermle Black Forest Clocks	P.O. Box 670	Amherst	VA	24521		434-946-7751		#http://www.hermleclock.com/index.html#
Sligh Furniture Company	Sligh Furniture Company	1201 Industrial Avenue	Holland	MI	49423				#http://sligh.com/home.asp#

A query is, of course, not relegated to returning a single column or all columns. It can specify multiple columns to be returned by separating them with a comma. The foregoing example can be modified thus:

"SELECT NAME,CITY,STATE FROM Suppliers"

These examples show how to control the number and name of columns returned from a database query; however, all of them still return *all* the records in the database. The WHERE clause allows a user to specify in a query which subset of the database's records will be returned. Example:

"Select * From Suppliers Where STATE = 'MA'"

This query will only return a single entry in the database — Chelsea Clock Company. Chelsea Clock Company is the only Supplier in the Suppliers Table that has an address in 'MA.' Notice that, when making textual comparisons as in strings, the string must be enclosed in single quotation marks "''". In addition to string comparisons, numerical comparisons are also possible. Sample:

"Select * From Employees Where SSN = 123456789"

This query would retrieve all employees whose Social Security Number (SSN) is 123456789 from the EMPLOYEES Table in the MCS Database. In this case, it returns "James Jerry Colegrove" from the Employees Table in the MCS.mdb database. A quick word about numeric data — it can only have numbers in it. If the user wished to store the Social Security Number with dashes "123-45-6789," then the SSN field would have to be of type TEXT. Similarly, in the case of storing a zip code where a leading zero may be required, the field must be stored as type text. This is because a Number field will delete a leading zero. For these reasons, the Suppliers Table does not contain *any* Numeric fields.

Chances are a single limiting factor for a query will not be enough, and most queries will require two or more conditions to be satisfied. Such queries are said to have *compound conditions*. Compound conditions are added to a query with the use of the logical operators AND and OR. The AND operator requires both conditions on either side of the AND operator to be TRUE in order for the query to return TRUE. The OR operator requires only one condition on either side of the OR operator to be True in order to return True. Complex queries can be built utilizing any number of compound conditions required to return the desired results. An example of a simple compound condition query:

"Select * From Suppliers Where STATE = 'MA' Or STATE = 'VA'"

This example query on the MCS.mdb database will return the following entries from the Suppliers Table: Chelsea Clock Company, Hermle Black Forest Clocks. The former is located in Massachusetts and the latter is located in Virginia.

All of these examples have tested for conditions of equality. Table 7.3 shows the most common comparison operators utilized in SQL queries. Note that, in most cases, the operators can be utilized on most data types: number, text, date, etc.

The final topic for discussion with respect to SQL queries is that of the LIKE construct. Often it is desirable to search for records in a database that match a certain pattern. For example, perhaps a subset of records within a table where the zip code begins with "063" is desired. Such a query would return all of the addresses in cities in New London County, CT. Example zip codes: New London, CT 06320; Gales Ferry, CT 06335; Waterford, CT 06385; Groton, CT 06340.

TABLE 7.3
Common SQL Comparison Operators

Operator	Description
=	Equal
<>	Not equal to
!=	
>	Greater than
<	Less than
>=	Greater than or equal to
<=	Less than or equal to

Such a query can be constructed utilizing the LIKE operator and the '%' wildcard character. The '%' can be utilized in a text query to indicate *any* character. Thus, a search for zip codes beginning with "063" would utilize "063%%," or, in the case of a nine-digit zip code, "063%%-%%%%". Here is a sample LIKE query for this example:

> "Select * From Suppliers Where ZIP LIKE '063%%'"

Another Example that will return records from the sample database is

> "Select * From Suppliers Where STATE Like 'M%'"

This query ends up returning the following companies: Howard Miller, Chelsea Clock Company, and Sligh Furniture Company. All of these companies are located in either Massachusetts ("MA") or Michigan ("MI"), which matches the query request.

7.5 USING THE MICROSOFT ODBC ADD-IN FOR MICROSOFT EXCEL (XLODBC.XLA)

The connection examples earlier in the chapter utilized the QueryTables Collection Object to return information queried from a Microsoft Access database by means of ODBC. The QueryTables Collection Object is somewhat limited in its capabilities. Fortunately, Microsoft has constructed an add-in for Excel that allows developers to harness the power of ODBC utilizing a few powerful functions. The XLODBC.XLA ODBC add-in allows developers to create macros that interact directly with ODBC through the ODBC driver manager.

In order to use these functions, the XLODBC.XLA must be added as a library reference within the project. If it is not added as a library reference, the error message "Sub or function not defined" will occur when an XLODBC.XLA function is called. To add it as a library reference, do the following (Figure 7.10):

1. In a *Visual Basic Module,* click References in the Tools menu.
2. In the References dialog box, click to select the XLODBC.XLA checkbox.

NOTE: If XLODBC.XLA is not present in the References dialog box, it can be added by clicking the Browse button and locating the XLODBC.XLA file. The XLODBC.XLA add-in should be located in the EXCEL\LIBRARY\MSQUERY or \PROGRAM FILES\MICROSOFT OFFICE\OFFICE\LIBRARY\MSQUERY folder.

A properly installed XLODBC.XLA Reference is shown in REF _Ref112138066 \h * MERGEFORMAT (Figure 7.11). Notice that, under References, XLODBC.XLA is checked, and

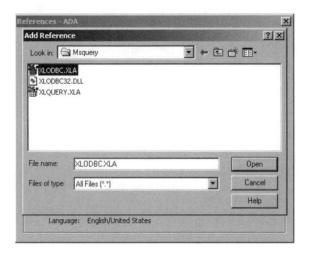

FIGURE 7.10 Adding the XLODBC.XLA Reference.

that a Folder named References appears in the Project Explorer that contains "Reference to XLODBC.XLA."

The XLODBC.XLA Reference makes a wide range of functions available for use in ODBC connectivity to databases using Excel VBA. The available functions and the action they each perform are listed in Table 7.4.

The XLODBC Reference is important because it allows Excel VBA programmers to access information in Microsoft Access using its predefined functions with relative ease. When extracting information from an Access database, the user should have the ability to gather information about

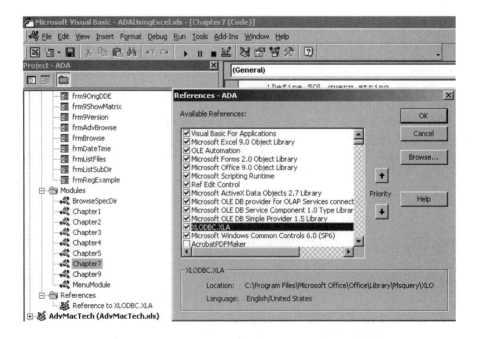

FIGURE 7.11 A properly installed XLODBC.XLA Reference.

TABLE 7.4
The XLODBC.XLA Visual Basic Functions

Function	Action
SQLBind	Specifies storage for a result column
SQLClose	Closes a data source connection
SQLError	Returns detailed error information
SQLExecQuery	Sends a query to a data source
SQLGetSchema	Gets information about a connected data source
SQLOpen	Establishes a connection to a data source
SQLRetrieve	Retrieves query results
SQLRetrieveToFile	Retrieves query results and places them in a file
SQLRequest	Connects with an external data source, performs the query, and returns the result set

the database with which a connection has been established. At a minimum, the developer may need to access the following information about the database on the fly:

1. The number of tables in the database
2. The names of the tables in the database
3. The ability to determine if a table is a system table
4. The number of fields in a particular table
5. The names of the fields in a particular table
6. The data type of a particular field in a particular table

VBA Code will be created, which accomplishes the tasks just listed by utilizing the XLODBC functions. The first task in dealing with any database would be to determine the number of tables in the database and their names. A single function can accomplish this task with an array as a passed parameter (passed by reference) to hold the table names. The total number of tables is the value that will be returned to the function.

```
Function GetTableNames(SQLOpenString As String, TableNames())
As Integer
'Returns the Total Number of Tables and their Names in
TableNames
Dim Chan As Variant
Dim TName As Variant, GetSchema() As Variant
Dim DatabaseName As Variant
On Error GoTo No_Database:
'Establish a connection to the NWind data source.
Chan = SQLOpen(SQLOpenString)
'Get the name of the database on the connection
DatabaseName = SQLGetSchema(Chan, 7)
'Get the table names for the connected database
GetSchema = SQLGetSchema(Chan, 4, DatabaseName & ".")
For ii = 1 To UBound(GetSchema)
  ReDim Preserve TableNames(ii)
  TableNames(ii) = GetSchema(ii, 1)
```

```
Next ii
GetTableNames = UBound(GetSchema)
' Close the connection to the NWind data source.
SQLClose (Chan)
Exit Function

No_Database:
MsgBox SQLOpenString, vbOKOnly + vbCritical, "Error: Cannot
Open Database"
GetTableNames = 0
' Close the connection to the NWind data source.
SQLClose (Chan)
End Function
```

This function can be tested quite easily on the sample MCS.mdb database by placing it in the folder "C:\temp\ADA\MCS.mdb" and utilizing the following code snippet:

```
Sub TestGetTableNames()
Dim Tables()
Debug.Print GetTableNames("DRIVER={Microsoft Access Driver
(*.mdb)};DBQ=C:\temp\ADA\MCS.mdb", Tables())
For i = 1 To UBound(Tables)
 Debug.Print i; Tables(i)
Next i
End Sub
```

Doing so will result in the following Table Names printed to the screen:

1. MSysAccessObjects
2. MSysACEs
3. MSysObjects
4. MSysQueries
5. MSysRelationships
6. Employees
7. Products
8. Suppliers

The first five table names are system tables. *System tables* hold the names of forms, reports, queries, and tables within a particular database. System tables are readily identifiable as they will be prefixed with "MSys."

The names of forms, reports, queries, and tables within a database can be accessed using the MSysObjects. For example, the following query will select all the forms:

select Name, DateUpdate from MSysObjects   where Type=−32768;

Queries and tables are a little problematic because they can be generated by the system! Often it is useful to limit returned values to just those created by a developer and suppress entities that are self generated. A query such as this will accomplish this task:

Select Name, DateUpdate from MSysObjects where Type=5 and Name not Like "~*." Like many self-generated operating system files, self-generated tables and queries in an Access database are preceded by a tilde "~," and this is how the query just presented identifies and suppresses them.

When all the Table Names are returned utilizing the above `GetTableNames` function, it is important to have a means of identifying system tables, as the developer would most likely want to prohibit access to them. The next function accomplishes this:

```
Function IsSystemTable(TableName As String) As Boolean
'Identifies MS Access System Tables
Select Case TableName
  Case "MSysAccessObjects"
    IsSystemTable = True
  Case "MSysACEs"
    IsSystemTable = True
  Case "MSysObjects"
    IsSystemTable = True
  Case "MSysQueries"
    IsSystemTable = True
  Case "MSysRelationships"
    IsSystemTable = True
End Select
End Function
```

Once the names of the tables that reside within the database have been determined, a method of determining the names of the fields that reside within those tables must be developed. The function `GetFieldNames` is similar to its predecessor function, except that it operates on a single specified table identified in the passed parameter `TableName`. The `GetFieldNames` function returns an integer number of the total fields that reside within the specified table. The passed-by reference parameter `FieldNames` is an array that is updated within the function to contain the names of the fields that reside within the given table.

```
Function GetFieldNames(SQLOpenString As String, TableName As
String, FieldNames()) As Integer
'Returns Total Number of Fields in Table
Dim Chan As Variant
Dim Fields As Variant
Dim i As Integer
On Error GoTo No_Table:
'Establish a DSN-less connection to the MCS Sample Data Base
Chan = SQLOpen(SQLOpenString)
'Get the field names for the TableName passed parameter
Fields = SQLGetSchema(Chan, 5, TableName)
'Fields is a two-dimension Variant array; (name, type)
For i = 1 To UBound(Fields)
  TotalFieldNames = TotalFieldNames + 1
  ReDim Preserve FieldNames(1 To TotalFieldNames)
  FieldNames(TotalFieldNames) = Fields(i, 1)
```

```
    GetFieldNames = TotalFieldNames
Next i
'Close the connection to the database
SQLClose (Chan)
Exit Function
No_Table:
MsgBox SQLOpenString, vbOKOnly + vbCritical, "Error: The
Referenced Table Does Not Exist in "
GetFieldNames = 0
'Close the connection to the database
SQLClose (Chan)
End Function
```

This function can be easily tested utilizing the following code snippet:

```
Sub TestGetFieldNames()
Dim Names()
Debug.Print GetFieldNames("DRIVER={Microsoft Access Driver
(*.mdb)};DBQ=C:\temp\ADA\MCS.mdb", "Suppliers", Names())
For i = 1 To UBound(Names)
  Debug.Print i; Names(i)
Next i
End Sub
```

Doing so will produce the following output using the MCS.mdb test database included on the CD-ROM. The output is a comprehensive listing of all the Field names contained in the Suppliers Table.

1. NAME
2. MANUFACTURER
3. ADDRESS
4. CITY
5. STATE
6. ZIP
7. ZIPPLUS4
8. PHONE
9. FAX
10. WEBSITE

7.6 CONSTRUCTING A DATABASE QUERY TOOL

Using the functions presented and the XLODBC add-in, it is possible to construct a generic database query tool that will illustrate complex query use within the Excel environment for data mining purposes. The sample application can be run by selecting ADA->Chapter 7->Access SQL Query, which will bring up the GUI shown in Figure 7.12.

The basic elements of the sample database query tool can be adapted for any custom application the reader has in mind. The sample tool allows the user to point to a particular Access file (*.mdb) of interest. Once a file has been selected, a valid query must be constructed. The GUI aids in this endeavor by first instructing the user to select a Table of interest within the chosen database. From there the user chooses the Fields that are to be operated on within the database. The user may also choose to add a Where clause to the given query by selecting the Condition 1 checkbox. If the user

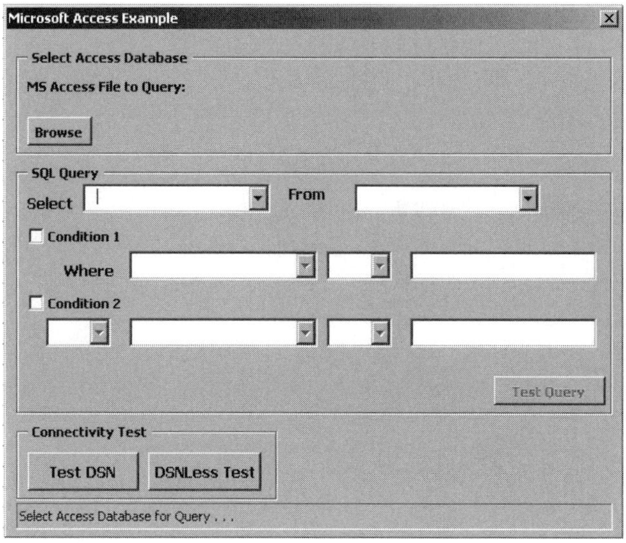

FIGURE 7.12 Sample database query tool.

wishes to add a compound condition using the AND/OR operators, the Condition 2 checkbox can also be selected. Once a valid query has been built, it can be tested by pushing the "Test Query" button.

The first step toward constructing such an application is to have the ability to point to a file and have its full path returned. This is a slightly different methodology than loading or saving a file that was covered in Chapter 2. The GetFileInfo function returns information related to a particular file chosen by means of a built-in dialog box.

By default, this function returns a file's full path (the filename and path). The user can also select to return just the filename, or just the path to the selected file by means of the optional Boolean passed parameters. The function allows the default directory for the dialog box to be specified, as well as a caption for the dialog box. The function also allows file filters to be specified for use in the dialog box. Filters must be three characters long and a default filter of "*.*" is always added to allow the user to choose any file.

```
Function GetFileInfo(ByVal filters$, ByVal defaultdir$, ByVal
WindowCaption$, _
Optional PathOnly As Boolean = False, Optional FileOnly As
Boolean = False) As String
'Function to Return File Information Using Dialog Box
'default returns full path
'may also return path without filename, or filename without path
Dim vFileName As String
Dim nochoice As Integer
Dim fillen As Integer
Dim strike As Integer
Dim ErrorBoxCaption$, ErrorBoxMsg$

On Error GoTo FileOpenErr

If PathOnly = True And FileOnly = True Then
```

```
      MsgBox "PathOnly and FileOnly are Mutually Exclusive",
      vbOKOnly + vbCritical, "Function Call Error"
  End If
  'Set Up Captions and Messages
  ErrorBoxCaption$ = "You must choose a File"

  filters$ = LCase(filters$)
  fillen = Len(filters$)
  If fillen Mod 3 = 0 Then
      'if valid filters then process
      For ii = 1 To fillen Step 3
        Search$ = Mid(filters, ii, 3)
        'Stop
        GoSub Build_Filters
    Next ii
  End If
  'Add default
  FileFilter = FileFilter & "All Files (*.*),*.*,"
  'remove trailing ','
  FileFilter = Left$(FileFilter, Len(FileFilter) - 1)

  'If the passes default directory is valid change the current
  directory to it
  If DirectoryExists(defaultdir$) = True Then
      ChDir defaultdir$
  End If
  Try_Again:
  vFileName = Application.GetOpenFilename(FileFilter, ,
  WindowCaption$)
  'Bug - Tries to Open "False.xls" on 2nd Pass Only
  If vFileName = "False" Then GoTo FileOpenErr
  If vFileName <> "" Then
      'File Name Captured with DialogBox
      If PathOnly = True Then
         GetFileInfo = Extract_Path(vFileName)
         Exit Function
         ElseIf FileOnly = False Then
         GetFileInfo = vFileName
         Exit Function
      End If
      If FileOnly = True Then
         GetFileInfo = Extract_File(vFileName)
         Exit Function
         ElseIf PathOnly = False Then
         GetFileInfo = vFileName
         Exit Function
```

```
      End If
  End If

  Build_Filters:
  Select Case Search$
    Case "xls"
      FileFilter = FileFilter & "Excel Workbooks (*.xls), *.xls,"
    Case "xlt"
      FileFilter = FileFilter & "Excel Templates (*.xlt), *.xlt,"
    Case "csv"
      FileFilter = FileFilter & "Comma Delimited (*.csv), *.csv,"
    Case "txt"
      FileFilter = FileFilter & "Text Files (*.txt), *.txt,"
    Case "mdb"
      FileFilter = FileFilter & "Access Files (*.mdb), *.mdb,"
    Case Else
      FileFilter = FileFilter & "File of Type (*." & Search$ & "),
      *." & Search$ & ","
  End Select
  Return

  FileOpenErr:
  'If no file is chosen you end up here
  strike = strike + 1
  Select Case strike
    Case 1
    ErrorBoxMsg$ = "You will get one more chance!"
    nochoice = MsgBox(ErrorBoxMsg$, vbOKOnly + vbExclamation,
    ErrorBoxCaption$)
          GoTo Try_Again
    Case 2
    ErrorBoxMsg$ = "Macro will terminate"
    nochoice = MsgBox(ErrorBoxMsg$, vbOKOnly + vbExclamation,
    ErrorBoxCaption$)
          End
  End Select
  End Function
```

Figure 7.13 shows an example test of the Get File Information that can be duplicated by typing the following into the debug window.

Debug.Print GetFileInfo("xlsmdbunk","c:\temp","Test File Info Function")

Notice that three filters are made available in the dialog box: "xls," "mdb," and "unk." The fourth filter is the default all inclusive "*.*." For file types that are not recognized, in this case the filter "unk," the function will simply add the text "File of Type" followed by the filter suffix (*.unk).

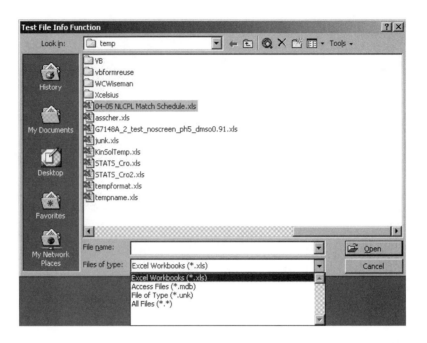

FIGURE 7.13 Testing the file information function.

Upon startup of the Database Query Tool, the system scans the Windows Registry for the last-selected Microsoft Access database file, and if one exists, it is selected as the default database file. After the selection of a valid database file, the tables corresponding to that database are populated in the GUI Combo Box immediately following the "From" label by means of the `PopulateTables` subroutine.

```
Sub PopulateTables()
'Populate the Database Tables Combobox
Dim Tables() As Variant
Call GetTableNames("DRIVER={Microsoft Access Driver
(*.mdb)};DBQ=" & frm7Access.Label_AFile.Caption, Tables())
For i = 1 To UBound(Tables)
  If IsSystemTable((Tables(i))) = False Or Tables(i) =
  "MSysAccessObjects" Then
     frm7Access.ComboBox_Tables.AddItem Tables(i)
  End If
  Debug.Print Tables(i)
Next i
End Sub
```

Notice that this subroutine utilizes the Access database file specified in the GUI, and that the only nonsystem tables are added to the tables Combo Box, with the exception of the "MsysAccess-Objects" Table, which is unique in that it always contains index values and is never empty.

The first thing that a user must do to create a query is select a table from the database and notice at this juncture a large red arrow pointing to the table Combo Box after the From statement is displayed, indicating that the user needs to select a value from here first (Figure 7.14). A statement to this effect is also shown in the status bar.

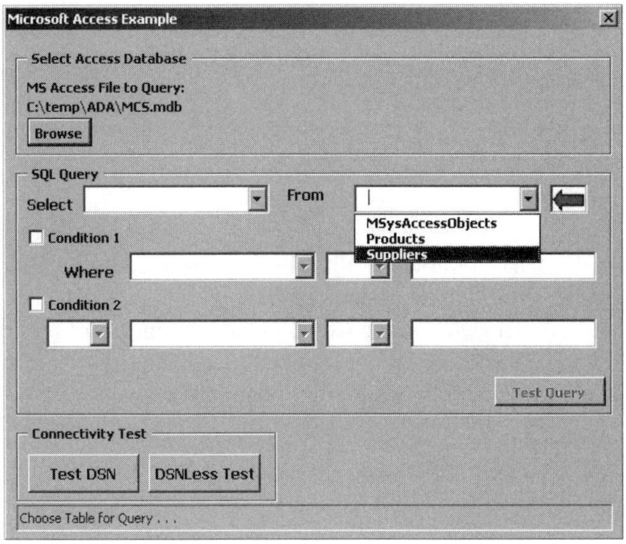

FIGURE 7.14 Step one in query creation — select a table.

Once a table has been selected, the fields available for query within the table are populated into the various Combo Boxes within the GUI. (Notice also that the large red arrow pointing at the table Combo Box disappears once a table has been selected.) The wildcard character "*" appears at the top of the Fields Combo Box, allowing the user to return information related to all the fields in the table. For example, the settings shown in Figure 7.15 will produce the following query:

Select * From Suppliers

If this query is tested, the entire contents of the Suppliers Table will be returned as shown in Figure 7.16. Notice that the sample application first writes the actual query transmitted to Access

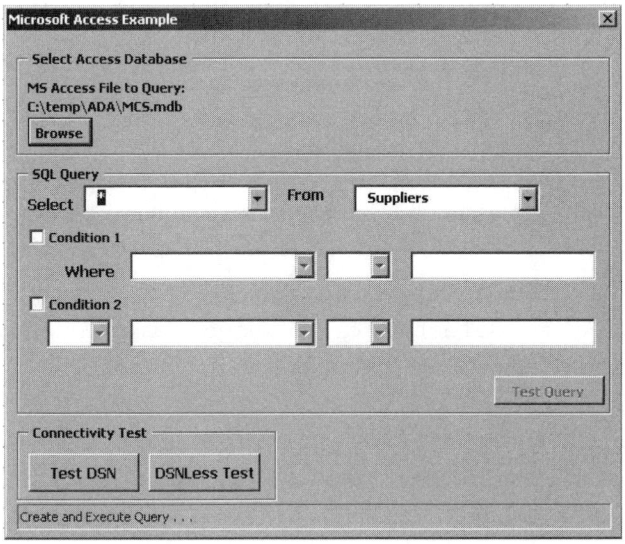

FIGURE 7.15 Creating a wildcard query.

	A	B	C	D	E	F	G	H	I	J
1	Select * From Suppliers									
2	NAME	MANUFACTURER	ADDRESS	CITY	STATE	ZIP	ZIPPLUS4	PHONE	FAX	WEBSITE
3	Howard Miller	Howard Miller	860 East Main Avenue	Zeeland	MI	49464	1300	616-772-7277		#http://www.howardmiller.com/#
4	Chelsea Clock Company	Chelsea Clock Company	284 Everett Avenue	Chelsea	MA	02150	1598	800-284-1778		#http://www.chelseaclock.com/#
5	Bulova Corporation	Bulova Corporation	One Bulova Avenue	Woodside	NY	11377		800-228-5682		#http://www.bulova.com/brands/office/office.aspx#
6	Hentschel	Hentschel	564 Weber St. N. - Unit 2	Waterloo	Ontario, Canada	N2L 5C6		866-565-1972		#http://www.hentschelclock.com#
7	Hermle Black Forest Clocks	Hermle Black Forest Clocks	P.O. Box 670	Amherst	VA	24521		434-946-7751		#http://www.hermleclock.com/index.html#
8	Sligh Furniture Company	Sligh Furniture Company	1201 Industrial Avenue	Holland	MI	49423				#http://sligh.com/home.asp#

FIGURE 7.16 Returned results of wildcard query.

in cell A1 of the worksheet. The returned results always start in cell A2 and occupy as many rows and columns necessary to return the requested results.

The available fields are populated in the Combo Boxes by means of the following subroutine. Notice that the wildcard character "*" is written prior to adding the Field results, and that if a condition is not enabled, its Combo Boxes are not populated.

```
Sub PopulateFields(TableName As String)
'Populate the Fields (Columns) Combobox
Dim Fields() As Variant
Call GetFieldNames("DRIVER={Microsoft Access Driver
(*.mdb)};DBQ=" & frm7Access.Label_AFile.Caption, TableName,
Fields())
frm7Access.ComboBox_Fields.AddItem "*"
If frm7Access.ComboBox_CFields1e.Enabled = True Then
frm7Access.ComboBox_CFields1e.AddItem "*"
If frm7Access.ComboBox_CFields2e.Enabled = True Then
frm7Access.ComboBox_CFields2e.AddItem "*"
For i = 1 To UBound(Fields)
   frm7Access.ComboBox_Fields.AddItem Fields(i)
   If frm7Access.ComboBox_CFields1e.Enabled = True Then
frm7Access.ComboBox_CFields1e.AddItem Fields(i)
   If frm7Access.ComboBox_CFields2e.Enabled = True Then
frm7Access.ComboBox_CFields2e.AddItem Fields(i)
Next i
End Sub
```

Note that this subroutine must be invoked whenever the Combo Box holding the Table Names is changed, as the Field Names must be updated whenever the Table of interest changes. The subroutine must also be called when the state of a compound query changes, such as when condition 1 or condition 2 are enabled by means of the checkboxes. The following subroutines accomplish the updating of the sample GUI information when a Table or compound query condition changes.

```
Private Sub ComboBox_Tables_Change()
    'When Table is Selected or Changed Names Must be Updated
    'First Clear Existing Entries
    Call ClearFields
    'Now Repopulate with Selected Table Fields
    Call PopulateFields(frm7Access.ComboBox_Tables.Text)
    frm7Access.StatusBar.SimpleText = "Create and Execute Query . . ."
    frm7Access.Image_Arrow.Visible = False
End Sub
```

The next two subroutines update the fields should a condition be enabled. They also enable/disable all conditional controls pertaining to a specific condition depending on the status change of the condition. This code works because the form was designed such that all objects associated with condition 1 have "1e" within their name, and all objects associated with condition 2 have "2e" within their name.

```
Private Sub Check_Cond1_Click()
Dim C As Control, State As Boolean
If frm7Access.Check_Cond1.value = True Then State = True Else
State = False

    For Each C In Controls
     If C.Name Like "*1e*" Then
        C.Enabled = State
    End If
Next C
If frm7Access.Check_Cond1.value = True Then Call
PopulateFields(frm7Access.ComboBox_Tables.Text)
End Sub

Private Sub Check_Cond2_Click()
Dim C As Control, State As Boolean
If frm7Access.Check_Cond2.value = True Then State = True Else
State = False

    For Each C In Controls
            If C.Name Like "*2e*" Then
                C.Enabled = State
            End If
      Next C
If frm7Access.Check_Cond2.value = True Then Call
PopulateFields(frm7Access.ComboBox_Tables.Text)

End Sub
```

The next subroutine will clear every Combo Box within the form that holds Field information about the selected Table of interest. This subroutine is possible because the forethought was put into designing the form such that all Combo Boxes holding Field names contained the word Field within the name.

```
Sub ClearFields()
'This will clear every ComboBox in the active GUI which holds
Table Field information
Dim C As Control
For Each C In Controls
  If TypeOf C Is ComboBox And C.Name Like "*Field*" Then
     C.Clear
  End If
Next C
End Sub
```

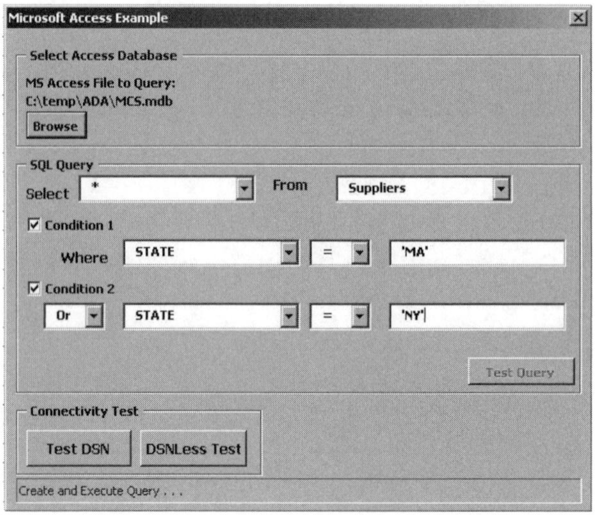

FIGURE 7.17 A Compound Query built using the Database Query Tool.

The concept of the compound query is demonstrated in Figure 7.17. Here the following query is built:

Select * From Suppliers Where STATE = 'MA' Or STATE = 'NY' which will return all suppliers whose addresses are listed in New York or Massachusetts.

Figure 7.18 shows the results for the query on the included sample database.

The actual query that is sent to Access is built from the GUI by means of the BuildSQL function. Keep in mind that, because the Excel 2000 implementation of OLE DB does not support background refresh, the form must be closed prior to the execution of the query, and therefore the SQL string must be built and stored in a string variable prior to closing the form.

```
Function BuildSQL() As String
'Build a SQL Query from Form Object Selections
BuildSQL = "Select " & frm7Access.ComboBox_Fields & " From "
& frm7Access.ComboBox_Tables & " "
If frm7Access.Check_Cond1.value = False Then Exit Function
BuildSQL = BuildSQL & "Where " &
frm7Access.ComboBox_CFields1e.Text & " " & _
frm7Access.ComboBox_CondEq1e.Text & " " &
frm7Access.TextBox_Cond1e.Text & " "

If frm7Access.Check_Cond2.value = False Then Exit Function
BuildSQL = BuildSQL & frm7Access.ComboBox_LogOps2e.Text & " "
& frm7Access.ComboBox_CFields2e.Text & _
```

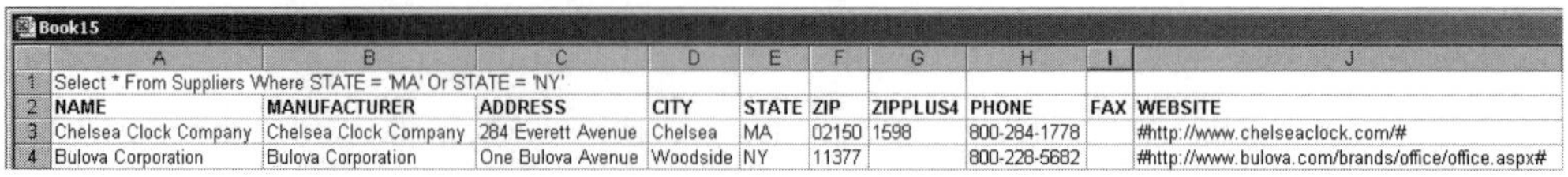

Book15

	A	B	C	D	E	F	G	H	I	J
1	Select * From Suppliers Where STATE = 'MA' Or STATE = 'NY'									
2	NAME	MANUFACTURER	ADDRESS	CITY	STATE	ZIP	ZIPPLUS4	PHONE	FAX	WEBSITE
3	Chelsea Clock Company	Chelsea Clock Company	284 Everett Avenue	Chelsea	MA	02150	1598	800-284-1778		#http://www.chelseaclock.com/#
4	Bulova Corporation	Bulova Corporation	One Bulova Avenue	Woodside	NY	11377		800-228-5682		#http://www.bulova.com/brands/office/office.aspx#

FIGURE 7.18 Complex Query Results on Sample MCS.mdb Database.

```
" " & frm7Access.ComboBox_CondEq2e.Text & " " &
frm7Access.TextBox_Cond2e.Text

End Function
```

The actual query is performed and returned utilizing the Test_Query subroutine.

```
Sub Test_Query()
'Perform Query Built in GUI
Dim sqlstring As String
'Build SQL String prior to closing form!
sqlstring = BuildSQL
'Unload Form - The Excel 2000 implementation of OLE DB does
not support background refresh

Unload frm7Access
'Will work when called from Form button press
DoEvents
'Define a DSN-less Connection String to Database Specified in GUI.
connstring = "ODBC;DRIVER={Microsoft Access Driver
(*.mdb)};DBQ=" & frm7Access.Label_AFile.Caption

'Write Query at Top of Worksheet
ActiveSheet.Range("A1") = sqlstring
'Create a QueryTable in worksheet beginning with cell A2.
With
Workbooks(ActiveWorkbook.Name).Worksheets(ActiveSheet.Name).Qu
eryTables.Add(Connection:=connstring, _
  Destination:=Range("A2"), Sql:=sqlstring)
  DoEvents
  .Refresh
End With
End Sub
```

7.7 USING DATA ACCESS OBJECTS — DAO

Thus far the discussion has centered around *reading* information from existing databases. Chances are good that, at some point, manipulations will be performed on data, and the reader will have a need to either write the analyzed data back to the database it originated from or create an entirely new database to store the analyzed parameters.

Writing data into Microsoft Access requires the use of Data Access Objects, termed DAO. This is simply a programming interface to access and manipulate database objects (Figure 7.19). At the top of the hierarchy of DAO resides the DBEngine object. It contains and controls all other objects within the DAO. Additional DBEngine objects cannot be created, and the DBEngine object is not an element of any other collection. Within the DBEngine, Workspaces are created from which databases can be manipulated.

In order to utilize DAO, a reference needs to be created to the DAO object within the VBA project. To add the DAO object as a library reference, select Tools->References from the menu and check Microsoft DAO *.* Object Library (where *.* is the version number; the author utilized version 3.6). Then click the OK button, and the reference will be installed (Figure 7.20).

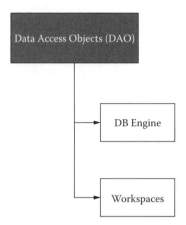

FIGURE 7.19 Data Access Objects — DAO hierarchy.

Once the project references the DAO, a small library of subroutines and functions can be built to allow Excel VBA to write information to Access utilizing DAO. The first function created allows the user to create a *completely new* database in Access from Excel. If the function is able to create the database, it returns True; if it fails, it returns False. The user must specify a name for the new database (FileName), and the path to which the database should be written to. Boolean passed parameters allow the user to determine if existing databases should be overwritten, or if paths should be created if they do not exist.

```
Function CreateMDBFile(FileName As String, FilePath As String, _
Optional DeleteExisting As Boolean, Optional CreatePath As
Boolean) As Boolean

'Creates Access Database at Location FilePath named FileName
'Returns True if successful

Dim ws As Workspace
```

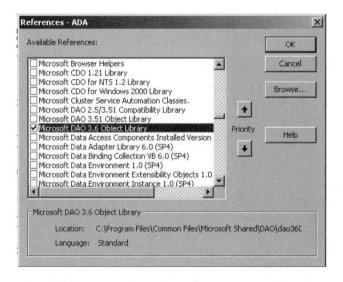

FIGURE 7.20 Installing a reference to the DAO.

```
Dim db As Database
Dim FullPath As String

'Ensure Path Ends with "\"
If Right(FilePath, 1) <> "\" Then FilePath = FilePath & "\"
'Ensure FileName Ends with ".mdb"
If Right(FileName, 4) <> ".mdb" Then FileName = FileName & ".mdb"

'Construct FullPath
FullPath = FilePath & FileName

'Get default Workspace
Set ws = DBEngine.Workspaces(0)

'Does this Access File Exist?
If Dir(FullPath) <> "" Then
    If DeleteExisting = True Then
        'Erase Existing File
        Kill FullPath
        Else
        'Preserve Existing File - Abort Return False
        Exit Function
     End If
End If

If DirectoryExists(FilePath) = False Then
    If CreatePath = True Then
        'Create the new Path Structure
        CreateDataDir (FilePath)
        Else
        'Do Not Create New Directories - Abort Return False
        Exit Function
    End If
End If

'Create a new mdb file of name FileName at Location FilePath
Set db = ws.CreateDatabase(FullPath, dbLangGeneral)

db.Close
Set db = Nothing
CreateMDBFile = True
End Function
```

Notice that this function makes use of the `CreateDataDir` subroutine discussed in Chapter 2 to create the specified subdirectory should it not exist (provided the user set the `CreatePath` passed parameter to True). It is the `CreateDatabase` method that actually creates the Microsoft Access database file.

The function can readily be tested for the creation of a new database file in an existing directory structure by typing

Debug.Print CreateMDBFile("Test","c:\temp",False ,False)

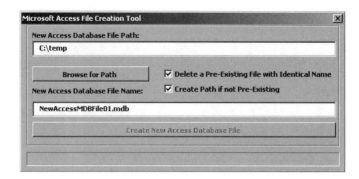

FIGURE 7.21 Sample application allows creation of a new Access database within Excel.

which will in turn result in True being printed to the immediate window (provided C:\temp exists on the user's machine). The execution of this statement will result in the creation of the file "Test.mdb" at location c:\temp.

The function can be tested for the case where the directory structure does not exist and must be created for the database file by typing

```
Debug.Print CreateMDBFile("Test","c:\temp\a\b\c\d",True ,True )
```

which will in turn result in True being printed to the immediate window. The execution of this statement will result in the creation of the file "Test.mdb" at location C:\temp\a\b\c\d.

A sample application has been packaged and deployed, which allows a user to create a new Microsoft Access database from within Excel. It can be run by selecting from the menu ADA->Chapter 7->Create Database. Doing so will bring up the GUI shown in Figure 7.21.

7.8 ELEMENTS IN THE DAO ARCHITECTURE

The reader will no doubt have some confusion as to some of the object variable declarations present in the CreateMDBFile subroutine example given in the previous section. The following section will explain the elements utilized in the DAO Architecture as they are applied to the code examples provided in the text. For a comprehensive discussion on DAO Architecture, the reader is invited to refer to Microsoft's Knowledge Base article entitled "Understanding the DAO Object Model," which can be referenced on Microsoft's website.

The developer should be aware of an oversight in DAO object declaration within Excel VBA. It should be enough to declare a DAO object as in the following manner, and often this is how such variables are declared in text and sample code:

```
Dim MCSdb As Database
```

VBA has the nasty tendency to confuse ADO with DAO database objects! If this occurs, a Type 13 Mismatch Error will occur, triggered by the declared variable in question. The solution to this problem is to prefix all declared DAO variables with (DAO). When a Recordset is not Prefixed with DAO, then a Type 13 Mismatch Error may occur. Therefore, it would be much safer to make the foregoing declaration as follows:

```
Dim MCSdb As DAO.Database
```

This Database object declared is utilized to represent a database with at least one open connection. It can be a Microsoft Access database as used in the examples in the text or an external data source such as an Oracle database.

The DBEngine object is at the top of the DAO hierarchy. The DAO holds all other objects and maintains all the database engine options. The common coding practice is to set a Workspace object equal to the desired DBEngine. The default DBEngine Workspace is always indexed at zero, as is shown in the following example:

```
Dim ws As DAO.Workspace

Set ws = DBEngine.Workspaces(0)
```

The Workspace object defines and manages the current user session. This object contains information on open databases and provides mechanisms for simultaneous transactions.

The Recordset object is utilized for the extraction of a subset of data from the database. Recordset objects represent a set of records defined by a query result.

```
Dim rs As DAO.Recordset
```

The TableDef object contains both Field and Index objects to describe the tables that reside in the database. It is within the fields of the tables that all the information in a database is contained.

```
Dim tdef As DAO.TableDef
```

The QueryDef object represents a stored SQL query statement, with zero or more parameters, maintained in a Microsoft database.

```
Dim qdf As DAO.QueryDef
```

The DAO model contains other objects that allow further manipulations to take place within a database. Among them are

The Container object represents a particular set of objects in a database for which permissions can be assigned in a secure workgroup. In addition to the Container objects provided by DAO, an application may define its own Container objects (such as saved forms, modules, reports, or script macros).

The Relation object represents a relationship between fields in tables and queries. The Relation object can be used to create, delete, or change the type of relationship and to determine which tables supply the fields that participate, whether to enforce referential integrity, and whether to perform cascading updates and deletes.

The Field object represents a field in a table, query, index, relation, or record set. A Field object contains data, and it can be used to read data from a record or write data to a record.

The Index object represents an index on a table in the database.

The Parameter object represents a value associated with a QueryDef object. Query parameters can be input, output, or both.

The Document object contains information about individual objects in the database (such as tables, queries, or relationships).

The User object represents a user account with particular access permissions.

The Group object represents a group of user accounts that have common access permissions in a particular workspace.

The Error object stores information about an error that occurred during a DAO operation. When more than one error occurs during a single DAO operation, each individual error is represented by a separate Error object.

7.9 SUMMARY

This chapter has focused on static operations to a database — static meaning that no elements of the database were altered and no new elements were added to the database. Static operations to a database are relatively easy to code because the elements in the database never change their location. They are simply read from their existing locations within the database file, and the database file structure remains unchanged.

Applications of any sophistication, however, will be dynamic in nature, meaning that elements within the database will be added, deleted, or altered depending upon the coding algorithm applied to the database. Performing dynamic actions on a database poses special coding challenges because the number of elements that an action is to be performed on is constantly changing as items are added, altered, and deleted from the database. Algorithms that address the challenges of dynamic coding structures on databases will be covered in detail in Chapter 8.

The concepts of using both DAO and SQL to access database information were introduced in this section and will be covered more in more detail in the subsequent chapter.

8 From Excel to Access and Back Again

8.1 INTRODUCTION

This chapter is dedicated to dynamic actions on databases, meaning the algorithms run in Excel will actually change the content within the database. Two methods will be covered to perform dynamic actions upon databases within Excel VBA. The first method is to utilize SQL commands, and the second method is to utilize the DAO model. Both methods were introduced in Chapter 7.

The advantage of using SQL is that both queries and manipulations on databases will occur much faster than if they had been coded using DAO. SQL commands also have the advantage of being more powerful, as they can be made as complex as the developer's understanding of the SQL language. When using the DAO model, the developer is limited to utilizing the properties and methods contained within the DAO framework. DAO is also self-limiting in that if a developer learns DAO, database manipulations can only be performed in environments that support DAO. SQL is the superior method to manipulate databases, its only downside being that one must learn the SQL language in order to utilize it. SQL is nearly platform independent for basic queries, but differences do exist between Microsoft Access SQL statements and Oracle SQL statements (in some instances). In general, learning SQL is usually well worth the time and effort put in up front.

The numerous examples contained within this chapter show how to perform the indicated operations using *both* SQL and DAO. It would not be appropriate to exclude either method, and for the sake of writing nearly platform-independent methods, SQL coverage is a must. DAO is also worth covering because its simplicity allows developers to put together applications quickly when time is of the essence.

8.2 USING POINTERS IN DYNAMIC DATABASE ALGORITHMS

Any C programmer is familiar with pointers. In C, a pointer is simply a variable that holds a memory address rather than a value. Because the values of all variables are stored in memory, if the address of a variable is known, its value can be read from memory without actually accessing the variable itself. Such a form of access is known as indirect referencing.

Although C formally supports pointers as operators, the fact is that pointers are utilized in many different algorithms and applications to perform a variety of operations. For all practical purposes, a pointer can be generically defined as a variable that defines the location of a value of interest. The location pointed to in the case of a database will be an index or a record number, and the value of interest will be a single or plurality of values defined by a query.

Writing code to access database elements poses some special programming challenges. Because databases are dynamic and always changing, the number of records (or rows) in them is in a constant state of fluctuation. To further complicate matters, individual elements in databases can change position. The combination of these two factors make using hard-coded For/Next loop structures impractical.

Figure 8.1 illustrates the challenges that can be encountered when accessing records from a dynamic database. In this example, a sample database includes five elements (a,b,c,d,e) that are located in the database in alphabetical order, and thus their record set (or index) numbers are (1,2,3,4,5), respectively.

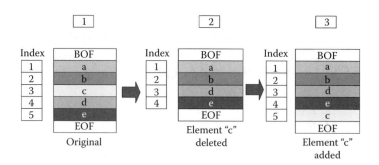

FIGURE 8.1 Dynamic indexing challenges in database Record sets.

At the top and bottom of each database are the terms BOF and EOF, which stand for Beginning Of File and End Of File, respectively. BOF and EOF are Properties that can be accessed using Excel VBA. The definition of these two properties is:

BOF — Indicates that the current record position is before the first record in a Record set object.

EOF — Indicates that the current record position is after the last record in a Record set object.

It will be shown later in this chapter how to utilize the BOF and EOF properties to create loop structures that allow the developer to create routines that loop through every element or a chosen subset of elements in a database.

Looking at Figure 8.1, let us suppose the element (c), which has an index of 3, is deleted from the database. (This scenario is shown in Section 2.) This change causes two profound effects that should be of concern to the developer. First, the number of records in the database is decreased by 1 from 5 to 4. Second, the indexes of elements (d and e) are now reduced by 1.

Upon inspection, an answer that comes to mind is to simply set the number of loops to an infinitely high maximum that the database should never attain and simply loop through the elements until a NULL is encountered, which indicates the end of the records. The problem with such an approach is that when a record is accessed beyond the end of the last index in the record set, it does not return a null but triggers an error. Thus, such an approach is not practical.

Another thought that comes to mind is to keep a certain *unique* element as the last record in the database, thus marking the end of the records in the database (in this example "e"). As soon as the unique element is encountered, an exit for statement could terminate the looping structure in the database. This solution is also impractical for the reasons shown in Figure 8.1, section (3).

Suppose another element is now added to the database — in this example, "c" is put back into the database. Now "e," the supposed "last record" in the database, precedes the record "c." Such a looping structure would now terminate prematurely before accessing every element in the database. Although it is true that a dynamic loop counter could be incremented with each record added, and decremented with each record deleted, such a system is hardly a model of efficiency. That is why using the BOF and EOF properties when dealing with databases is so important for maintaining clean and easy-to-understand code.

Also notice that, in Figure 8.1, section (3), the indexes of records (d,e,c) are now different from the original database. To complicate matters even further, if the user implements complex SQL commands to manipulate the database, it will not be possible to count the number of records added or deleted utilizing a counter. Thus, pointers for individual database records or elements are impractical and should not be used.

One situation in which pointers *must* be utilized is when populating dynamic arrays. Oftentimes when coding, it is convenient to store the results of a query in an array or to use an array to populate a list box or combo box on a GUI. Looking at Figure 8.2, if elements (a,c,d) of the example

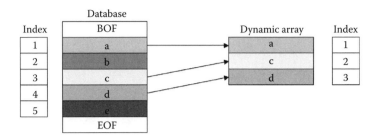

FIGURE 8.2 Using pointers for populating dynamic arrays.

database are to be extracted to an array because they meet some specified criteria, the indexes of the database and the indexes of the array will not correspond because some items in the database will not be extracted to the array, depending upon the criteria of the query. In such an instance, it is necessary to utilize a pointer to point to the next available index in a dynamically sized array. Also keep in mind that if a combo box or list box object is to be populated, the first index is 0. The first index of an array will be dependent upon the setting of the Option Base command. In the absence of the Option Base statement, the lower limit index of an array is set to 0.

When looking at the coding examples in this chapter of the text, it is necessary for the reader to keep these concepts in mind. It is also very useful when debugging code in database applications where unexpected results have occurred, to print the elements extracted and their corresponding indexes in both the database and the array they have been populated into. This makes finding errors much easier.

8.3 CONCEPTS IN DATABASE ALTERATION AND MANAGEMENT

Certain concepts in database management and alteration transcend the various means of implementation. In this text, database management will be implemented using examples in SQL and DAO, and certain terms and commands will be utilized using both schemes, although the syntax may differ.

The three most important concepts in database management and alteration are that of commit, update, and rollback. Whenever a change is made to a record in a database, the record must be updated for the change to be recorded. Updating is nothing more than changing the data in a database table.

When a series of changes is made to a database, they are not implemented until the commit command is issued. Commit finalizes all the changes made in the database since the last commit command was issued.

The rollback command is essentially the opposite of the commit command. Issuing the rollback command undoes any changes made to the database using the update command since the last commit command was issued.

These concepts are utilized when dealing with the Workspace object introduced in Chapter 7.

Examples:
BeginTrans *begins a new transaction.*
CommitTrans *ends the current transaction and saves the changes.*
Rollback *ends the current transaction and restores the databases in the Workspace object to the state they were in when the current transaction began.*

8.4 CREATING NEW TABLES IN ACCESS FROM EXCEL

Effective database alteration and maintenance begin at the table level. To effectively maintain a database, a programmer must be able to accomplish the following from Excel: (1) Determine which Tables already exist, (2) determine which Tables are System Tables (that is, tables created and utilized by Access), (3) create New Tables in an existing database using either DAO or SQL.

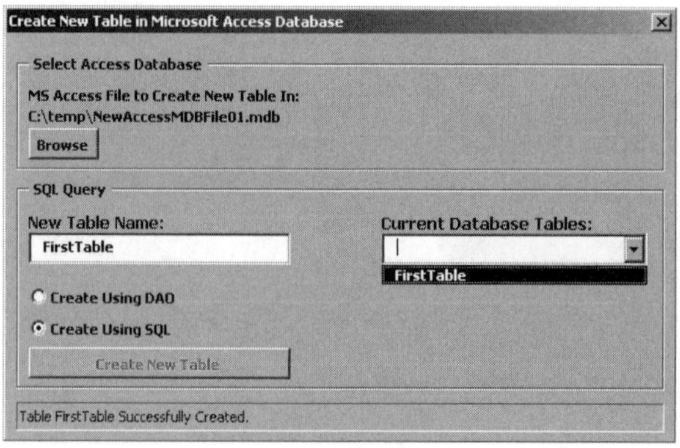

FIGURE 8.3 Sample application to create database tables in Excel.

Figure 8.3 shows the sample application that can be run by selecting ADA->Chapter 8->Add Table to Database from the Excel Menu. If a database has been selected in the past, the application will remember this by utilizing the Windows Registry. If a previous database is stored in the Windows Registry when the application is run, the program will populate the "Current Database Tables" combo box with the names of all the tables in the database that *are not* system tables. This process begins when the application is initialized. The color-coding scheme should help the reader comprehend the series of events that take place.

```
Private Sub UserForm_Initialize()
'Retrieve Registry Settings
If GetSetting("AccessCT", "File", "FullPath") <> "" Then
    frm7NewTable.Label_AFile.Caption = GetSetting("AccessCT",
    "File", "FullPath")
    Call PopulateTablesNT
    frm7NewTable.StatusBar.SimpleText = "Add a New Table to the
    Database . . ."
    Else
    frm7NewTable.StatusBar.SimpleText = "Select Access Database
    for Query . . ."
End If
End Sub
```

Once initialized, the PopulateTablesNT subroutine populates the combobox in the GUI with the names of the tables in the database selected by the user. The subroutine is dependent on two functions: GetTableNames and IsSystemTable.

```
Sub PopulateTablesNT ()
'Populate the Database Tables Combobox
Dim Tables() As Variant
Call GetTableNames ("DRIVER={Microsoft Access Driver
(*.mdb)};DBQ=" & frm7NewTable.Label_AFile.Caption, Tables())
For i = 1 To UBound(Tables)
  If IsSystemTable ((Tables(i))) = False Or Tables(i) =
  "MSysAccessObjects" Then
  frm7NewTable.ComboBox_Tables.AddItem Tables(i)
```

```
End If
'Debug.Print Tables(i)
Next i
End Sub
```

The GetTableNames function returns an integer that is the total number of tables present in the selected database (including system tables). The names of the tables are returned by means of a passed by reference array variable TableNames(). It is the XLODBC function named SQLGetSchema that actually returns the table names from the selected database.

```
Function GetTableNames (SQLOpenString As String, TableNames())
As Integer
'Returns the Total Number of Tables and their Names in
TableNames
Dim Chan As Variant
Dim TName As Variant, GetSchema() As Variant
Dim DatabaseName As Variant
On Error GoTo No_Database:
'Establish a connection to the data source.
Chan = SQLOpen(SQLOpenString)
'Get the name of the database on the connection
DatabaseName = SQLGetSchema(Chan, 7)
'Get the table names for the connected database
GetSchema = SQLGetSchema(Chan, 4, DatabaseName & ".")
For ii = 1 To UBound(GetSchema)
  ReDim Preserve TableNames(ii)
  TableNames(ii) = GetSchema(ii, 1)
Next ii
GetTableNames = UBound(GetSchema)
' Close the connection to the data source.
SQLClose (Chan)
Exit Function
No_Database:
MsgBox SQLOpenString, vbOKOnly + vbCritical, "Error: Cannot
Open Database"
GetTableNames = 0
' Close the connection to the NWind data source.
SQLClose (Chan)
End Function
```

The IsSystemTable function simply takes a table name as a passed parameter and compares it to a list of known system table names. If the table is a system table, the function returns the Boolean value True, or else it returns False.

```
Function IsSystemTable (TableName As String) As Boolean
'Identifies MS Access System Tables
Select Case TableName
  Case "MSysAccessObjects"
```

```
      IsSystemTable = True
    Case "MSysACEs"
      IsSystemTable = True
    Case "MSysObjects"
      IsSystemTable = True
    Case "MSysQueries"
      IsSystemTable = True
    Case "MSysRelationships"
      IsSystemTable = True
  End Select
  End Function
```

Now, the more difficult task of actually creating an Access table from within Excel is covered. Two functions have been created named `CreateTableSQL` and `CreateTableDAO`, which utilize SQL and DAO, respectively, to create new tables in Access. Both functions return a Boolean value indicative of whether the function was successful in creating the table. The `CreateTableDAO` function is presented first.

```
Function CreateTableDAO(fullmdbpath As String, TableName As
String, Optional ColumnName As String = "ID") As Boolean
'Creates a Single Table named TableName in the Specified
Database with one Column named ColumnName
'DAO Cannot Create a Table without any fields in MS Access!
Dim MCSdb As Database
Dim tdef As TableDef
Dim ws As Workspace
Dim ConnectString As String

On Error GoTo Table_Exists:
'Get default Workspace
Set ws = DBEngine.Workspaces(0)
'Establish a DSN-less connection to the MCS Sample Data Base
ConnectString = "ODBC;DRIVER={Microsoft Access Driver
(*.mdb)};DBQ=" & fullmdbpath
Set MCSdb = DBEngine.OpenDatabase(fullmdbpath, , ,
ConnectString)
'Begin the Transaction
ws.BeginTrans
Set tdef = MCSdb.CreateTableDef(TableName)
With tdef
 .Fields.Append .CreateField(ColumnName, dbText)
End With

MCSdb.TableDefs.Append tdef
MCSdb.TableDefs.Refresh
ws.CommitTrans

' Refresh the cache to ensure that the latest data is available.
```

```
DBEngine.Idle dbRefreshCache
MCSdb.Close: ws.Close
Set MCSdb = Nothing: Set tdef = Nothing: Set ws = Nothing
CreateTableDAO = True
Exit Function

Table_Exists:
MsgBox "Error: " & Err & ": " & Error$, vbOKOnly + vbCritical,
"Error Accessing: " & fullmdbpath
CreateTableDAO = False
End Function
```

One idiosyncrasy of DAO is that it is not capable of creating a table without any columns. The preceding function gets around this limitation by creating a default column named "ID" should the user not specify one. (Nearly all tables have a field named "ID" anyway.) The next function creates a table using SQL and XLODBC functions, and SQL can create a table without any columns.

```
Function CreateTableSQL(fullmdbpath As String, TableName As
String) As Boolean
'Creates a Single Table named TableName in the Specified Database
'Only Known Method to Create a Table Without Fields
Dim Chan As Variant
Dim SQLStatement As String
Dim SQLOpenString As String
On Error GoTo Table_Exists:

'Establish a DSN-less connection to the Data Base
SQLOpenString = "DRIVER={Microsoft Access Driver (*.mdb)};DBQ="
& fullmdbpath
Chan = SQLOpen(SQLOpenString)
'Build SQL to Create a new Table
SQLStatement = "CREATE TABLE " & TableName & ";"
'Create the Table
SQLExecQuery Chan, SQLStatement
'Close the connection to the database
SQLClose (Chan)
CreateTableSQL = True
Exit Function

Table_Exists:
Debug.Print Error$, Err
MsgBox SQLOpenString, vbOKOnly + vbCritical, "Error: The Table
Name Already Exists!"
CreateTableSQL = False
'Close the connection to the database
SQLClose (Chan)
End Function
```

Notice how much more compact the SQL function is as opposed to the DAO function. This is due largely to the fact that a number of operations can be explicitly declared right from within an SQL statement. DAO proceeds on an element-by-element basis, whereas SQL can operate *en masse*, which explains why SQL has a performance advantage over DAO.

8.5 ADDING AND REMOVING FIELDS IN ACCESS TABLES FROM EXCEL

Tables are, of course, only as good as the information put in them. It is imperative that the developer has the ability to add or delete fields from a database table at will. A sample application has been constructed to show the reader how this can be accomplished using both DAO and SQL. Although the sample was written for use with a Microsoft Access Database, the routines can be adapted with some minor syntax changes to whatever database the reader happens to be utilizing.

Figure 8.4 shows the sample application to add or delete fields in an Access Database that can be run by selecting ADA->Chapter 8->Database Fields from the Excel Menu. Notice that the application permits the user to either add or remove fields from a specified table, the action to be taken specified by means of an option button. As in the earlier application, another option button allows the user to select if DAO or SQL should be used to perform the operation.

As in the previous example, upon startup, a Microsoft Access database file must be selected, and the existing Table Names corresponding to the selected database are placed in a combo box for selection by the user.

First take the case where the user wishes to add a field to an existing table. Upon selection of a table by the user, a textbox to the left of the combo box holding the table to be acted upon is utilized for the user to specify the new field name for the table. The user then selects whether to create the table using DAO or SQL. In the case of DAO, the following function is utilized to create a field in the specified table.

```
Function CreateFieldDAO(fullmdbpath As String, TableName As
String, ColumnName As String) As Boolean
'Adds a Column named ColumnName to Table named TableName in
the Specified Database
Dim MCSdb As Database
Dim ws As Workspace
Dim tbl As TableDef
```

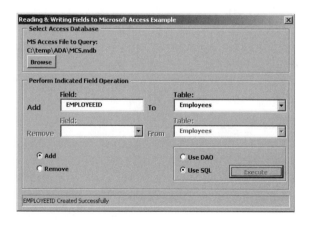

FIGURE 8.4 Adding fields (columns) to Microsoft Access.

```
Dim ConnectString As String

On Error GoTo Table_Exists:
'Get default Workspace
Set ws = DBEngine.Workspaces(0)
'Establish a DSN-less connection to the MCS Sample Data Base
ConnectString = "ODBC;DRIVER={Microsoft Access Driver
(*.mdb)};DBQ=" & fullmdbpath
Set MCSdb = DBEngine.OpenDatabase(fullmdbpath, , , ConnectString)
'Begin the Transaction
ws.BeginTrans
' Add a Column to the Specified Table
Set tbl = MCSdb.TableDefs(TableName)
'Create a Reference to the Table to be Operated On
tbl.Fields.Append tbl.CreateField(ColumnName, dbText)
'Commit the Changes
ws.CommitTrans
' Refresh the cache to ensure that the latest data is available.
DBEngine.Idle dbRefreshCache
MCSdb.Close: ws.Close
Set MCSdb = Nothing: Set tbl = Nothing: Set ws = Nothing
CreateFieldDAO = True
Exit Function

Table_Exists:
Debug.Print Error$, Err
MsgBox "Error: " & Err & ": " & Error$, vbOKOnly + vbCritical,
"Error Accessing: " & fullmdbpath
CreateFieldDAO = False
End Function
```

The function returns a Boolean value of True if the field was successfully generated in the table. In the case shown in Figure 8.4, the field EMPLOYEEID was created in the table named Employees. A similar function was constructed to allow fields to be created using SQL.

```
Function CreateFieldSQL(fullmdbpath As String, TableName As
String, _
      FieldName As String) As Boolean
'Creates a Single Field named FieldName in the Specified Table
Dim Chan As Variant
Dim SQLStatement As String
Dim SQLOpenString As String

On Error GoTo Table_Exists:

'Establish a DSN-less connection to the Data Base
SQLOpenString = "DRIVER={Microsoft Access Driver (*.mdb)};DBQ="
& fullmdbpath
```

```
Chan = SQLOpen(SQLOpenString)

'Build SQL to Create a new Table
SQLStatement = "ALTER TABLE " & TableName & " ADD COLUMN ["
& FieldName & "] TEXT;"
'Create the Table
SQLExecQuery Chan, SQLStatement
'Close the connection to the database
SQLClose (Chan)
CreateFieldSQL = True
Exit Function

Table_Exists:
Debug.Print Error$, Err
MsgBox SQLOpenString, vbOKOnly + vbCritical, "Error: The Table
Name Already Exists!"

CreateFieldSQL = False
'Close the connection to the database
SQLClose (Chan)
End Function
```

Notice that the SQL Statement generated and utilized in the above function makes use of the ALTER command to generate new columns (or fields) in an existing table. Further notice that the type of column to be added (data type) is specified at the end of the SQL statement. The data type at the end of this sample statement can be altered to reflect whatever data type the user wishes for the field to be created.

Removing fields from an existing table adds one more degree of complexity to the operation in that the existing field names must be identified and presented to the user. In this instance, a second combo box is utilized to hold all the field names for the table of interest. The function GetFieldNames was created to return the number of fields in a specified table and pass a string array back that contains the names of all the fields in the specified table. If an error occurs, the specified table does not exist, or the specified table cannot be accessed, the function returns 0 and the string array is returned empty.

```
Function GetFieldNames(SQLOpenString As String, TableName As
String, FieldNames()) As Integer
'Returns Total Number of Fields in Table
Dim Chan As Variant
Dim Fields As Variant
Dim i As Integer

On Error GoTo No_Table:
'Establish a DSN-less connection to the MCS Sample Data Base
Chan = SQLOpen(SQLOpenString)

'Get the field names for the TableName passed parameter
Fields = SQLGetSchema(Chan, 5, TableName)
'Fields is a two-dimension Variant array; (name, type)
```

```
For i = 1 To UBound(Fields)
  TotalFieldNames = TotalFieldNames + 1
  ReDim Preserve FieldNames(1 To TotalFieldNames)
  FieldNames(TotalFieldNames) = Fields(i, 1)
  GetFieldNames = TotalFieldNames
Next i
'Close the connection to the database
SQLClose (Chan)
Exit Function

No_Table:
MsgBox SQLOpenString, vbOKOnly + vbCritical, "Error: The
Referenced Table Does Not Exist in "
GetFieldNames = 0
'Close the connection to the database
SQLClose (Chan)
End Function
```

It is the `PopulateFieldsNF` subroutine that actually takes the array passed back from the preceding function and utilizes it to populate the combo box on the form. The subroutine is triggered in the program any time the table name changes in the Table Name combo box in the Remove Field section of the GUI.

```
Sub PopulateFieldsNF(TableName As String)
'Populate the Fields (Columns) Combobox
Dim Fields() As Variant
Call GetFieldNames("DRIVER={Microsoft Access Driver
(*.mdb)};DBQ=" & frm7NewField.Label_AFile.Caption, TableName,
Fields())
frm7NewField.ComboBox_Fields.AddItem "*"
For i = 1 To UBound(Fields)
 frm7NewField.ComboBox_Fields.AddItem Fields(i)
Next i
End Sub
```

Once all the field names have been populated to the combo box, the user may select one at will to be deleted. The function `DeleteFieldDAO` will delete a specified field in a specified table using DAO. If the deletion is successful, the function will return the Boolean value of True.

```
Function DeleteFieldDAO(fullmdbpath As String, TableName As
String, ColumnName As String) As Boolean
'Adds a Column named ColumnName to Table named TableName in
the Specified Database
Dim MCSdb As Database
Dim ws As Workspace
Dim tbl As TableDef
Dim ConnectString As String
```

```
On Error GoTo Table_Exists:
'Get default Workspace
Set ws = DBEngine.Workspaces(0)
'Establish a DSN-less connection to the MCS Sample Data Base
ConnectString = "ODBC;DRIVER={Microsoft Access Driver
(*.mdb)};DBQ=" & fullmdbpath
Set MCSdb = DBEngine.OpenDatabase(fullmdbpath, , , ConnectString)

'Begin the Transaction
ws.BeginTrans
' Add a Column to the Specified Table
Set tbl = MCSdb.TableDefs(TableName)
'Create a Reference to the Table to be Operated On
tbl.Fields.Delete ColumnName
'Commit the Changes
ws.CommitTrans
' Refresh the cache to ensure that the latest data is available.
DBEngine.Idle dbRefreshCache
MCSdb.Close: ws.Close
Set MCSdb = Nothing: Set tbl = Nothing: Set ws = Nothing
DeleteFieldDAO = True
Exit Function

Table_Exists:
Debug.Print Error$, Err
MsgBox "Error: " & Err & ": " & Error$, vbOKOnly + vbCritical,
"Error Accessing: " & fullmdbpath
DeleteFieldDAO = False
End Function
```

A similar function has been constructed to allow fields to be deleted using SQL. Notice that the ALTER command is again used in SQL along with the DROP command to delete an existing field in a database. This function also returns TRUE upon success.

```
Function DeleteFieldSQL(fullmdbpath As String, TableName As
String, _

       FieldName As String) As Boolean
'Deletes a Single Field named FieldName in the Specified Table
Using SQL
Dim Chan As Variant
Dim SQLStatement As String
Dim SQLOpenString As String

On Error GoTo Table_Exists:
'Establish a DSN-less connection to the Data Base
SQLOpenString = "DRIVER={Microsoft Access Driver (*.mdb)};DBQ="
& fullmdbpath
Chan = SQLOpen(SQLOpenString)
```

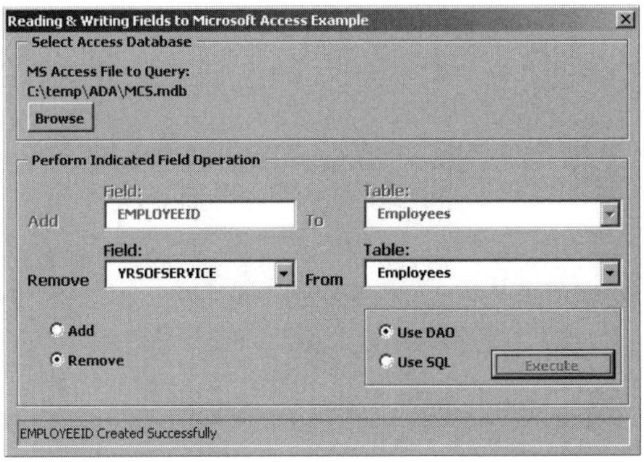

FIGURE 8.5 Deleting field names in tables.

```
'Build SQL to Create a new Table
SQLStatement = "ALTER TABLE " & TableName & " DROP " & FieldName
& ";"
'Create the Table
SQLExecQuery Chan, SQLStatement
'Close the connection to the database
SQLClose (Chan)
DeleteFieldSQL = True
Exit Function

Table_Exists:
Debug.Print Error$, Err
MsgBox SQLOpenString, vbOKOnly + vbCritical, "Error: The Table
Name Already Exists!"
DeleteFieldSQL = False
'Close the connection to the database
SQLClose (Chan)
End Function
```

Figure 8.5 shows the example program ready to delete the field YRSOFSERVICE in a table named Employees in the sample MCS.mdb database using DAO.

8.6   ADDING RECORDS TO SPECIFIC FIELDS IN DATABASE TABLES

Having a database with specific tables and fields means nothing unless something is stored within them. That "something" to be stored within the fields of the tables are individual records.

A sample application was constructed to show how records can be added to a database using Excel VBA. Figure 8.6 shows a sample record consisting of a first, middle, and last names being added to the sample MCS.mdb database, which is included on the accompanying CD-ROM.

Keep in mind that a record may contain far more than three fields; in fact, it may contain as many fields as there are columns in the database. Also note that not every field in a record must contain data; it is perfectly acceptable to write values to only a subset of the given fields in a database.

SQL is by far the most efficient means of adding records into a database. This is because using SQL allows the developer to populate a multitude of fields with a single SQL statement.

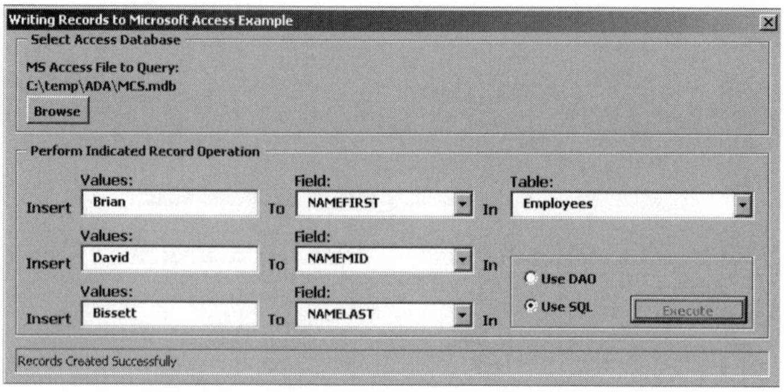

FIGURE 8.6 Adding a record to an Access database.

The `CreateRecordSQL` function is utilized in the sample application to add records to the chosen database using SQL.

```
Function CreateRecordSQL(fullmdbpath As String, TableName As
String, _
  FieldName As String, FieldName1 As String, FieldName2 As String, _
  Value As String, Value1 As String, Value2 As String) As Boolean
'Writes 3 Fields as a Record to the Specified Table Using SQL
Dim Chan As Variant
Dim SQLStatement As String
Dim SQLOpenString As String
On Error GoTo Table_Exists:

'Establish a DSN-less connection to the Data Base
SQLOpenString = "DRIVER={Microsoft Access Driver (*.mdb)};DBQ="
& fullmdbpath
Chan = SQLOpen(SQLOpenString)

'Build SQL to Add Record to Specified Table
SQLStatement = "INSERT  INTO " & TableName & "(" & FieldName
& "," & _
        FieldName1 & "," & FieldName2 & ")" & " VALUES ('" & _
        Value & "','" & Value1 & "','" & Value2 & "');"
'Create the Record
SQLExecQuery Chan, SQLStatement
'Close the connection to the database
SQLClose (Chan)
CreateRecordSQL = True
Exit Function

Table_Exists:
Debug.Print Error$, Err
MsgBox SQLOpenString, vbOKOnly + vbCritical, "Error: The Table
Name Already Exists!"
```

```
CreateRecordSQL = False
'Close the connection to the database
SQLClose (Chan)
End Function
```

Notice that, in the above function, it is the INSERT command that allows records to be added using SQL. Further notice that the fields to have data inserted into them are specified and surrounded by parentheses, with each element separated by a comma. The VALUES to be inserted into the specified fields are also surrounded by parentheses with each element separated by a comma.

The function `CreateRecordDAO` performs the exact same operations as the predecessor function; however, it utilizes DAO to perform the indicated operations. Notice how, in the following function, each database field must be operated upon independently as opposed to *en masse* as is done using SQL.

```
Function CreateRecordDAO(fullmdbpath As String, TableName As
String, _
  FieldName As String, FieldName1 As String, FieldName2 As String, _
  Value As String, Value1 As String, Value2 As String) As Boolean
'Writes 3 Fields as a Record to the Specified Table Using DAO

Dim MCSdb As Database, r As Long
'If Recordset not Prefixed with DAO. then Type 13 Mismatch
Error may occur
'VBA may confuse ADO with DAO database objects!
Dim rs As DAO.Recordset

On Error GoTo Record_Error:
'Establish a DSN-less connection to the MCS Sample Data Base
ConnectString = "ODBC;DRIVER={Microsoft Access Driver
(*.mdb)};DBQ=" & fullmdbpath
Set MCSdb = DBEngine.OpenDatabase(fullmdbpath, , , ConnectString)
'Define Recordset to Table
'If Recordset Object Variable not Prefixed with DAO Type 13
Mismatch Error may occur here:
Set rs = MCSdb.OpenRecordset(TableName, dbOpenTable)
With rs
  .AddNew 'Create a New Record
  'Add Values to Specific Fields
  .Fields(FieldName) = Value
  .Fields(FieldName1) = Value1
  .Fields(FieldName2) = Value2
  .Update ' stores the new record(s)
End With
rs.Close
MCSdb.Close
Set MCSdb = Nothing: Set rs = Nothing
CreateRecordDAO = True
Exit Function
```

```
Record_Error:
Debug.Print Error$, Err
MsgBox "Error: " & Err & ": " & Error$, vbOKOnly + vbCritical,
"Error Accessing: " & fullmdbpath
CreateRecordDAO = False
End Function
```

Both the preceding functions return a Boolean value of True if the write operation was successful in writing the specified record to the database. The sample application can be run by selecting ADA->Chapter 8->Records in Databases from the Excel menu.

8.7 DELETING RECORDS IN DATABASES USING BOUND CONTROLS

Creating a tool to delete records in a database is more problematic than creating a tool to add them. Prior to deleting a record, it is imperative for the user to have the ability to see all the records in the database and select those that should be deleted. The tool of choice for accomplishing this is the bound control.

Bound controls are simply controls that populate themselves with the contents of a selected database. List boxes and combo boxes are commonly bound to databases, from which they derive their content.

In this example, two list boxes will be utilized to allow the user to delete selected items in a chosen database. The first list box will show the contents of the entire database. The second list box will show the elements of the database that the user has selected to delete. For this example, the Microsoft Northwind Example Access database will be utilized. (Usually present at location: C:\Program Files\Microsoft Office\Office\Samples\Northwind.mdb.)

Some special settings exist for the list box that needs to be invoked in order to have a list box function as it does in this example. For example, list boxes have the capability of allowing a user to select either a single item (by means of an option button) or a plurality of items (by means of check boxes). This functionality can be set at design time by using the `Multiselect` property.

List boxes can also be set up to have column headers. This function can be particularly useful when using them to display the contents of databases, as all databases have names for the columns or fields within them. To enable column headers in a List box, the user must set the `Columnheads` property equal to TRUE.

Just because list boxes can be set up to utilize column headers does not mean any type of character can be put in them. Ordinarily, the `ListFillRange` property allows the user to populate the column headers with the values of choice, usually field names in database tables or spreadsheet headers. *Excel, however, does not support this property.*

Excel does have two properties that should allow the developer to set the data source for headers in a list box. Those properties are: `RowSourceType` and `RowSource`. The problem is that these properties do not function properly in Excel. The author has tried the following sample code and found that it does not work as advertised. Using the two properties, it should be possible to set the headers in a list box using a variety of formats including table references, SQL query results, and arrays. Sadly, none of the examples below function properly. They do not trigger errors when run; however, they do not function either. Various postings on Usenet groups confirm that this is not an isolated problem. The following code snippets are provided to avoid duplication of effort by the reader.

```
'BUG -> Adding Headers to Listbox Fails in Excel
'Using a Table Name
ListBox_All.RowSourceType = 2
```

```
ListBox_All.RowSource = frm7Bind.ComboBox_Tables.Text

'Using SQL Statement Type
'ListBox_All.RowSourceType = 3
'ListBox_All.RowSource = "SELECT * FROM " &
frm7Bind.ComboBox_Tables.Text

'Using Array
'ConnectString = "DRIVER={Microsoft Access Driver
(*.mdb)};DBQ=" & strDB
'Call GetFieldNames((ConnectString),
frm7Bind.ComboBox_Tables.Text, Fields())
'ListBox_All.RowSourceType = 5
'ListBox_All.RowSource = "Fields"
```

Looking at Figure 8.7, here the contents of the table named Customers is loaded into the list box on the left-hand side of the GUI. Because list boxes may not support some "special" characters that databases may contain (i.e., ©, ±, ~, §, £, etc.), a mechanism must be built in place to handle errors that will inevitably occur when such characters are encountered and to allow the application to continue loading data that is in an acceptable format.

When a new table is selected from the combo box, the following things need to happen. Any existing entries in the list boxes must be removed. The database table to be utilized must be set, a connection to the database must be made, and the list box must be populated with the chosen database records. The next subroutine triggers these items to occur.

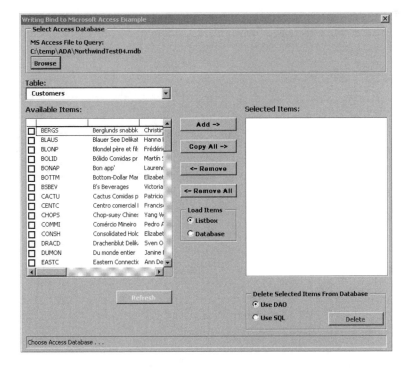

FIGURE 8.7 Sample application using bound controls to delete database items.

```
Private Sub ComboBox_Tables_Change()
  'When Table is Selected or Changed ListBoxes be Updated
  'First Clear Existing Entries
  Call ClearListBoxes
  'Now Repopulate with Selected Table Fields
  ' Database Table to Use.
  strDBTable = ComboBox_Tables.Text
  Call DBConnect
  Call PopulateListBox_AllData
End Sub
```

The subroutines that accomplish the tasks discussed are now presented in chronological order. Although clearing a list box is not really an intellectual feat, there is a bug in Excel VBA that has caused the author aggravation on numerous occasions. The bug is that, when utilizing the TypeOf statement, the 'ListBox' object is not recognized. This makes it impossible to clear all list boxes on a form by just looking at the object type. Some sample code to demonstrate this bug has been commented out and is presented to the readers so that they may save time and be aware of this pitfall.

```
Sub ClearListBoxes()
'This will clear every ListBox in the active GUI
Dim C As Control
For Each C In Controls
 'VBA Bug - TypeOf 'ListBox' Not Recognized!
 'Next Line of Code Will Not Work
 'If TypeOf C Is ListBox And C.Name Like "*List*" Then
 'Replace with:
 If TypeName(C) = "ListBox" And C.Name Like "*List*" Then
 '*** TEST ***
 'This NEVER Triggered!
 'If TypeOf C Is ListBox Then Debug.Print "I am a ListBox"
 'This Triggered when Object is a ListBox!
 'If TypeName(C) = "ListBox" Then Debug.Print "I am also a  ListBox"
 'Debug.Print TypeName(C), C.Name
 'The Clear Method is not Present in Autocomplete!
 C.Clear
  End If
Next C
End Sub
```

A connection is then established to the selected database.
```
Sub DBConnect()
'Get default Workspace
Set oWorkSpace = DBEngine.Workspaces(0)
'Establish a DSN-less connection to the Data Base
ConnectString = "ODBC;DRIVER={Microsoft Access Driver
(*.mdb)};DBQ=" & strDB
```

```
Set oDataBase = DBEngine.OpenDatabase(strDB, , , ConnectString)
'Begin the Transaction
oWorkSpace.BeginTrans
' Set the record set to the specified table.
'Set oRecordSet = oDataBase.OpenRecordset(strDBTable)
' Set the record set to return all records using SQL
Set oRecordSet = oDataBase.OpenRecordset("SELECT * FROM " &
strDBTable & ";")
End Sub
```

With a connection established to the database, and the contents of the table of interest queried via SQL, it is now possible to populate the list box. The On Error Resume Next statement allows the subroutine to skip over invalid or null characters read in from the database. The number of columns in the list box must also be set to the number of columns present in the database in order for all the field information to be shown in the list box. Finally, a separate subroutine (Clean-UpQuery) is called at the end of the subroutine to release the objects utilized during the query.

```
Sub PopulateListBox_AllData()
'Populate the ListBox with every Record in the Selected Table
Dim row As Integer, col As Integer
'Skips Errors When Null Fields Encountered
On Error Resume Next
oRecordSet.MoveFirst
ListBox_All.ColumnCount = oRecordSet.Fields.Count
Do Until oRecordSet.EOF
 ListBox_All.AddItem 'Add next row to list
 For col = 0 To oRecordSet.Fields.Count
   ListBox_All.List(row, col) = oRecordSet.Fields(col)
 Next col
 row = row + 1 'rows in listboxes begin @ 0!
 oRecordSet.MoveNext ' Move to next record in table.
Loop
Call CleanUpQuery
End Sub
```

If the commands in the CleanUpQuery subroutine are not called, at some point the application will end up locking up. This is because resources will be constantly allocated, and never freed with each successive query. Such actions will result in a "memory leak," where memory will be utilized and not returned to the system, and at some point the system will run out of usable memory and the program will either crash or not function properly. (Usually, first the program will not function properly and will, shortly thereafter, crash.)

```
Sub CleanUpQuery( )
'Clean Up Object Variables
'Commit Any Changes
oWorkSpace.CommitTrans
'Free Up Object Variables
Set oRecordSet = Nothing
```

```
Set oDataBase = Nothing
Set oWorkSpace = Nothing
End Sub
```

With the current database records now contained in the first list box, the next task to be tackled is how to select the records that should be deleted. Notice in Figure 8.7 that four buttons are present on the GUI. They are Add, Copy All, Remove, and Remove All. Each can be utilized to select items to be removed from the current database. Items that are selected for removal can be populated to the second list box by one of two means. They can be added from the list box that contains all the database items, or they can be added directly from the database itself. An option button on the GUI allows the user to choose which scheme to utilize. The advantage of reading items in directly from the database is that order is always preserved in the listing of records. The advantage of reading items in directly from the list box is speed, because the database only needs to be queried once until such time it has been updated.

Keep in mind that these four buttons really constitute a "Shell Game," where individual elements of the database are selected and moved around at will. The concepts covered earlier in this chapter in the "Using Pointers in Dynamic Database Algorithms" section are put to work in this example.

Taking the case first of copying over only the SELECTED elements in the left-hand list box:

```
Private Sub Button_Add_Click()
'Copy over only SELECTED Items to Chosen Listbox FROM Database
Dim index As Integer, DeletedItems As Integer
Dim lbrow As Integer, dbrow As Integer, col As Integer
On Error Resume Next
Select Case OB_db.Value
  'Load From Database
  Case True
  Call DBConnect
  oRecordSet.MoveFirst
  ListBox_Chosen.ColumnCount = oRecordSet.Fields.Count
  Do Until oRecordSet.EOF '*** Use For Loop to .Fields of All
  listbox instead?
   If frm7Bind.ListBox_All.Selected(dbrow) = True Then
   ListBox_Chosen.AddItem 'Add next row to list
   'Track Items that have moved, 0 = not selected, else set
   to position + 1
   ReDim Preserve dBptr(dbrow)
   dBptr(dbrow) = lbrow + 1
   For col = 0 To oRecordSet.Fields.Count
    frm7Bind.ListBox_Chosen.List(lbrow, col) =
    oRecordSet.Fields(col)
   Next col
   'Increment Listbox row
   lbrow = lbrow + 1 'rows in listboxes begin @ 0!
  End If
  'Increment Database Row
  dbrow = dbrow + 1 'rows in listboxes begin @ 0!
```

```
    oRecordSet.MoveNext ' Move to next record in table.
  Loop
  Call CleanUpQuery
  'Load from Listbox
  Case False
  'Copy over only SELECTED Items to Chosen Listbox
  'Set Number of Columns
  frm7Bind.ListBox_Chosen.ColumnCount =
  frm7Bind.ListBox_All.ColumnCount
  For index = 0 To frm7Bind.ListBox_All.ListCount - 1
    'Check for Error Condition induced by indexes
    If (index - DeletedItems) > frm7Bind.ListBox_All.ListCount
    - 1 Then Exit For
  If frm7Bind.ListBoxs_All.Selected(index - DeletedItems) = True Then
     ListBox_Chosen.AddItem 'Add next row to list
     'Track Items that have moved, 0 = not selected, else set
     to position + 1
     ReDim Preserve dBptr(index)
     dBptr(index) = lbrow + 1
     'This item is selected, add to Chosen ListBox
     For col = 0 To frm7Bind.ListBox_All.ColumnCount - 1
      'ListBox_All.AddItem oRecordSet.Fields(strDBField)
      frm7Bind.ListBox_Chosen.List(lbrow, col) =
     frm7Bind.ListBox_All.List(index - DeletedItems, col)
     Next col
     'Debug.Print index, lbrow, col,
  frm7Bind.ListBox_Chosen.List(lbrow, 0),
  frm7Bind.ListBox_All.List(index - DeletedItems, 0)
        'Increment Listbox row
        lbrow = lbrow + 1 'rows in listboxes begin @ 0!
        'No () Here! frm7Bind.ListBox_All.RemoveItem (index - 1
        - DeletedItems) = ERROR!
        frm7Bind.ListBox_All.RemoveItem index - DeletedItems
        DeletedItems = DeletedItems + 1
    End If
  Next index
  End Select
  End Sub
```

Some comments on the preceding subroutine. It will copy elements either directly from the database or from the preexisting list box of database elements. Notice that elements are not copied to the second list box unless they are checked (the selected property of the corresponding index of the list box is set to TRUE). Most important, notice that both the list box and the database have separate pointers to reference their data. This is imperative because the records in the database may not necessarily have a one-to-one correspondence (indexwise) with the items present in the list box. There will never be identical indexes between the database and list boxes if an element in the database fails to load into the list box due to an error (special or null characters, etc.). Each time such a behavior occurs, the index of the database element will be one greater than the current list box element.

The case of copying all the items over from the first list box is even easier because no checking needs to be done.

```
Private Sub Button_CopyAll_Click()
'Copy over ALL Items to Chosen Listbox FROM Database
Dim index As Integer, DeletedItems As Integer
Dim lbrow As Integer, dbrow As Integer, col As Integer
On Error Resume Next
Select Case OB_db.Value
  'Load From Database
  Case True
  Call DBConnect
  oRecordSet.MoveFirst
  ListBox_Chosen.ColumnCount = oRecordSet.Fields.Count
  Do Until oRecordSet.EOF '*** Use For Loop to .Fields of All
  listbox instead?

   ListBox_Chosen.AddItem 'Add next row to list
   'Track Items that have moved, 0 = not selected, else set
   to position + 1

   ReDim Preserve dBptr(dbrow)
   dBptr(dbrow) = lbrow + 1
     For col = 0 To oRecordSet.Fields.Count
     'ListBox_All.AddItem oRecordSet.Fields(strDBField)
     'Debug.Print oRecordSet.Fields(2)
     frm7Bind.ListBox_Chosen.List(lbrow, col) =
     oRecordSet.Fields(col)
     Next col
   'Increment Listbox row
   lbrow = lbrow + 1 'rows in listboxes begin @ 0!
 'Increment Database Row
 dbrow = dbrow + 1 'rows in listboxes begin @ 0!
 oRecordSet.MoveNext ' Move to next record in table.
 Loop
   Call CleanUpQuery
 'Load from Listbox
 Case False
 'Copy over ALL Items to Chosen Listbox
 'Set Number of Columns
 frm7Bind.ListBox_Chosen.ColumnCount =
 frm7Bind.ListBox_All.ColumnCount
 For index = 0 To frm7Bind.ListBox_All.ListCount - 1
   ListBox_Chosen.AddItem 'Add next row to list
```

```
     'Track Items that have moved, 0 = not selected, else set to
     position + 1
     ReDim Preserve dBptr(index)
     dBptr(index) = index + 1
     'Add to Chosen ListBox
     For col = 0 To frm7Bind.ListBox_All.ColumnCount - 1
     frm7Bind.ListBox_Chosen.List(index, col) =
     frm7Bind.ListBox_All.List(index - DeletedItems, col)
     Next col
     frm7Bind.ListBox_All.RemoveItem 0
     DeletedItems = DeletedItems + 1
    Next index
  End Select
End Sub
```

Figure 8.8 and Figure 8.9 show the example macro in action. In Figure 8.8, notice how each item in the list box has a corresponding check box allowing the user to select individual records. In this instance, ALFKI, ANTON, and BERGS are selected from the customer's table. Figure 8.9 shows what happens after the user presses the Add button, which adds the checked items to the second list box on the right-hand side of the GUI.

It is conceivable, however, that the user will accidentally add items or want to remove items that have been previously slated for deletion. The next subroutine removes only those items selected in the Selected Items list box.

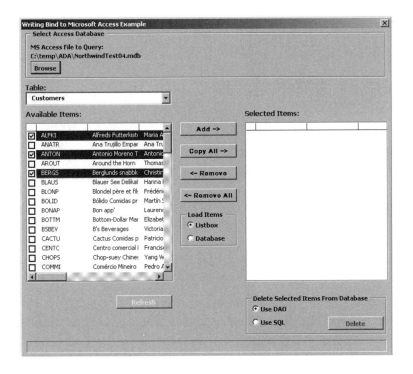

FIGURE 8.8 Selecting records for deletion.

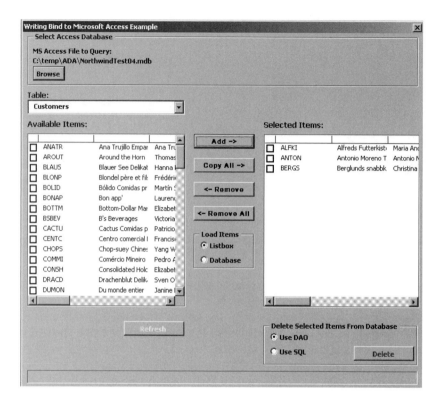

FIGURE 8.9 Selected records for deletion.

```
Private Sub Button_Remove_Click()
'Remove Selected Items from Chosen Listbox
Dim index As Integer, DeletedItems As Integer
Dim allrow As Integer
On Error Resume Next
For index = 0 To frm7Bind.ListBox_Chosen.ListCount - 1
    If frm7Bind.ListBox_Chosen.Selected(index - DeletedItems )
    = True Then
        ListBox_All.AddItem 'Add next row to list
        'Track Items that have moved, 0 = not selected, else
        set to position + 1
        ReDim Preserve dBptr(index)
        dBptr(index) = index + 1
        'Next added Row Number
        allrow = ListBox_All.ListCount - 1
        If allrow < 0 Then allrow = 0
        For col = 0 To frm7Bind.ListBox_Chosen.ColumnCount - 1
         'ListBox_All.AddItem oRecordSet.Fields(strDBField)
         'Debug.Print oRecordSet.Fields(2)
         frm7Bind.ListBox_All.List(allrow, col) =
         frm7Bind.ListBox_Chosen.List(index - DeletedItems, col)
```

```
            Next col
        frm7Bind.ListBox_Chosen.RemoveItem index - DeletedItems
        DeletedItems = DeletedItems + 1
        End If
    Next index
End Sub
```

Notice that it is necessary to keep track of the number of deleted items. Obviously, as items are deleted, the number of items in the list box decreases. The index is only valid until the first item is deleted and removed from the list box. From then forward, the index must be adjusted to reflect the number of items deleted from the list box. Notice also that each list box has its own pointer indexes, as was true in the previous subroutines.

As in the previous example, it is somewhat easier to remove all of the selected items than simply a subset of the selected items. The Remove All button triggers the following code.

```
Private Sub Button_RemoveAll_Click()
'Remove All Items from Chosen Listbox
Dim index As Integer, allrow As Integer
On Error Resume Next
    For index = 0 To frm7Bind.ListBox_Chosen.ListCount - 1
    ListBox_All.AddItem 'Add next row to list
    'Track Items that have moved, 0 = not selected, else set
    to position + 1
    ReDim Preserve dBptr(index)
    dBptr(index) = index + 1
    'Next added Row Number
    allrow = ListBox_All.ListCount - 1
    'Add to Chosen ListBox
    For col = 0 To frm7Bind.ListBox_All.ColumnCount - 1
      frm7Bind.ListBox_All.List(allrow, col) =
      frm7Bind.ListBox_Chosen.List(0, col)
    Next col
    frm7Bind.ListBox_Chosen.RemoveItem 0
    allrow = allrow + 1
  Next index
End Sub
```

Once a subset of items has been chosen for deletion, a mechanism must be utilized to actually delete the selected items from the database. As in previous examples, methods for deleting items using both SQL and DAO will be presented. When the delete button is pushed, the following subroutine is invoked.

```
Private Sub Button_Delete_Click()
Dim MsgBoxRet
Select Case frm7Bind.OptionButton_SQL.Value
  Case True
  'Use SQL To Delete Selected Records
  MsgBoxRet = MsgBox("This Action will Alter the Original
  Database! Continue?", vbYesNoCancel + vbExclamation, _
```

```
        "Warning - This Action will Delete All items in the
        Selected Items Listbox!")
  If MsgBoxRet = vbYes Then
    'Call DBConnect 'Not Necessary for XLODBC SQL Statement
    Call SQLDelete
    Else
    Exit Sub
  End If
  Case False
  'Use DAO
  MsgBoxRet = MsgBox("This Action will Alter the Original
  Database! Continue?", vbYesNoCancel + vbExclamation, _
        "Warning - This Action will Delete All items in the
        Selected Items Listbox!")
  If MsgBoxRet = vbYes Then
    Call DBConnect
    Call DAODelete
    Else
    Exit Sub
  End If
End Select
End Sub
```

Two subroutines are responsible for deleting the records in the chosen database. The first, SQLDelete, does so using SQL. Notice that this is a three-step process that consists of (1) deleting the actual record, (2) removing the deleted records from the Selected Items Listbox, and (3) refreshing the Available Items Listbox to reflect the changes made within the database.

```
Sub SQLDelete()
Dim dBindex As Integer, ptrindex As Integer
Dim DeletedItems As Integer
'Dim SQLStr As String
'Delete Chosen Records Using SQL
On Error Resume Next
For index = 0 To UBound(dBptr())
  If dBptr(index) <> 0 Then
      Call DeleteRecordSQL(strDB, strDBTable, Trim(str(index
      - DeletedItems)))
      DeletedItems = DeletedItems + 1
  End If
Next index
'Remove All Items from Selected Listbox
For index = 0 To frm7Bind.ListBox_Chosen.ListCount - 1
  frm7Bind.ListBox_Chosen.RemoveItem (0)
Next index
Call Button_Refresh_Click
End Sub
```

Notice that the actual deletion is performed by the function `DeleteRecordSQL`, which returns a Boolean value indicating whether the operation was successful or not.

```
Function DeleteRecordSQL(fullmdbpath As String, TableName As
String, _
     ID As String) As Boolean
'Deletes a Single Record in the Specified Table Using SQL
Dim Chan As Variant
Dim SQLStatement As String
Dim SQLOpenString As String
On Error GoTo Rec_Exists:
'Establish a DSN-less connection to the Data Base
SQLOpenString = "DRIVER={Microsoft Access Driver (*.mdb)};DBQ="
& fullmdbpath
Chan = SQLOpen(SQLOpenString)
'Delete Chosen Row
'Works with Oracle But Now MS Access
SQLStatement = "DELETE FROM " & TableName & " WHERE rowid =
" & Trim(ID) & ";"
'Delete the Record
SQLExecQuery Chan, SQLStatement
SQLExecQuery Chan, "COMMIT;"
'Close the connection to the database
SQLClose (Chan)
DeleteRecordSQL = True
Debug.Print SQLStatement
Exit Function
Rec_Exists:
Debug.Print Error$, Err
MsgBox SQLOpenString, vbOKOnly + vbCritical, "Error: Unable
to Delete: " & SQLStatement
DeleteRecordSQL = False
'Close the connection to the database
SQLClose (Chan)
End Function
```

The preceding function is completely dependent on accurate pointers to delete the items that the user selected. If the pointers are inaccurate, the algorithm fails. Hence the discussion at the start of the chapter. Also be aware that, for the SQL command to make a permanent change in the database, the COMMIT command must be invoked.

Records can also be deleted using DAO as shown in the next subroutine.

```
Sub DAODelete()
Dim dBindex As Integer, ptrindex As Integer
'Delete Chosen Records Using DAO
'On Error Resume Next
oRecordSet.MoveFirst
```

```
Do Until oRecordSet.EOF
  If dBptr(ptrindex) <> 0 Then
      'Iff ptr <> 0 then item selected to be deleted
      Debug.Print ptrindex, dBptr(ptrindex),
      oRecordSet.Fields(0)
      oRecordSet.Delete
  End If
  oRecordSet.MoveNext ' Move to next record in table.
  ptrindex = ptrindex + 1
Loop
'Remove All Items from Selected Listbox
For index = 0 To frm7Bind.ListBox_Chosen.ListCount - 1
  frm7Bind.ListBox_Chosen.RemoveItem (0)
Next index
Call CleanUpQuery
Call Button_Refresh_Click
End Sub
```

Threads common to both methodologies include (1) reliance upon pointers to delete the selected records, (2) the need to "clean up" after each successive database operation, closing sessions, freeing object variables, etc., (3) warning the user prior to permanently altering the database, and (4) refreshing the current database information to reflect the most recent changes.

8.8 COMPACTING DATABASES USING VBA

As data is changed in a database, the database file can become fragmented and use more disk space than is necessary (much like a computer hard disk). Periodically, a database should be compacted to defragment the database file. The compacted database is usually smaller and often runs faster. Also, the compacting process will repair any inconsistencies and help fix or prevent corruption of the database file.

A routine has been created to allow developers to compact database utilizing Excel VBA code. The advantage here is that, if the user makes numerous changes, the database can be automatically compacted at specified periods.

The subroutine functions in the following manner. First, the original database is compacted, but under a different temporary name. Once the database has been compacted, the original database file is deleted. The compacted database is then renamed to the original database name, which was previously deleted.

```
Sub DAOCompactDatabase(fullmdbpath As String)
'Compacts and Cleans Up an Access Database
Dim dBpath As String
Dim fulltempdBpath As String
dBpath = StriptoPath(fullmdbpath)
fulltempdBpath = dBpath & "TempDBCompact.mdb"
'Use a scratch file TempDBCompact.mdb to write the compacted dB to.
'Make sure this file (TempDBCompact.mdb) does not already exist
If Dir(fulltempdBpath) <> "" Then Kill fulltempdBpath
'Compact Database to New FileName
DBEngine.CompactDatabase fullmdbpath, fulltempdBpath
```

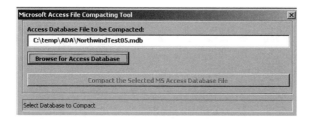

FIGURE 8.10 A sample database compacting application.

```
frm8Compact.StatusBar.SimpleText = "Compacting Selected
Database .    .    ."
' Delete the original database
Kill fullmdbpath
frm8Compact.StatusBar.SimpleText = "Deleting Original
Uncompacted Database .    .    ."
' Rename the file back to the original name
Name fulltempdBpath As fullmdbpath
frm8Compact.StatusBar.SimpleText = "Renaming Compacted Database
to Original Filename .    .    ."
End Sub
```

The preceding subroutine could be modified to create another layer of safety by not deleting the original filename but rather archiving it under a time–date stamp filename. When a sufficient period of time has passed, archived files older than a certain interval could be automatically deleted.

The sample compacting application shown in Figure 8.10 can be run by selecting ADA->Chapter 8->Compact Database from the Excel Menu.

8.9 RETURNING THE RESULTS OF A REMOTE ACCESS DATABASE QUERY TO AN EXCEL WORKSHEET

The last topic of this chapter may well be the most important, and that is, how to get information out of a database and into an Excel Worksheet. The most efficient way to obtain a subset of data from a database is through the use of an SQL query. This brings to light another limitation in the (Excel–Access) paradigm, which is that SQL queries are not supported by DAO. SQL queries, however, are supported by ADO. The coding differences between DAO and ADO are minor and will be covered in the next sample application.

Although the user could use DAO or even ADO without SQL to make a series of comparisons on the database, such a method would be highly inefficient and will not be covered.

Figure 8.11 shows the sample application that can be run by selecting ADA->Chapter 8->Extract from Database from the Excel Menu. The GUI shown in Figure 8.11 allows the user to construct a sample SQL query, the results of which are returned to a new workbook. The user can execute the SQL query using either ADO or the XLODBC functions, and this sample application shows how to accomplish the task either way. The query created in Figure 8.11 will return all the CustomerID fields when the CustomerID field contains a "B." Everything is set into motion when the "Extract Database Information to New Workbook" button is pressed.

```
Private Sub Button_Extract_Click()
SaveSetting "AccessSamples", "ExtractDB", "File",
frm8Extract.TextBox_File
```

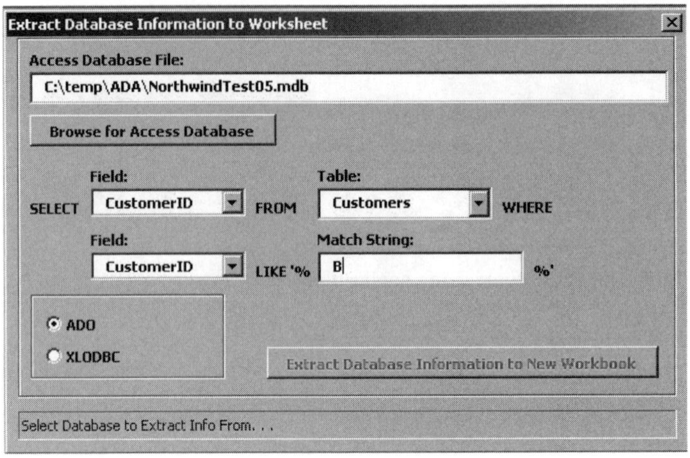

FIGURE 8.11 SQL-based data extractions to Excel.

```
SQLExtract = "SELECT " & frm8Extract.ComboBox_Fields & " FROM
" & frm8Extract.ComboBox_Tables & _
        " WHERE " & frm8Extract.ComboBox_Fields2 & " LIKE '%"
        & frm8Extract.TextBox_Query & "%';"
'Debug.Print SQLExtract
If SQLExtract = "" Or frm8Extract.TextBox_File = "" Then Exit Sub
Workbooks.Add
Select Case frm8Extract.OB_ADO
  Case True 'USE ADO
    Call QueryToWorksheet (frm8Extract.TextBox_File, SQLExtract)
  Case False 'USE XLODBC
    Call RetrieveDataXLODBC (frm8Extract.TextBox_File,
    SQLExtract, Range("A1"))
End Select
End Sub
```

Two subroutines are utilized, each of which will execute the SQL query and return its results in a different manner. The QueryToWorksheet subroutine utilizes ADO.

```
Sub QueryToWorksheet(mdbfile As String, SQLString)
'Query Results, Return Results to Worksheet Using ADO
Dim wksrow As Integer, col As Integer
Dim cn As ADODB.Connection
Dim rs As ADODB.Recordset
Dim Fields()
On Error Resume Next
'Set ADO Connection
Set cn = New ADODB.Connection

'Establish a DSN-less connection to the MCS Sample Data Base
```

```
'ConnectString = "ODBC;DRIVER={Microsoft Access Driver
(*.mdb)};DBQ=" & mdbfile
'REMOVE "ODBC;" PREFIX WHEN USING ADO CONNECTIONS
ConnectString = "DRIVER={Microsoft Access Driver (*.mdb)};DBQ="
& mdbfile
cn.Open ConnectString
Set rs = New ADODB.Recordset
'Return Recordset limited by Query
rs.Open SQLString, cn, adOpenForwardOnly, adLockReadOnly,
adCmdText
'Write in Headers (Done Automatically with XLODBC )
Call GetFieldNames((ConnectString),
frm8Extract.ComboBox_Tables.Text, Fields())
For col = 0 To rs.Fields.Count - 1
 If frm8Extract.ComboBox_Fields.Text = "*" Then
    ActiveSheet.Cells(1, col + 1).Value = Fields(col + 1)
 Else
   ActiveSheet.Cells(1, col + 1).Value =
 Fields(frm8Extract.ComboBox_Fields.ListIndex + col)
 End If
Next col
'Now Write Information to Worksheet
wksrow = 2 'Start at Row 2, Row 1 = Column Headers
rs.MoveFirst
Do Until rs.EOF '*** Use For Loop to .Fields of All listbox instead?
   For col = 0 To rs.Fields.Count - 1
     ActiveSheet.Cells(wksrow, col + 1).Value = rs.Fields(col)
   Next col
   'Increment Worksheet Row
   wksrow = wksrow + 1
   rs.MoveNext ' Move to next record in table.
Loop
'Release Objects!
rs.Close
Set rs = Nothing
cn.Close
Set cn = Nothing
End Sub
```

Notice that the ADO connection syntax is nearly identical to the DAO connection syntax. One notable exception is that the connection string using ADO must not have the "ODBC"; prefix. Doing so will result in a failed connection. Speaking of failed connections, a word of warning is in order with respect to use of the On Error Resume Next syntax. Although including this statement allows a macro to jump over database entries that trigger errors, it also allows the code to continue running when ANY error results. This can be problematic when debugging subroutines

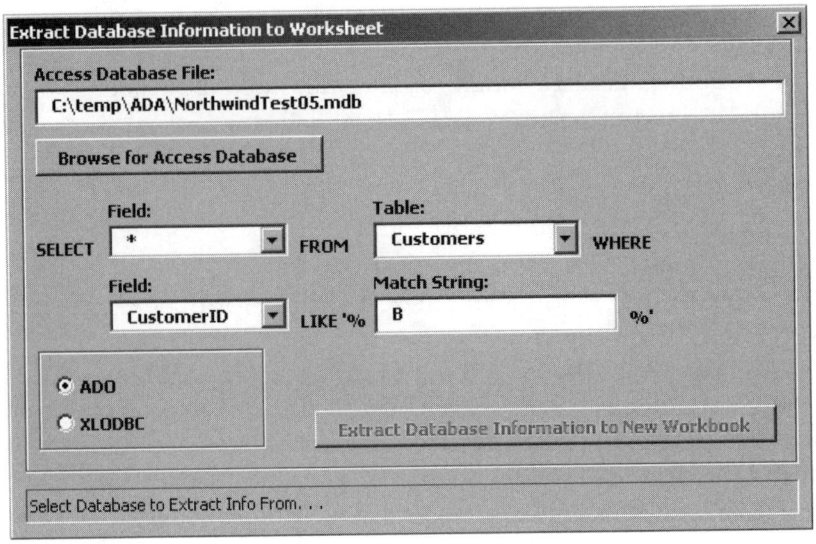

FIGURE 8.12 Returning all fields in an SQL query with.'*'

because, if a database connection fails to connect, the code will keep running, no results will be returned, and the developer will not have the vaguest idea why. It is best to comment out this statement once a subroutine has been fully tested.

Notice also that the user must manually write in the field names (column headers) if they are desired in the output. The XLODBC functions automatically include them in the returned query as the first record, but ADO does not. The XLODBC subroutine (RetrieveDataXLODBC) is much shorter because much of the work has already been done in the XLODBC function library.

```
Sub RetrieveDataXLODBC(mdbfile As String, SQLString, Position
As Range)
'Query Results, Return to Results to Worksheet Using XLODBC
Functions

Dim Chan As Variant
'Establish a DSN-less connection to the Data Base
SQLOpenString = "DRIVER={Microsoft Access Driver (*.mdb)};DBQ="
& mdbfile
Chan = SQLOpen(SQLOpenString)
```

	A	B	C	D	E	F	G	H	I	J	K
1	CustomerID	CompanyName	ContactName	ContactTitle	Address	City	Region	PostalCode	Country	Phone	Fax
2	BERGS	Berglunds snabbköp	Christina Berglund	Order Administrator	Berguvsvägen 8	Luleå		S-958 22	Sweden	0921-12 34 65	0921-12 34 67
3	BLAUS	Blauer See Delikatessen	Hanna Moos	Sales Representative	Forsterstr. 57	Mannheim		68306	Germany	0621-08460	0621-08924
4	BLONP	Blondel père et fils	Frédérique Citeaux	Marketing Manager	24, place Kléber	Strasbourg		67000	France	88.60.15.31	88.60.15.32
5	BOLID	Bólido Comidas preparadas	Martín Sommer	Owner	C/ Araquil, 67	Madrid		28023	Spain	(91) 555 22 82	(91) 555 91 99
6	BONAP	Bon app'	Laurence Lebihan	Owner	12, rue des Bouchers	Marseille		13008	France	91.24.45.40	91.24.45.41
7	BOTTM	Bottom-Dollar Markets	Elizabeth Lincoln	Accounting Manager	23 Tsawassen Blvd.	Tsawassen	BC	T2F 8M4	Canada	(604) 555-4729	(604) 555-3745
8	BSBEV	B's Beverages	Victoria Ashworth	Sales Representative	Fauntleroy Circus	London		EC2 5NT	UK	(171) 555-1212	
9	FURIB	Furia Bacalhau e Frutos do Mar	Lino Rodriguez	Sales Manager	Jardim das rosas n. 32	Lisboa		1675	Portugal	(1) 354-2534	(1) 354-2535
10	LAUGB	Laughing Bacchus Wine Cellars	Yoshi Tannamuri	Marketing Assistant	1900 Oak St.	Vancouver	BC	V3F 2K1	Canada	(604) 555-3392	(604) 555-7293
11	SIMOB	Simons bistro	Jytte Petersen	Owner	Vinbæltet 34	København		1734	Denmark	31 12 34 56	31 13 35 57
12	THEBI	The Big Cheese	Liz Nixon	Marketing Manager	89 Jefferson Way□ Suite 2	Portland	OR	97201	USA	(503) 555-3612	

FIGURE 8.13 Query results for query setup in Figure 8.12.

```
SQLExecQuery Chan, SQLString
' Return the starting at cell range specified in Position
variable on the active sheet.
SQLRetrieve Chan, Position, , , True
SQLClose (Chan)
End Sub
```

In Figure 8.11, a single field (CustomerID) was returned using an SQL query. However, it is possible to return every field in the records returned by an SQL query by utilizing the '*' operator (Figure 8.12). Figure 8.13 shows the results written to the Worksheet when the query shown in Figure 8.12 is executed.

8.10 SUMMARY

Chapters 7 and 8 provide the reader with the tools required to manipulate database information. Chapter 7 focused on how to extract information from databases, whereas this chapter focused on writing information to and modifying information in databases. Concepts in database alteration and management are covered with explicit attention given toward the concepts of utilizing pointers in database manipulations. Utilizing bound controls to display database information is also covered with specific instruction given as to utilizing list boxes to display and manipulate database information.

9 Analyses Via External Applications

9.1 INTRODUCTION

Although Excel has a number of built-in functions, the reality is that its ability to perform computations of any degree of sophistication internally is somewhat limited. There are many possible solutions to Excel's limitations. Custom functionality can be written by the user utilizing the VBA macro programming language. Custom software libraries or Excel Add-Ins can be purchased that will perform the desired operations within Excel. The last and most flexible option is to harness the engine of another computational tool the user is intimate with and have it return its computations to Excel. In this section the reader will be shown how to utilize the computational power of Matlab and Origin from within the Excel environment. The techniques employed here could be modified and used with other computational tools such as SAS, Mathematica, JMP, or any other tool that accepts COM, OLE, ActiveX, or DDE Commands.

9.2 SETTING UP A MATLAB ACTIVEX SERVER FROM EXCEL

The first step toward harnessing the power of the Matlab computational engine (or any other software tool for that matter) is to establish a reference from Excel to the tool to be controlled. The software that executes the function calls (in this case Excel) is referred to as the *client*. The software that performs the operations and returns the results (in this case Matlab) is referred to as the *server*. Client server communication is accomplished by means of an Object variable. An object variable should be declared at the top of the module that will contain the ActiveX server control subroutines from Excel.

```
'Declare a Matlab Object Variable to reference Matlab from Excel
Dim MatLab As Object
```

This object variable can be utilized to create a reference to Matlab from Excel. A very practical way to think about this is that the object variable creates a "tunnel" between Excel (the client application) and Matlab (the server application). Simply declaring the object variable at the top of the module does not establish the connection (or tunnel) between Excel and Matlab. This must be done via the Set command.

```
'Initiate a reference to Matlab
Set MatLab = CreateObject("Matlab.Application")
```

As soon as this Set command is executed, an instance of Matlab is started (just like running the program from the start menu), and a connection is established between Excel and Matlab via the MatLab Object variable. With the establishment of this connection, a number of COM server functions can be invoked using the Matlab Object variable (Table 9.1).

Equally important as establishing the connection is the termination of a connection. Leaving a connection to the server open after the client no longer requires its use is not only bad form but wasteful of system resources as well. Connections are terminated via the `Quit` command. In addition to terminating the connection, the object reference should be released. This can be easily done by setting the Matlab object variable equal to nothing.

TABLE 9.1
COM Server Functions Available in VBA

Function Name	Description
enableservice	Enable DDE or COM Automation server
Execute	Execute MATLAB command in server
Feval	Evaluate MATLAB function in server
GetCharArray	Get character array from server
GetFullMatrix	Get matrix from server
GetVariable	Returns data from variable in server workspace
GetWorkspaceData	Get data from server workspace
MaximizeCommandWindow	Display server window on Windows desktop
MinimizeCommandWindow	Minimize size of server window
PutCharArray	Store character array in server
PutFullMatrix	Store matrix in server
PutWorkspaceData	Store data in server workspace
Quit	Terminate MATLAB server

```
MatLab.Quit      ' When you finish, use the Quit method to
close the application
Set MatLab = Nothing     'Release the reference by setting it
equal to Nothing.
```

Execute is the most versatile COM server function. The Execute function executes the MATLAB statement specified by the command string. The server returns any output generated from the command in a string. The result string usually contains the results of a calculation, but may also contain any warning or error messages that might have been generated by MATLAB as a result of the command. An empty result might be returned if the MATLAB command string is terminated with a semicolon (the ";" in Matlab suppresses printing the answer to the screen) and there are no warnings or error messages. The Execute COM server function is of the following form:

```
Execute(command As String) As String
An example in Visual Basic of using the Execute function:
Sub VersionCheck()
'Return the Version Number of Matlab on this Machine
Dim MLVer As String
'Initiate a reference to Matlab
Set MatLab = CreateObject("Matlab.Application")
'Query Version
MLVer = MatLab.Execute("version")
Debug.Print MLVer
'Quit and release the reference
MatLab.Quit
Set MatLab = Nothing
End Sub
```

Here, the version of the current Matlab server is queried and returned using the "version" command. In this instance, the following value is returned to the MLVer string variable and printed in the immediate (debug) window.

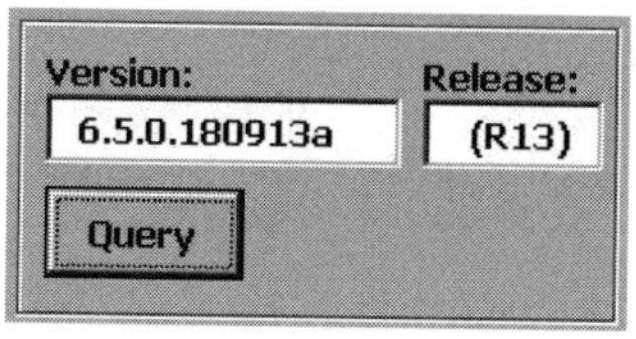

FIGURE 9.1 Query the Matlab version and release number.

ans =
6.5.0.180913a (R13)

If the result string were returned to a textbox, the following would be displayed:

¶ans =¶¶6.5.0.180913a (R13)¶¶

Here, the backwards script P "¶" is indicative of a new paragraph or line. The version and release numbers can very easily be extracted from this string by means of the InStr function.

```
MLVer = MatLab.Execute("version")
frm9ShowMatrix.TextBox_Version = Mid(MLVer, InStr(MLVer, ".")
- 1, InStr(MLVer, "(") - InStr(MLVer, "."))
frm9ShowMatrix.TextBox_Release = Mid(MLVer, InStr(MLVer, "(")
- 1, InStr(MLVer, ")") + 2 - InStr(MLVer, "("))
```

Because the Execute function is capable of executing any Matlab statement remotely, it obviously provides the ultimate in flexibility in terms or remotely harnessing the Matlab computational engine. Matlab is a matrix- and vector-driven language, meaning that all computations are set up by using matrices and vectors to carry out calculations. In the next section, it will be shown how to set up matrices and vectors from Worksheet values. The version query tool can be run by selecting ADA->Chapter 9->Show Matlab Version, which will bring up the GUI shown in Figure 9.1.

9.3 MATRIX AND VECTOR BUILDING

Because all calculations in Matlab are accomplished by means of vectors and matrices, methods must be developed for building vectors and matrices within the Excel environment. Once built, the vectors and matrices can be sent to Matlab for computation, and the results will be returned to Excel.

Vectors come in two forms: the row vector and the column vector. A row vector is denoted like this:

[2 0 1]

A row vector is created simply by separating its elements by commas or spaces. Thus the Matlab commands

c = [2,0,1]
c = [2 0 1]

will produce the shown row vector.

A column vector is denoted like this:

$$\begin{bmatrix} 2 \\ 0 \\ 1 \end{bmatrix}$$

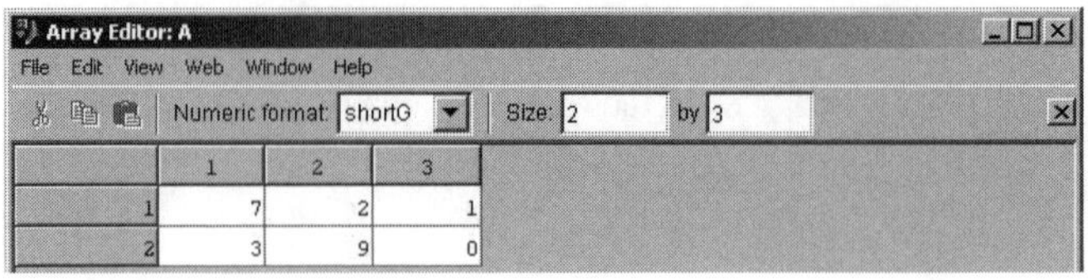

FIGURE 9.2 The array editor in Matlab.

A column vector is created by separating its elements by semicolons. Thus, the Matlab command

c = [2;0;1] will produce the previous column vector.

The Transpose notation, indicated by a "'", will convert a row vector into a column vector and vice versa. Thus, the Matlab command

c = [2,0,1]'

will also produce the earlier-described column vector. Conversely, the Matlab command

c = [2;0;1]' would produce the earlier-described row vector.

Although vectors have a single row or column, matrices have multiple rows and/or columns. (A vector is also considered a $1 \times n$ matrix; thus, the term matrix or matrices is also representative of vectors.) One example of a matrix is

$$\begin{bmatrix} 1 & 3 & 6 \\ 2 & 4 & 7 \\ 9 & 8 & 7 \end{bmatrix}$$

This matrix is called a 3×3 matrix. It is also a square matrix. A matrix is called square if the number of rows equals the number of columns. Matrices are always referred to in row-by-column format. The "3×3" in the above example stands for 3 rows by 3 columns. Here are two examples of a 2×3 matrix and a 4×2 matrix, respectively.

$$\begin{bmatrix} 7 & 2 & 1 \\ 3 & 9 & 0 \end{bmatrix} \quad \begin{bmatrix} 1 & 9 \\ 8 & 2 \\ 7 & 3 \\ 6 & 4 \end{bmatrix}$$

Matrices are written using commas (or spaces) to separate row elements, and semicolons to indicate the start of a new row. Typical mathematical matrix notation uses a capital boldface letter to denote a matrix, whereas a vector utilizes a boldface lowercase letter. The Matlab commands to enter the matrices shown could be

A = [7,2,1; 3,9,0]
B = [1,9;8,2;7,3; 6,4]

or alternatively, using spaces to separate row elements:

A = [7 2 1; 3 9 0]
B = [1 9;8 2;7 3; 6 4]

Notice how the ";" denotes the beginning of a new row.

Any element within a matrix or vector can be referenced using (row, column) notation. For example, in matrix A, 9 could be replaced with 8 by typing the following command:

A(2,2) = 8

Recall that in Excel VBA, Worksheet cells are referenced by the Cells form of the Range method, which also denotes cell location in the row, column format. Worksheets can easily be set up to hold vector or matrix elements and transfer them to Matlab upon command. Matlab, in fact, has a tool called the Array Editor that allows users to view matrix and vector elements in an Excel Worksheet type of format. (See Figure 9.2) A sample application will be set up to show the reader how to define matrix and vector objects in Matlab remotely from the Excel environment.

9.4 DEFINING MATRICES AND VECTORS IN MATLAB FROM EXCEL

A sample application can now be constructed that will allow the user to define any matrix or vector entity in Matlab remotely using an ActiveX server. Figure 9.3 shows the GUI for the sample application that allows matrices and vectors to be defined from within Excel. The sample matrix building tool can be run by selecting ADA->Chapter 9->Matlab Matrix Builder, which will bring up the GUI shown in Figure 9.3.

The GUI in Figure 9.3 allows the user to select what is to be defined — a matrix or a vector. If a matrix is selected, the mapping variable is converted to uppercase. If a vector is selected, the mapping variable is converted to lowercase, and the user must choose between a column or row

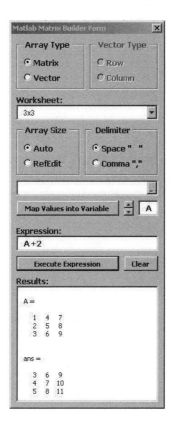

FIGURE 9.3 The matrix builder sample application.

vector using the specified option buttons. A Worksheet must also be selected from which the data will be extracted to create the matrix or vector. The drop-down box on the GUI contains the names of all the Worksheets within the active Workbook. Once the location and type of object to be created has been determined, the span of cells it occupies must be defined. This can be done manually by means of the RefEdit control on the GUI, or automatically. To automatically read data into an object, some assumptions must be made. In this instance, the application assumes that the data begins in cell A1 and is an $(n \times m)$-sized matrix, where: n = the bottom-most row that contains data, and m = the right-most column that contains data. The user may also specify the column delimiter to be utilized. Recall that, in Matlab, either a space or a comma may be used to delineate row elements in a matrix or vector. Regardless of which delimiter the user chooses, Matlab always returns its messages delineated with spaces (column elements) and carriage returns (row elements).

The user must select a character identifier to write each matrix or vector to. This is done by means of a textbox with a spin button control. The textbox can scroll through the alphabet (A–Z) by pressing the spin buttons. When a vector is chosen, the textbox reverts to lowercase letters, and when a matrix is chosen, the textbox returns to uppercase letters. This allows the application to maintain standard matrix- and vector-naming conventions.

The code that accomplishes this task is contained within four subroutines. The first two subroutines set the starting point for the alphabetic matrix identifier (A = ASCII code 65, a = ASCII code 97). The starting point is stored in a module-level variable (UorLCase) at the top of the form.

```
Private Sub OB_Matrix_Click()
frm9ShowMatrix.OB_Row.Enabled = False
frm9ShowMatrix.OB_Column.Enabled = False
frm9ShowMatrix.Frame_VType.Enabled = False
UorLCase = 65 'Use Upper Case Identifier for Matrix
If Asc(TextBox_Array.Text) > 90 Then
    TextBox_Array.Text = Chr(Asc(TextBox_Array.Text) - 32)
End If
End Sub

Private Sub OB_Vector_Click()
frm9ShowMatrix.OB_Row.Enabled = True
frm9ShowMatrix.OB_Column.Enabled = True
frm9ShowMatrix.Frame_VType.Enabled = True
UorLCase = 97 'Use Lower Case Identifier for Vector
If Asc(TextBox_Array.Text) < 97 Then
    TextBox_Array.Text = Chr(Asc(TextBox_Array.Text) + 32)
End If
End Sub
```

When a spin button is pushed, the ASCII value of the current character in the box is extracted to the variable `MatRef`. If `MatRef` is greater than the starting point `UorLCase`, then the value is allowed to decrement as the current letter must be greater than {A or a}. If `MatRef` is less than (`UorLCase + 25`), then the value is allowed to increment as the current letter must be less than {Z or z}.

```
Private Sub SpinButton_SpinDown()
Dim MatRef As Integer
MatRef = Asc(TextBox_Array.Text)
```

```
If MatRef > UorLCase Then
   MatRef = MatRef - 1
End If
TextBox_Array.Text = Chr(MatRef)
End Sub

Private Sub SpinButton_SpinUp()
Dim MatRef As Integer
MatRef = Asc(TextBox_Array.Text)
If MatRef < (UorLCase + 25) Then
   MatRef = MatRef + 1
End If
TextBox_Array.Text = Chr(MatRef)
End Sub
```

Once a matrix or vector object has been chosen and a letter has been selected to represent the chosen matrix or vector, the user can write the matrix or vector to Matlab by pressing the "Map Values into Variable" button. When this button is pressed, the Matrix will be written into Matlab using the specified identifier with the following code:

```
Private Sub Button_SetMatrix_Click()
If frm9ShowMatrix.ComboBox_Worksheets.Text <> "" Then
   Call RunMatrixBuilder
   Else
   MsgBox "You must choose a Worksheet!", vbOKOnly +
   vbExclamation, "Missing Parameter"
End If
End Sub
```

If a Worksheet has not been specified in the GUI from which to pull the matrix data, the process will be halted. Otherwise, the RunMatrixBuilder subroutine will be called.

```
Sub RunMatrixBuilder()
'Build Matrices in Excel -> Export to Matlab
Dim startcol As Integer, endcol As Integer
Dim startrow As Integer, endrow As Integer
Dim row As Integer, Col As Integer
Dim ReturnMsg As String, SendMsg As String
'ReEstablish Connection to Matlab Object
Set MatLab = CreateObject("Matlab.Application")
'Determine Array Size
Select Case frm9ShowMatrix.OB_Auto
Case True    'Auto Range Size
  'Matrix Always starts at A1
  startcol = 1: startrow = 1
  endcol = LastCol(ActiveWorkbook.Name,
  frm9ShowMatrix.ComboBox_Worksheets.Text)
```

```
   endrow = LastRow(ActiveWorkbook.Name,
   frm9ShowMatrix.ComboBox_Worksheets.Text)

Case False    'Determine Size from RefEdit
   'Get the address, or reference, from the RefEdit control.
   Addr = frm9ShowMatrix.RefEdit.Value
   'Set the SelRange Range object to the range specified in
   the RefEdit control.
   Set CoeffMatrix = Range(Addr)
   'Create "A" Matrix Based on Cell Values
   CoeffRng = CoeffMatrix.Address(, , xlR1C1)
   Call SetRefEditMatrixRanges(CoeffRng, startrow, startcol,
   endrow, endcol)

End Select
SendMsg = CreateMatrix (frm9ShowMatrix.TextBox_Array,
ActiveWorkbook.Name, _

  frm9ShowMatrix.ComboBox_Worksheets.Text, startrow, startcol,
  endrow, endcol, True, frm9ShowMatrix.OB_Space.Value)
'Debug.Print SendMsg

'If this Matrix or Vector Already Exists in Memory Clear it out!
MatLab.Execute ("clear " & frm9ShowMatrix.TextBox_Array & ";")
'Assign Matrix or Vector in Matlab
ReturnMsg = MatLab.Execute(SendMsg)
'Debug.Print ReturnMsg
frm9ShowMatrix.TextBox_Results.Text =
frm9ShowMatrix.TextBox_Results.Text & ReturnMsg
End Sub
```

Notice that the preceding subroutine accomplishes several things. It reestablishes its connection to the Matlab server, defines the limits of the matrix, calls a function to create the matrix, clears out the matrix should it already exist in Matlab's memory, writes the matrix to Matlab, and displays any results returned by Matlab to Excel to the Results textbox. Notice that it is the CreateMatrix function that actually constructs the matrix passed to Matlab. The function has the following passed parameters that determine how the elements that form the matrix will be assigned and written:

ID — a unique letter assignment for the matrix
wkbook — the Workbook the matrix data resides in
wksheet — the Worksheet the matrix data resides in
startrow — the starting row for the matrix data
startcol — the starting column for the matrix data
endrow — the ending row for the matrix data
endcol — the ending column for the matrix data
ReturnMsgs — return messages from Matlab to Excel (T/F)?
SpaceDelim — should spaces be utilized as the column delimiters (T/F)?

```
Function CreateMatrix(ID As String, wkbook, wksheet, startrow,
startcol, endrow, endcol, ReturnMsgs As Boolean, SpaceDelim
As Boolean) As String
```

```
'Build a matrix or vector string
Dim delim$
Select Case SpaceDelim
  Case True
  delim$ = Chr(32)
  Case False
  delim$ = Chr(44)
End Select
CreateMatrix = ID & " = ["
For row = startrow To endrow
  For Col = startcol To endcol
     If Col <> endcol Then
        CreateMatrix = CreateMatrix &
        Workbooks(wkbook).Worksheets(wksheet).Cells(row,
        Col).Value & delim$

         Else
         CreateMatrix = CreateMatrix &
         Workbooks(wkbook).Worksheets(wksheet).Cells(row,
         Col).Value & "; "
         End If
     Next Col
Next row
Select Case ReturnMsgs
  Case True
  'Remove last ";" and add "]"
  CreateMatrix = Mid(CreateMatrix, 1, Len(CreateMatrix) - 2) & "]"
  Case False
  'Remove last ";" and add "];"
  CreateMatrix = Mid(CreateMatrix, 1, Len(CreateMatrix) - 2) & "];"
End Select
Debug.Print CreateMatrix
End Function
```

Although the function requires that many parameters be specified up front, it basically is an engine for creating a text (or string) representation of a matrix using specified Worksheet elements. The column delimiter is set using standard ASCII elements (32 = {space}, 44 = {,}). The string is preceded by the identifier to be assigned to the matrix. The elements of the array to fill the matrix are then extracted by means of a dual looping structure. Finally, should the user not wish to have messages returned to the user, a ";" is appended to the end of the matrix.

This application also has the capability to execute any Matlab statement remotely from Excel. The expression to be executed is simply written into the expression textbox and the "Execute Expression" button is pressed. Looking at Figure 9.3, the integer 2 can be added to the Matrix A by simply typing "A+2" in the expression textbox and executing the expression. The answer is then returned to the results window. This is accomplished using the following subroutine:

```
Private Sub Button_Execute_Click()
'Execute Expression in Matlab
```

```
ReturnMsg =
MatLab.Execute(frm9ShowMatrix.TextBox_Expression.Text)
'Add Results to Textbox Log
frm9ShowMatrix.TextBox_Results.Text =
frm9ShowMatrix.TextBox_Results.Text & ReturnMsg
End Sub
```

Notice that no error checking is done on the string to be executed; therefore, if it is invalid, an error message will be returned by Matlab and printed to the results window. Also, the results written to the results window can be cleared using the following code:

```
Private Sub Button_Clear_Click()
frm9ShowMatrix.TextBox_Results.Text = ""
End Sub
```

The results window is simply a textbox with the vertical scrollbars enabled (ScrollBars property set to 2). The messages returned from Matlab are simply appended to the end of the existing text within the textbox.

When the form is closed and the application is terminated, it is imperative to shut down the Matlab server and release the reference.

```
Private Sub UserForm_Terminate()
'Quit and release the reference
MatLab.Quit
Set MatLab = Nothing
End Sub
```

Although the preceding example is somewhat rudimentary in nature, it sets the stage for building more complex projects that perform more sophisticated analyses. With the ability to define matrices and vectors at will, execute expressions upon those definitions, and return the results generated by Matlab on the preceding actions, any calculations possible within Matlab can now be harnessed by an Excel client.

9.5 USING MATLAB TO PERFORM MORE ADVANCED FORMS OF REGRESSION

Building upon the methods discussed in the previous section, an application will be created to solve a linear system of equations of the form $Ax = b,$ where:

A = $n \times n$ matrix of known coefficients
x = a column vector of n unknowns
b = a column vector of n known coefficients

In this instance, the number of equations is equal to the number of unknowns. In such a case, $\mathbf{Ax = b}$ is said to be consistent and will possess a unique solution. For $\mathbf{Ax = b}$ to be consistent, the following must be true:

1. $\mathbf{Ax = 0}$ has only the trivial solution x = 0.
2. $\mathbf{A}$ is nonsingular and its inverse exists.
3. **Rank A** = n.
4. **Det(A)** $\neq$ **0.**
5. The Reduced Row Echelon Form (RREF) of $\mathbf{A}$ is the identity matrix.
6. $\mathbf{A}$ has n linearly independent rows and columns.

When such a case exists, **A** can be solved for a single unique solution. Three other cases are, of course, possible. Those are when the number of equations equals the number of unknowns and does not satisfy the conditions just set forth, in which case it will be underdetermined and have more than one solution. Similarly, an underdetermined case also exists when there are less equations than unknown variables. In such an instance, the system will either have an infinite number of solutions or lack a solution entirely.

The last case would be that of the overdetermined system. Overdetermined systems can often be inconsistent, but the inconsistency is often due to the inaccuracies of the coefficients. This is especially true when the coefficients have been determined experimentally. In such an instance, it is desirable to solve for a solution that best approximates the given system. A method for doing so will be covered later in this chapter.

Consider a system of equations of the form: $y = a_0 + a_1 x_1 + a_2 x_2 + a_3 x_3$ where the constant a_0 term can be omitted from the solution at will. A tool can be created within Excel to solve just such a system of equations using a Matlab server. The GUI for such a tool is shown in Figure 9.4.

The tool allows the user to select the appropriate region of cells for the construction of the **A** matrix and the **b** solution vector. The user must then pick the appropriate pivot points for the x solution vector, residuals, and ill-conditioning terms. The Matlab server solves for these parameters that can be written down a column or across a row in relation to the selected pivot points.

The sample multiple linear regression tool can be run by selecting ADA->Chapter 9->Matlab Multiple Linear Regression, which will bring up the GUI shown in Figure 9.4. This sample application will be run on various systems to demonstrate its utility and functionality. If the user

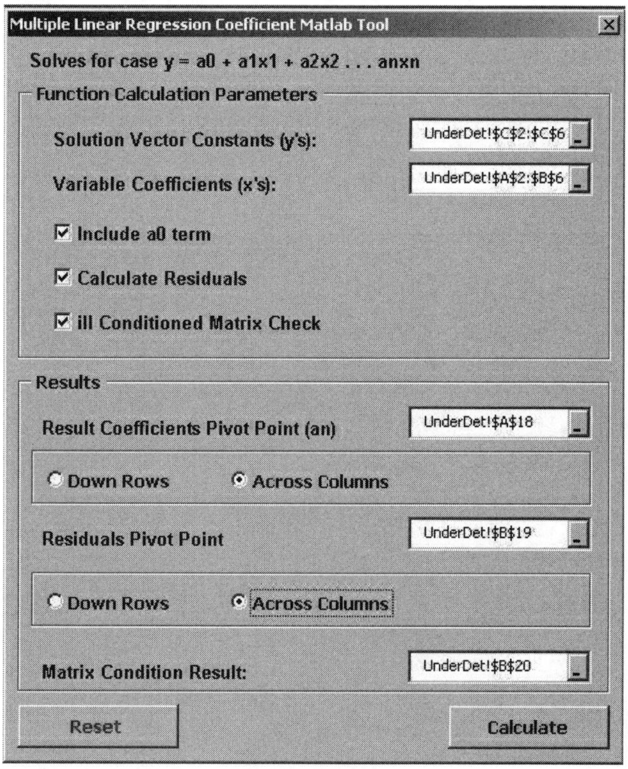

FIGURE 9.4 A multiple linear regression solution tool.

	A	B	C	D	E
1	x_1	x_2	y		
2	1.00	1.00	2.00		
3	2.50	-1.00	0.95		
4	3.05	1.50	3.95		
5	-1.05	2.00	0.95		
6	4.00	-1.00	3.00		
7					
8					
9					
10	Omitting a_0 term:				
11	x_1	x_2			
12					
13	Residuals:				
14	Condition:				
15					
16	Including a_0 term:				
17	a_0	x_1	x_2		
18					
19	Residuals:				
20	Condition:				
21					

FIGURE 9.5 Overdetermined sample MLR data set report.

loads the "MatlabMLRDataOver.XLT" sample file from the CD-ROM in location Chapter 9\Samples, the Worksheet template shown in Figure 9.5 will appear.

Here, an overdetermined system is presented with two areas reserved for results — the top area for the case where the a_0 term is omitted from the solution, and the bottom area for the case where the a_0 term is included in the returned solution. Both result areas have designated cells for the coefficients, residuals, and conditioning parameters to be written in horizontally (across columns). The sample application can easily solve this overdetermined system with an a_0 term by setting up the GUI as shown in Figure 9.6.

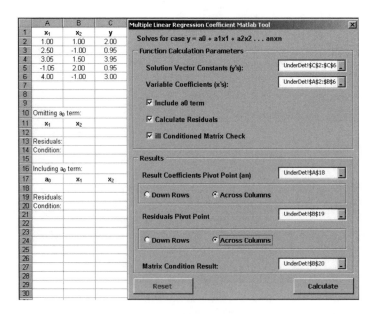

FIGURE 9.6 Solving the overdetermined case with an a_0 term.

	A	B	C	D	E	F	G
1	x_1	x_2	y				
2	1.00	1.00	2.00				
3	2.50	-1.00	0.95				
4	3.05	1.50	3.95				
5	-1.05	2.00	0.95				
6	4.00	-1.00	3.00				
7							
8							
9							
10	Omitting a_0 term:						
11	x_1	x_2					
12							
13	Residuals:						
14	Condition:						
15							
16	Including a_0 term:						
17	a_0	x_1	x_2				
18	-0.007	0.900	0.931				
19	Residuals:	-0.175	0.363	0.187	-0.039	-0.336	
20	Condition:	6.584					
21							

FIGURE 9.7 Results of the solved overdetermined case with an a_0 term.

If the user fills in the GUI as shown in Figure 9.6 and presses the "Calculate" button, the results of the analysis will be calculated in Matlab and returned to the Worksheet as shown in Figure 9.7. Similarly, the solution can also be calculated without an a_0 term by setting up the GUI as shown in Figure 9.8.

If the user fills in the GUI as shown in Figure 9.8 and presses the "Calculate" button, the results of the analysis will be calculated in Matlab and returned to the Worksheet at the specified locations. Figure 9.9 shows the "MatlabMLRDataOver.xlt" template file with all calculations completed.

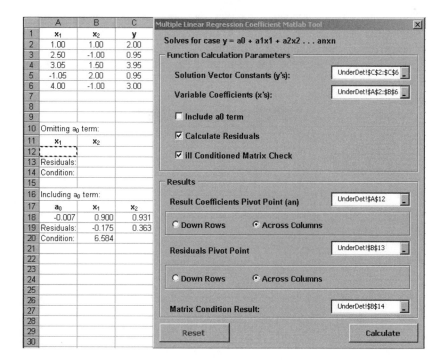

FIGURE 9.8 Solving the overdetermined case without an a_0 term.

	A	B	C	D	E	F	G
1	x_1	x_2	y				
2	1.00	1.00	2.00				
3	2.50	-1.00	0.95				
4	3.05	1.50	3.95				
5	-1.05	2.00	0.95				
6	4.00	-1.00	3.00				
7							
8							
9							
10	Omitting a_0 term:						
11	x_1	x_2					
12	0.898	0.929					
13	Residuals:	-0.173	0.367	0.184	-0.035	-0.335	
14	Condition:	1.957					
15							
16	Including a_0 term:						
17	a_0	x_1	x_2				
18	-0.007	0.900	0.931				
19	Residuals:	-0.175	0.363	0.187	-0.039	-0.336	
20	Condition:	6.584					
21							

FIGURE 9.9 Completed overdetermined case Worksheet.

The following are topics of interest with respect to the overdetermined solutions given in the sample. Notice that any solution to an overdetermined system can only approximately satisfy all of the equations in the system. This is self-evident in that the residuals are nonzero, indicative of deviation. Also notice that, for perfect conditioning, the function `rcond` will return unity (1). The higher the value rcond returns, the worse the conditioning of the matrix. Notice that, in this instance, omitting the a_0 term will produce a better-conditioned matrix to work with, and thus the residuals solved for without an a_0 term will generally be smaller than those solved for with an a_0 term.

This tool is also capable of solving a system of equations where the opposite condition exists — that of the underdetermined system. If the user loads the "MatlabMLRDataUnder.xlt" sample file from the CD-ROM in location Chapter 9\Samples, the Worksheet template shown in Figure 9.10 will appear. Here, an underdetermined system is presented with two areas reserved for results. The top area is for the case where the a_0 term is omitted from the solution, and the bottom area is for the case where the a_0 term is included in the returned solution. The result areas have designated places for

MatlabMLRDataUnder.XLT

	A	B	C	D
1	x_1	x_2	x_3	y
2	1.00	2.00	3.00	5.00
3	-3.50	1.75	-1.50	2.25
4				
5				
6	Omitting a_0 term:			
7	x_1	x_2	x_3	
8				
9	Residuals:			
10	Condition:			
11				
12	Including a_0 term:			
13	a_0	x_1	x_2	x_3
14				
15	Residuals:			
16	Condition:			
17				
18				

FIGURE 9.10 The underdetermined case example template.

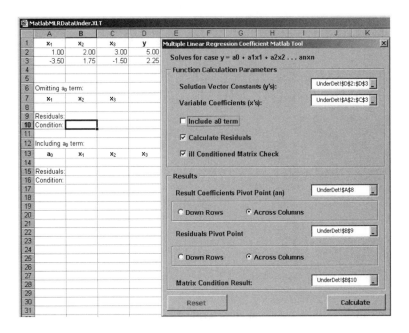

FIGURE 9.11 GUI set up to solve underdetermined system without an a_0 term.

the coefficients, residuals, and conditioning parameters to be written in horizontally (across columns). The sample MLR tool can easily solve this underdetermined system with a solution omitting an a_0 term by setting up the GUI as shown in Figure 9.11.

The sample MLR tool can also solve this underdetermined system with a solution including an a_0 term by setting up the GUI as shown in Figure 9.12.

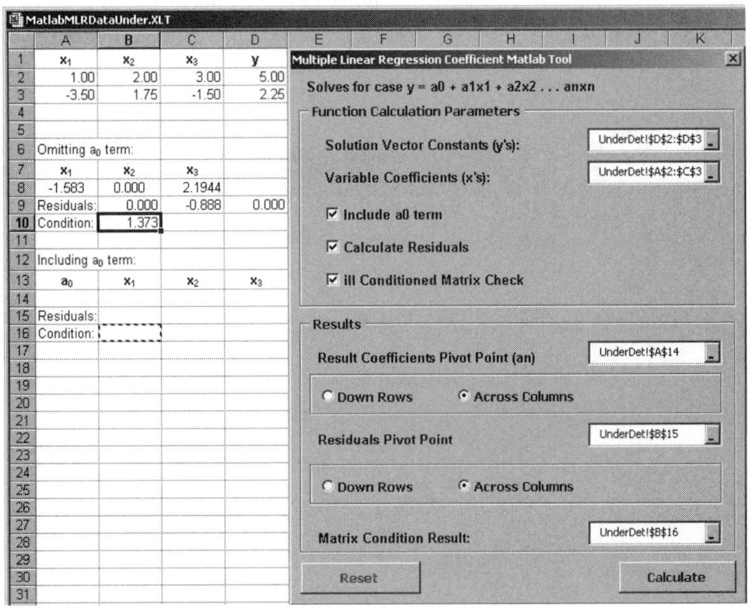

FIGURE 9.12 GUI set up to solve underdetermined system with an a_0 term.

	A	B	C	D
	MatlabMLRDataUnder.XLT			
	x_1	x_2	x_3	y
1	x_1	x_2	x_3	y
2	1.00	2.00	3.00	5.00
3	-3.50	1.75	-1.50	2.25
4				
5				
6	Omitting a_0 term:			
7	x_1	x_2	x_3	
8	-1.583	0.000	2.1944	
9	Residuals:	0.000	-0.888	0.000
10	Condition:	1.373		
11				
12	Including a_0 term:			
13	a_0	x_1	x_2	x_3
14	0.000	-1.583	0.000	2.1944
15	Residuals:	0.000	-0.888	0.000
16	Condition:	1.2693		
17				

FIGURE 9.13 Underdetermined case solutions.

The solutions to both cases for the underdetermined set of equations are shown in Figure 9.13. Some points are worth mentioning here. Underdetermined systems typically have an infinite number of solutions that can typically be described with a series of curves. Matlab will return only one solution to the underdetermined system, and no warning is given to the user to indicate that the returned solution is merely one of many.

Notice in this instance that Matlab randomly chose to assign x_2 to zero. It could have just as easily picked x_1 or x_3 to fix at zero, or any other constant for that matter. In an underdetermined case, arbitrary assignments must be made to some variables in order to converge on a solution for the other variables. Because this system of equations has 3 unknowns and 2 equations, only one of the variables to be solved for had to be arbitrarily assigned a value. If the system consisted of 4 unknowns and 2 equations, two of the variables would have to be fixed for convergence.

The last case for consideration is that of the consistent system, where the number of equations equals the number of unknowns. If the user loads the "MatlabMLRConsistent.xlt" sample file from the CD-ROM in location Chapter 9\Samples, the Worksheet template shown in Figure 9.14 will appear.

	A	B	C	D
	MatlabMLRConsistent.XLT			
	x_1	x_2	x_3	y
1	x_1	x_2	x_3	y
2	5	-2	3	10
3	3	-1	2	7
4	1	5	-3	2
5				
6				
7				
8				
9				
10	Omitting a_0 term:			
11	x_1	x_2	x_3	
12				
13	Residuals:			
14	Condition:			
15				
16	Including a_0 term:			
17	a_0	x_1	x_2	x_3
18				
19	Residuals:			
20	Condition:			
21				

FIGURE 9.14 Consistent system sample template.

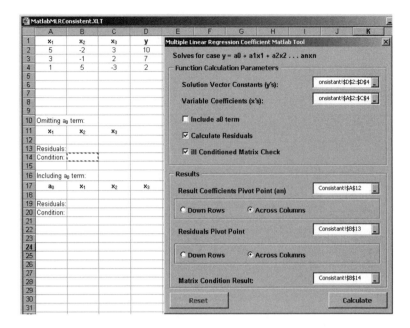

FIGURE 9.15 Solving the consistent system without a_0 for a perfect solution.

This system of equations can be solved for a perfect solution by many means, notably by substitution and Cramer's rule. Recall that the system of equations will be of the form: $y = a_0 + a_1 x_1 + a_2 x_2 + a_3 x_3$, where a_0 is an optional parameter for the model. In this instance, a perfect solution is $a_0 = 0$, $x_1 = 1$, $x_2 = 2$, $x_3 = 3$. Figure 9.15 illustrates the GUI setup to solve for a system solution without an a_0 term.

The system can also be solved by forcing a solution with an a_0 term as shown in Figure 9.16. The results of both solutions are shown in Figure 9.17.

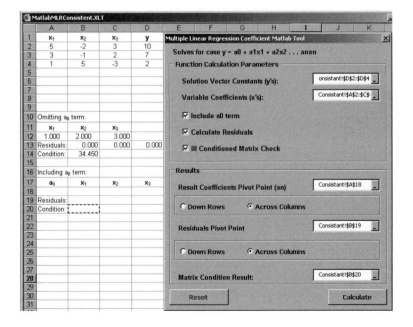

FIGURE 9.16 Solving the consistent system forcing the inclusion of an a_0 term.

MatlabMLRConsistent.XLT

	A	B	C	D	E
1	x_1	x_2	x_3	y	
2	5	-2	3	10	
3	3	-1	2	7	
4	1	5	-3	2	
5					
6					
7					
8					
9					
10	Omitting a_0 term:				
11	x_1	x_2	x_3		
12	1.000	2.000	3.000		
13	Residuals:	0.000	0.000	0.000	
14	Condition:	34.450			
15					
16	Including a_0 term:				
17	a_0	x_1	x_2	x_3	
18	2.700	1.300	-0.400	0	
19	Residuals:	0.000	-0.355	-0.178	0.000
20	Condition:	20.858			
21					

FIGURE 9.17 Comparison of consistent system solutions.

It is obvious upon inspection that the solution without the a_0 term is perfect, as the residuals are all 0, and indicate no deviation between the model and the data. The solution with the a_0 term, however, leaves more to be desired. The residuals show a deviation between the model and the actual data. Although the matrix in this solution is better conditioned than the matrix utilized in the former solution (20.858 vs. 34.450), the solution converged upon is less optimal than that of the solution with the a_0 term omitted. Thus, better conditioning is not always indicative of a superior solution set.

9.6 THE INNER WORKINGS OF THE MULTIPLE LINEAR REGRESSION EXAMPLE

The discussion so far has focused on the use of the MLR Matlab tool, but it is the mechanisms that the tool utilizes to work that will be of paramount interest to the reader. When the Calculate button is pressed, a series of actions take place that accomplish the following:

1. A Matlab ActiveX server is initiated.
2. The required data is extracted from the Worksheet.
3. The Extracted Data is formatted into Matlab-friendly strings.
4. The Formatted Data is sent to Matlab.
5. Commands are sent to Matlab to perform the required operations.
6. The results from the Matlab operations are returned to Excel.
7. The returned results from Matlab are parsed.
8. The parsed results are written to the designated area of the Worksheet.

It all begins with the code behind the calculate button:

```
Private Sub CalculateButton_Click()
'Save Settings to Windows Registry
```

```
'RefEdit Settings
SaveSetting "MLRCMatlab", "RefEdit", "YRefEdit",
frm9MLRC.YRefEdit.Text
SaveSetting "MLRCMatlab", "RefEdit", "XRefEdit",
frm9MLRC.XRefEdit.Text
SaveSetting "MLRCMatlab", "RefEdit", "RPPRefEdit",
frm9MLRC.RPPRefEdit.Text
SaveSetting "MLRCMatlab", "RefEdit", "APPRefEdit",
frm9MLRC.APPRefEdit.Text
SaveSetting "MLRCMatlab", "RefEdit", "IllCond",
frm9MLRC.RefEdit_Cond
Call RunMLRC
End Sub
```

Here, the settings to be utilized during the run are saved to the Windows Registry, where they will be recalled the next time the macro is executed. The RunMLRC subroutine is then called, which begins the execution of the just-described processes. The RunMLRC subroutine is quite extensive, so critical code segments will be listed and discussed in their order of appearance. The entire RunMLRC subroutine can be viewed on the included CD-ROM in the Chapter 9 module of the "ADAUsingExcel.xls" file.

This section of code sets up the Matlab ActiveX Server.

```
'Create the Matlab Object
Set MatLab = CreateObject("Matlab.Application")
This section of code builds the A matrix based on the location
of the x values specified in the RefEdit control.
'Get the address, or reference, from the RefEdit control.
Addr = frm9MLRC.XRefEdit.Value
'Set the SelRange Range object to the range specified in the
RefEdit control.
Set CoeffMatrix = Range(Addr)
'Create "A" Matrix Based on Cell Values
CoeffRng = CoeffMatrix.Address(, , xlR1C1)
Call SetRefEditMatrixRanges(CoeffRng, startrow, startcol,
endrow, endcol)
'Build the "a" matrix
Select Case frm9MLRC.a0CheckBox.Value
  Case True
    Amatrix = "a = [1 "
  Case False
    Amatrix = "a = ["
End Select
For row = startrow To endrow
    For Col = startcol To endcol
    If frm9MLRC.a0CheckBox.Value = True Then
        'Include a0 term, prefix a matrix with ones
        If Col <> endcol Then
```

```
      Amatrix = Amatrix &
      Workbooks(ActiveWorkbook.Name).Worksheets(ActiveSheet.Name).
      Cells(row, Col).Value & Chr(32)
      Else
      Amatrix = Amatrix &
      Workbooks(ActiveWorkbook.Name).Worksheets(ActiveSheet.Name).
      Cells(row, Col).Value & "; 1 "
      End If
      Else
      If Col <> endcol Then
      'Do not include a0 term, no prefix
      Amatrix = Amatrix &
      Workbooks(ActiveWorkbook.Name).Worksheets(ActiveSheet.Name).
      Cells(row, Col).Value & Chr(32)

      Else
      Amatrix = Amatrix &
      Workbooks(ActiveWorkbook.Name).Worksheets(ActiveSheet.Name).
      Cells(row, Col).Value & "; "

      End If
    End If
  Next Col
  Next row
  'Remove last ";" and add "];"
  Select Case frm9MLRC.a0CheckBox.Value
    Case True
      Amatrix = Mid(Amatrix, 1, Len(Amatrix) - 4) & "];"
    Case False
      Amatrix = Mid(Amatrix, 1, Len(Amatrix) - 2) & "];"
  End Select
```

The b matrix (in this case a column vector) is constructed in a similar manner.

```
  'Build "b" Matrix
  'Get the address, or reference, from the RefEdit control.
  Addr = frm9MLRC.YRefEdit.Value
  'Set the SelRange Range object to the range specified in the
  RefEdit control.
  Set SolMatrix = Range(Addr)
  SolRng = SolMatrix.Address(, , xlR1C1)
  Call SetMatrixRanges(SolRng, startrow, startcol, endrow, endcol)
  'Debug.Print startrow, startcol, endrow, endcol
  Bmatrix = "b = ["
  For row = startrow To endrow
    For Col = startcol To endcol
      If Col <> endcol Then
```

```
          Bmatrix = Bmatrix &
          Workbooks(ActiveWorkbook.Name).Worksheets(ActiveSheet.Name)
          .Cells(row, Col).Value & Chr(32)
            Else
          Bmatrix = Bmatrix &
          Workbooks(ActiveWorkbook.Name).Worksheets(ActiveSheet.Name).
          Cells(row, Col).Value & "; "
          End If
      Next Col
  Next row
  'Remove last ";" and add "];"
  Bmatrix = Mid(Bmatrix, 1, Len(Bmatrix) - 2) & "];"
```

Now that Matlab-"friendly" strings have been constructed, the command strings must be sent to the Matlab server. The following series of commands sends the strings to Matlab, which in turn assigns the values specified in the strings to the *a* and *b* matrices.

```
  'Execute Matlab Commands for Multiple Linear Regression
  SendMatlabCmd = MatLab.Execute(Amatrix)
  SendMatlabCmd = MatLab.Execute(Bmatrix)
```

With the matrices assigned, the system can easily be solved using the following series of commands. Here the commands are given to calculate the MLRC coefficients and the residuals. The results of the calculations are passed back to Excel in the form of a string. The string to capture the results of the calculations is set equal to the command executed. The returned string will contain all of the results separated by spaces (known as space delimiting). Unfortunately, the returned string will also contain other information such as matrix identification "x =," "r =," etc. The results must be separated out from the identification information. The first step in accomplishing this is to utilize the $\texttt{Split}$ function, which will tokenize the returned string using a delimiter specified by the user. The tokens will be returned into an array, in this case, MLRCArray() and ResidArray().

```
  'Calculate MLRC
  MLRC_Results = MatLab.Execute("x = a\b")
  MLRCArray() = Split(MLRC_Results, " ")
  'Calculate Residuals
  Residuals =  MatLab.Execute("r = (a*x-b)'")
  ResidArray() = Split(Residuals, " ")
```

The next section of code loops through each array of tokens in sequential order from the returned results and determines the start of valid data. This is done by means of a flag that is tripped (to True) when the valid results are about to begin. The flag is tripped to True when the "=" character is found within a token. Subsequent tokens are then taken as valid results, provided they are numeric. The rationale behind this algorithm is that results are returned in a format similar to this:

x =
   2.7000
   1.3000
 −0.4000
     0

The array is then resized and reindexed based on the number of True results found within the returned string.

```
'Now Extract "Real" Results from Split Generated Arrays
'For MLRCArray
n = 0
For index = LBound(MLRCArray, 1) To UBound(MLRCArray, 1)
  If trip = True And IsNumeric(MLRCArray(index)) = True Then
      'Write Results starting in Position 1 "Str(Val" -> Remove CR&LF
      MLRCArray(n) = str(Val(MLRCArray(index)))
      n = n + 1
  End If
  If InStr(1, MLRCArray(index), "=", vbTextCompare) > 0 Then
      'Start looking for results now!
      trip = True
  End If
Next index
'ReSize Array Based on True Number of Elements
ReDim Preserve MLRCArray(n - 1)
```

The array containing the results of the residuals is reindexed and resized in a similar fashion.

```
'For ResidArray
n = 0
For index = LBound(ResidArray, 1) To UBound(ResidArray, 1)
  If trip = True And IsNumeric(ResidArray(index)) = True Then
      'Write Results starting in Position 1 "Str(Val" -> Remove CR&LF
      ResidArray(n) = str(Val(ResidArray(index)))
      n = n + 1
End If
  If InStr(1, ResidArray(index), "=", vbTextCompare) > 0 Then
      'Start looking for results now!
      trip = True
  End If
Next index
'ReSize Array Based on True Number of Elements
ReDim Preserve ResidArray(n - 1)
```

Once the results have been parsed out of the returned strings from Matlab, it is time to write them into the Excel Worksheet at the specified pivot points.

```
'Write Out Results to Worksheet in Specified Position
'Write MLR Coefficient Results
'Get the address, or reference, from the RefEdit control.
Addr = frm9MLRC.APPRefEdit.Value
'Determine sheet to write MLRC to
For ii = 1 To Len(Addr)
```

```
   If Mid(Addr, ii, 1) = "!" Then
      resultsheet$ = Left$(Addr, ii - 1)
      resultsheet$ = ReformatSheetName (resultsheet$)
      Exit For
   End If
Next ii
'Set the ResultPos Range object to the range specified in the
RefEdit control.
Set ResultPos = Range(Addr)
'Convert to R1C1 format in string
ResultRng = ResultPos.Address(, , xlR1C1)
'Determine "Pivot Point of Results"
rrow = FindRow(ResultRng)
rcol = FindCol(ResultRng)
'Write Results out to Worksheet at Specified Location by row
or column

For Each Rslt In MLRCArray
  Workbooks(ActiveWorkbook.Name).Worksheets(resultsheet$).Cells
  (rrow, rcol).Value = Rslt
  Select Case frm9MLRC.ResultOBAcross.Value
    Case True
      rcol = rcol + 1
    Case False
      rrow = rrow + 1
    End Select
Next Rslt
```

If an ill-conditioned matrix check is to be performed, this section of code will implement the check and return the results.

```
'Check for ill conditioned matrix
If frm9MLRC.CheckBox_Ill = "True" Then
    IllCond = MatLab.Execute("cond(a)")
    IllCondArray = Split(IllCond, " ")
    n = 0
    For index = LBound(IllCondArray, 1) To UBound(IllCondArray, 1)
    If trip = True And IsNumeric(IllCondArray(index)) = True Then

      'Write Results starting in Position 1 "Str(Val" -> Remove CR&LF

      Addr = frm9MLRC.RefEdit_Cond
      Workbooks(ActiveWorkbook.Name).Worksheets(resultsheet$).Ra
      nge(Addr) = str(Val(IllCondArray(index)))

      Exit For
      End If
      If InStr(1, IllCondArray(index), "=", vbTextCompare) > 0 Then
```

```
        'Start looking for results now!
        trip = True
    End If
  Next index
End If
```

If the residuals are not to be calculated, this section of code will allow the macro to skip over processing the residual results.

```
'No Residuals to Calculate - Exit Sub
If frm9MLRC.ResidualCheckBox.Value = "False" Then GoTo
Quit_Now:
```

The final section of code terminates the Matlab ActiveX Server (very important as it frees up system resources), closes and unloads the GUI, and terminates the macro.

```
Quit_Now:
MatLab.Quit        ' Quit method to close the application
Set MatLab = Nothing      'this releases the reference.
frm9MLRC.Hide
Unload frm9MLRC
End Sub
```

9.7 INTERFACING EXCEL AND ORIGIN TO PERFORM MORE COMPLEX ANALYSES

Another very powerful computational tool is OriginLab's Origin. Although Matlab is a matrix- and vector-based calculation tool, Origin is a spreadsheet-based computational tool like Excel. Origin has the ability to open Excel spreadsheets in its native environment and, like Excel, has a macro language that allows automation. The similarities end there, however.

Unlike Excel, Origin has *presentation*-quality graphics. Origin's scripting language, termed Labtalk, is C-based, unlike Excel's scripting language that is a derivative of BASIC. As a matter of practicality, the learning curve to automate data analysis applications in Origin is several times more steep than learning how to automate data analysis applications in Excel. However, it is also equally true that Origin has much greater capabilities than Excel in terms of the complexity of the computational tools offered within the package, and the quality and flexibility of its graphics.

Origin's scripting abilities span across two domains. The first is Labtalk, the first and original macro language available from within Origin. Although Labtalk is an extremely robust language *modeled* on C, it is not as robust as a natively compiled language such as C. To help improve performance and increase the scripting capabilities, Origin C was developed. Origin C runs much faster than Labtalk but not quite as fast as a regular C compiler. In addition to regular C constructs, Origin C has unique commands specific for use in the Origin environment. Origin C can, in fact, utilize Labtalk commands within Origin C, and Origin C functions can be called from Labtalk within Origin. This interoperability provides a great deal of flexibility when it comes to constructing constructs and algorithms for automation purposes.

It is beyond the scope of this text to fully teach the reader all the intricacies of Labtalk and Origin C. However, to utilize the Origin computational engine, some knowledge of both languages is required. The text will delve into Labtalk and Origin C only to the extent necessary to produce routines that can be called by Excel to perform nonlinear curve fitting, filtering, or other higher-level analyses. For more information on Origin C and Labtalk syntax, the reader is encouraged to make use of the following freely available resources on the World Wide Web:

http://www.originlab.com/
http://g.webring.com/hub?ring=originwebring

Excel has two principal methods that can be utilized to harness the Origin computational engine. The first, DDE, will be covered briefly. Although DDE is no longer supported by Microsoft, it is still widely utilized. DDE remains extremely useful, especially when a method is required to perform a quick computation utilizing the Origin computational engine, and return only a number as opposed to returning a result and other ancillary information such as a graph. The second method requires utilizing an Automation Server. This approach utilizes the most current techniques and allows any client application (such as Microsoft Excel) to make calls to Origin, perform analyses, and return pertinent information such as fitted data, graphs, etc.

A typical Excel client application might involve some or all of the following actions:

1. Launching an instance of the Origin application.
2. Opening a previously created Origin template.
3. Sending data from an external source (the client) to Origin (the server) for analysis.
4. Sending Origin specific commands to analyze the data.
5. Requesting the processed results from Origin.
6. Returning images of graphs prepared in the Origin analysis.
7. Saving the results generated by the Origin project.
8. Closing the instance of the Origin application.

This list encompasses a wide range of processes and actions and is fairly comprehensive in terms of application to application interaction. In order to effectively demonstrate using DDE and Automation Server Technologies with Excel, a fairly complicated example will need to be constructed for the reader. The reader will be shown how to calculate pKa values for one and two pKa systems utilizing the Origin computational engine.

A pKa (also termed the Dissociation Constant), is a measure of the strength of an acid or a base. A pKa determination shows the charge on a molecule at any given pH (Figure 9.18). A pKa value is often useful in predicting if a specific compound does or does not have drug-like properties. Compounds screened in drug discovery settings typically have pKa values from 3 to 12.

To accomplish pKa curve fitting in Origin for Excel, a number of tasks must be completed. They are as follows in order of increasing complexity:

1. A Origin Worksheet template must be constructed.
2. pKa Curve Fitting functions must be defined in Origin.

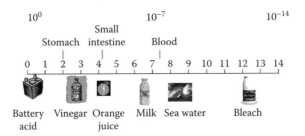

FIGURE 9.18 Relative pH scale indicators.

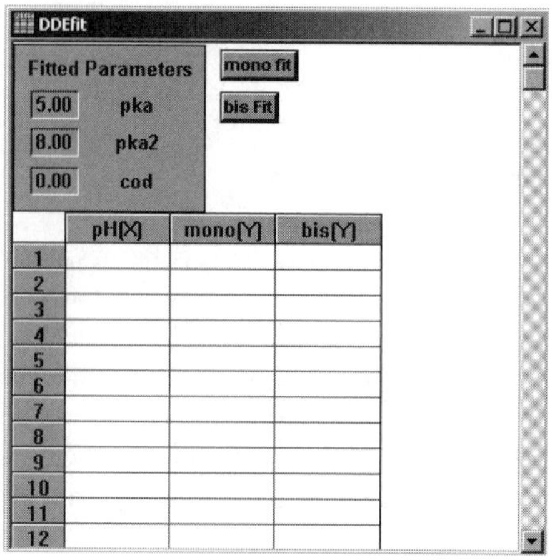

FIGURE 9.19 The Origin Worksheet utilized for pKa analysis.

3. Labtalk Scripts must be written to run the NLSF Fitter.
4. Origin C Scripts must be written to perform the curve fitting.

For this sample application, an Origin Worksheet Template has been created for the reader. It is located on the CD-ROM in the directory: (Chapter 9\Samples\Origin\DDEFit.ogw). The Worksheet is shown in Figure 9.19. Notice that it contains three columns. The first column is to hold the pH buffer values (the independent variables). The second and third columns hold the y-axis (dependent variables) for a single (monotonic) and a dual (bis) pKa system, respectively.

The Worksheet also contains two buttons labeled "mono fit" and "bis fit," respectively. Both of these buttons contain Labtalk scripts that will execute when pressed. It is possible to remotely press these buttons using automation server or DDE methods, and a Visual Basic script will be written in Excel VBA to do so. Both Labtalk scripts initialize the curve fitter and perform a curve fit on the specified data set. They return the results to the radio box in the upper left-hand corner of the Worksheet. These scripts each call unique curve fitting functions that must be predefined. The single (mono) pKa equation is

$$y = (10^{\wedge}pKa*A1+10^{\wedge}pH*A2)/(10^{\wedge}pKa+10^{\wedge}pH) \qquad (9.1)$$

parameters: y, pKa, A1, A2, pH

The dual (bis) pKa system is a more complicated system:

$$y = (ACAT*10^{\wedge}(-pH)*10^{\wedge}(-pH)+ANEUT*10^{\wedge}(-PK1)*10^{\wedge}(-pH)+AANION*10^{\wedge}(-PK1)*10^{\wedge}$$
$$(-PK2))/(10^{\wedge}(-pH)*10^{\wedge}(-pH)+10^{\wedge}(-pH)*10^{\wedge}(-PK1)+10^{\wedge}(-PK1)*10^{\wedge}(-PK2)) \qquad (9.2)$$

parameters: y, PK1, PK2, ACAT, ANEUT, AANION, pH

Origin Functions are stored in *.fdf files, which stands for function definition files. The functions for Equation 9.1 and Equation 9.2 are stored in the files monopka.fdf and bispka.fdf, respectively, which are contained on the CD-ROM in \Chapter 9\Samples\Origin. To utilize these fitting functions, they must be placed in the \FitFunc directory under the Origin program directory structure. The files must also be added to the nlsf.ini file for Origin to know they exist. This can be done by adding the following text to the bottom of the nlsf.ini file:

```
[pKa]
monopKa=monopKa
```

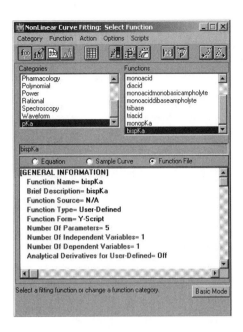

FIGURE 9.20 NLSF curve-fitting module with pKa functions.

bispKa=bispKa
Default Function=monopKa

Adding the preceding text to the nlsf.ini file will create a custom function category called pKa that contains both the mono and bis pKa curve-fitting functions.
(Note: multiple nlsf.ini files may exist under the Origin directory structure. Modify the one residing in the subfolder that is named after the username of the person logged into the Machine and will be using Origin.)

Figure 9.20 shows the nlsf curve-fitting GUI with a custom pKa category added and its respective fitting functions.

With the functions now defined and in place, the scripts that initialize and run the curve-fitting functions can be explained. The scripts associated with the buttons can be accessed by selecting Edit->Button Edit Mode from the menu. Right-clicking on the button (*which will be displayed as text in the button edit mode*) and selecting Label Control displays the script associated with the button. The mono pKa curve-fitting Labtalk script (associated with the mono fit button) follows:

```
// DDE Mono pKa Fitting Script
// Save the name of the current Worksheet
%W = %H;
dependataset=2;
// Create a plot window named "monopKa"
GetEnumWin monopKa;
// Select the Custom Fitting Function
nlsf.func$=monopKa;
// Initialize the nlsf engine
nlsf.init();
```

```
// initialize parameter values
A1=%(%W,dependataset,1); //.p1
A2=%(%W,dependataset,8); //.p2
pka = 5;                      //.p3
// Dependant Data (Y-Axis) to be Fit
nlsf.y$ = %(%W,dependataset);
// Independent Data (X-Axis) to be Fit
nlsf.x$ = %(%W,1);
// Total Number of Points for Fitted Curve
nlsf.xpoints = 60;
// Improve accuracy of fit
nlsf.tolerance = 1e-5;
// Perform Curve Fitting
nlsf.fit(20);
// Add the original Data to the plot
layer -i %(%W,2);
// Rescale X & Y Axis to show All data
layer -at;
// Update "FitPams" Status Object
%W!FitPams.v1 = $(nlsf.p3,.2);
%W!FitPams.v2$ = "";
%W!FitPams.v3 = $(nlsf.cod,.2);
// Rename Axis Labels
label -xb pH Buffer Values;
label -yl Absorbance;
// Close the plot Window
win -c %H;
// Delete NLSF Fitset Hidden Worksheet
win -c NLSF1;
// Clean Up all internal Objects Used in nlsf
nlsf.cleanupfitdata();
```

The preceding script, in essence, performs the following functions. To fully understand the script, an understanding of the Labtalk commands nlsf, win, label, layer, and the Labtalk reserved variables is essential. This information is readily available online and in the Origin Labtalk Manual. The script creates a plot window, chooses the mono pKa fitting function, and initializes the NLSF curve-fitting engine. Reasonable starting values for the undefined parameters A1, A2, and pKa are then chosen and set. The X (pH) and Y (absorbance) data to be fitted are then chosen (in this case, the Y-data are utilized from column 2), and a 20-iteration curve fit is performed. The fitting parameters are then assigned to the GUI in the Worksheet showing the pKa and the cod or r^2 value (an indication of the goodness of fit).

Finally, the windows are closed, and the NLSF fitting module is cleaned up by erasing old fitting parameters and freeing up allocated memory. A similar Labtalk script is created for the "bis fit" button:

```
// DDE Bis pKa Fitting Script
// Save the name of the current Worksheet
```

```
%W = %H;
// Set Dataset (Column) for y data to be fit
dependataset=3;
// Create a plot window named "bispKa"
GetEnumWin bispKa;
// Select the Custom Fitting Function
nlsf.func$=bispKa;
// Initialize the nlsf engine
nlsf.init();
// initialize parameter values
Acat=%(%W,dependataset,1); //.p1
Aneut=%(%W,dependataset,4); //.p2
Aanion=%(%W,dependataset,8); //.p3
pk1 = 5;            //.p4
pk2 = 8;            //.p5
// Do Not Allow Acat or Aanion to Vary!
nlsf.v1=0;
nlsf.v3=0;
// Dependant Data (Y-Axis) to be Fit
nlsf.y$ = %(%W,dependataset);
// Independent Data (X-Axis) to be Fit
nlsf.x$ = %(%W,1);
// Total Number of Points for Fitted Curve
nlsf.xpoints = 60;
// Improve Accuracy of Fit
nlsf.tolerance = 1e-5;
// Perform Curve Fitting
nlsf.fit(20);
// Add the original Data to the plot
layer -i %(%W,dependataset);
// Rescale X & Y Axis to show All data
rescale;
%W!FitPams.v1 = $(nlsf.p4,.2);
%W!FitPams.v2$ = $(nlsf.p5,.2);
%W!FitPams.v3 = $(nlsf.cod,.2);
// Rename Axis Labels
label -xb pH Buffer Values;
label -yl Absorbance;
// Close the plot Window
win -c %H;
// Delete NLSF Fitset Hidden Worksheet
win -c NLSF1;
// Clean Up all internal Objects Used in nlsf
nlsf.cleanupfitdata();
```

TABLE 9.2
Sample pKa Data

pH	mono01	mono02	bis01	bis02
3	0.07313	0.69146	0.07313	0.54821
4	0.07347	0.68217	0.07347	0.26453
5	0.07641	0.58883	0.13971	0.10114
6	0.1097	0.29511	0.25001	0.07217
7	0.29511	0.1097	0.3	0.07046
8	0.58883	0.07641	0.30001	0.08004
9	0.68217	0.07347	0.40787	0.14483
10	0.69146	0.07313	0.60013	0.30012

The preceding script is nearly identical to the mono pKa curve-fitting script given earlier, with a few important exceptions. The script obviously utilizes a different fitting function for a dual pKa system "bispka." Reasonable starting values must be assigned for more parameters: PK1, PK2, ACAT, ANEUT, AANION. In the bis pKa case, it is necessary to fix the values for ACAT and AANION at the extremes. Not allowing them to vary is necessary for a successful convergence and fit on certain data sets. The Y (absorbance) data to be fitted is then chosen (in this case, the Y data is utilized from column 3), and a 20-iteration curve fit is performed. The fitting parameters are then assigned to the GUI above the Worksheet showing the 1st pKa, 2nd pKa (if it exists), and the cod or r^2 value.

As in the previous instance, the windows are closed, and the NLSF fitting module is cleaned up by erasing old-fitting parameters and freeing up allocated memory.

To run the pKa curve-fitting example will require some sample data that is shown in Table 9.2. The first column contains the pH buffer data (X axis) followed by absorbance data for two single pKa and two bis pKa systems (Y axis).

Graphically, the pKa value is equal to the pH at the midpoint of a titration, where the titration is represented on a graph as a slope. In order to have two or more pKas, a data set must contain as many unique slopes as pKas. To provide the reader an idea as to what reasonable solutions are graphically, solution graphs of some of the example pKa data sets will be shown.

Looking at Figure 9.21, it is clear there is a single slope, and the midpoint of the sloping region is around 7.3. The crosshairs on the graph show the location of the pKa whose exact value is 7.2628. This example represents a classic well-defined pKa system and is the solution to the mono01 dataset (column B). The mono02 data set is the mirror image of the mono01 data set, and its solution is graphically quite similar.

Figure 9.22 shows a more complicated pKa system. Here, it could be argued that the data set in fact contains four distinct slopes over the intervals (3–4.5, 4.5–6.0, 6.0–8.0, 8.0–10.0). Clearly, however, it is the slopes between the intervals of (4.5–6.0, 8.0–10.0) that have the steepest slopes. Thus, the pKas are calculated to be pKa1 ≈ 5.35, pKa2 ≈ 9.37. Looking at the graph in Figure 9.22, the returned results seem reasonable.

Figure 9.23 shows another two-pKa system quite different from that of Figure 9.22. Here, the function is a U shape, with each side of the U having a distinct slope and thus a unique pKa. The calculated pKas are pKa1 ≈ 3.7, pKa2 ≈ 9.5. Looking at the graph in Figure 9.23, this seems reasonable.

The natural progression of this exercise is to find a way to have Excel harness the power of the Origin computational engine and utilize the Origin Worksheet templates and their associated Labtalk scripts.

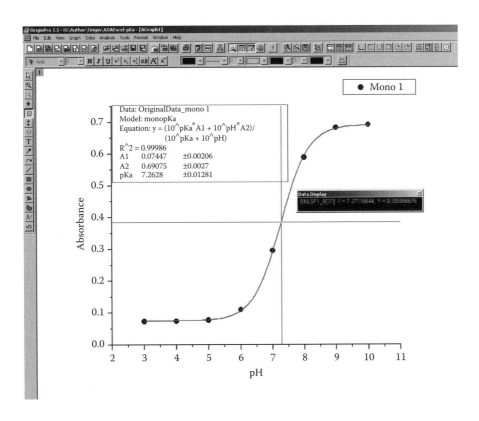

FIGURE 9.21 A single pKa system.

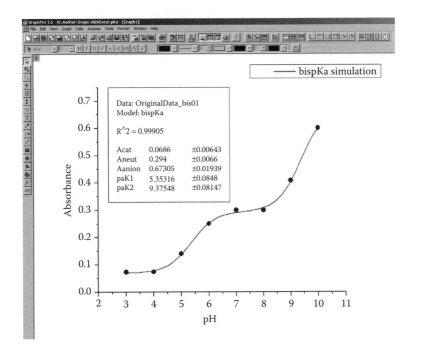

FIGURE 9.22 A two-pKa system (bis01).

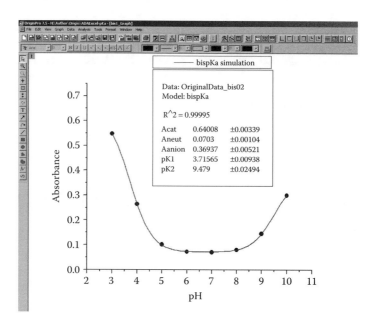

FIGURE 9.23 A two-pKa system (bis02).

9.8 EXCEL-TO-ORIGIN DDE EXAMPLE

Depending upon the requirements of the application, the user may wish to simply return raw computed values to Excel, or in more complex cases, return not just the raw values to Excel but ancillary material as well, such as graphs and calculation-specific information.

Dynamic Data Exchange (DDE) is the most straightforward way to utilize a computational engine to simply return results and their associated parameters. To demonstrate DDE use, a sample application will be constructed that allows the user to compute pKa values for a specified data set using the Origin computational engine.

Figure 9.24 shows the sample data sets utilized for this example that are contained in the Workbook "pKa_Sample_Data.xls," which can be found on the CD-ROM under \Chapter 9\ Samples\Origin. This Workbook contains two fitting examples for both single and dual pKas. What is desired is an application that will allow the user to pick the data set of interest to fit, trigger the Origin computational engine to perform the fit, and then retrieve the desired calculated parameters from Origin. Figure 9.25 shows the GUI that allows the user to specify the data set and the fitting function to be utilized on the data set (mono or bis). The user may choose to fit a single specified data set or all the given data sets, depending upon which button on the GUI is pressed. The internal constructs that allow the sample to function will now be examined.

First, some defined constants and variables declared at the beginning of Chapter 9 module are examined.

```
Dim Origin_TaskID As Variant
Dim Channel As Variant
Public pKaDataBook As String
'These paths are Machine Specific and Will Need to be Changed!
Public Const origin_path$ = "C:\Program
Files\OriginLab\OriginPro75\Origin75.exe"
```

	A	B	C	D	E
1	pH	mono01	mono02	bis01	bis02
2	3	0.07313	0.69146	0.07313	0.54821
3	4	0.07347	0.68217	0.07347	0.26453
4	5	0.07641	0.58883	0.13971	0.10114
5	6	0.1097	0.29511	0.25001	0.07217
6	7	0.29511	0.1097	0.3	0.07046
7	8	0.58883	0.07641	0.30001	0.08004
8	9	0.68217	0.07347	0.40787	0.14483
9	10	0.69146	0.07313	0.60013	0.30012
10					
11					
12					
13	pKa1				
14	pKa2				
15	cod				

FIGURE 9.24 Sample pKa data sets.

```
Public Const origin_sheet_path$ =
"C:\Author\Books\Excel\Chapter 9\Samples\Origin\DDEFit.ogw"
Public Const excel_sheet_path$ = "C:\Author\Books\Excel\Chapter
9\Samples\Origin\pKa_Sample_Data.xls"
```

The task ID is a unique number that identifies a running program in Windows. When a program is started in VBA by utilizing the `Shell` command, a unique task ID is assigned to the program VBA started. This task ID allows VBA to communicate with the program it started.

When a DDE conversation is initiated between two applications, the communications must take place over a channel, much like people who choose to communicate over small two-way radios must be on the same channel for communication to occur. When the DDE conversation is initiated, the channel for communication to take place over is returned. The `pKaDataBook` is the Workbook that will hold the pKa data and the fitted results. The declared constants are simply the paths to Origin, the `DDEFit` Origin Worksheet, and the Excel sample data Workbook. These constants will need to be changed to the paths specific to the machine they are running on. The following subroutine is run when the following is selected from the Menu ADA->Chapter 9->Origin DDE Sample.

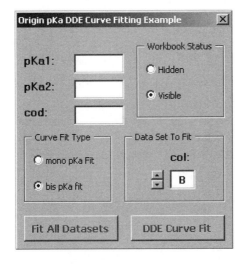

FIGURE 9.25 GUI for DDE fitting example.

```
Sub Origin_DDE_Sample()
 Application.DisplayAlerts = False
 pKaDataBook = Open_Template(excel_sheet_path$,
 "c:\temp\junk.xls")
 frm9OrigDDE.Show
End Sub
```

This subroutine suppresses any alerts Excel would generate, opens and saves the Excel Workbook with the pKa data to a different name "c:\temp\junk.xls" (to prevent corruption to the original Workbook), and opens the GUI shown in Figure 9.25.

The user must now select the type of function for fitting (mono or bis), a data set for fitting (if only fitting one data set), and press either the "Fit All Datasets" button to fit all data sets, or the "DDE Curve Fit" to fit a single data set. In the case of fitting a single data set, the following subroutine will be executed:

```
Sub Run_Sample()
Call Start_Origin(origin_path$, origin_sheet_path$,
vbMinimizedNoFocus)
Call Initiate_Conversation
Call Send_pH_Data
Call Send_Fitting_Data
Call Set_Initial_Parameters
Call Click_Button
Call Get_Parameters
Call Terminate_Conversation
End Sub
```

The Start_Origin subroutine opens up Origin using the Shell command, stores the task ID of the Origin process it started, and returns the operating system's focus back to Excel using the AppActivate command.

```
Sub Start_Origin(apppath$, projpath$, winstyle)
'Starts Origin Remotely, Stores its Task ID to Origin_TaskID
'apppath$ = Full path to Origin75.exe
'projpath$ = Full path of Project to Open when Origin Starts
(Optional)
'winstyle% - Window Focus and Style of Opened Object (Origin)
Dim pathname$
'A space must be placed between apppath$ and projpath$ and
'projectpath$ must be in "" (Chr$(34) = ")
pathname$ = apppath$ + " " + Chr(34) + projpath$ + Chr(34)
'WD98: WindowStyle Argument of Shell Function Ignored (Q178328)
'If the pathname argument is set to a document instead of a program,
'the windowstyle argument will have no effect.
Origin_TaskID = Shell(pathname$, winstyle)
'Since Shell windowstyle is ignored
'Use AppActivate to set focus back to Excel
AppActivate "Microsoft Excel"
End Sub
```

Just because Origin has been started does not mean that Excel can automatically communicate with it. The next subroutine establishes and stores a communication channel between Excel and Origin.

```
Sub Initiate_Conversation()
'This will initiate a DDE "Conversation" Channel from Excel
to Origin

  Channel = DDEInitiate("Origin", "Variable")
End Sub
```

The independent data set (pH–X-axis) will always be the same no matter which data set is to be fit. The pH data must be sent from Excel to the Origin Worksheet template. The following subroutine accomplishes this. Both subroutines that send fitting data to Origin utilize the Send_Data subroutine to actually transmit the data.

```
Sub Send_pH_Data()
'Send pH data to Origin - Excel Column 1
Dim ii As Integer
For ii = 1 To 8
  Call Send_Data ("DDEfit", "pH", (ii), ActiveSheet.Cells(ii +
  1, 1).value)

Next ii
End Sub
```

The dependent data set (absorbance–Y-axis) must also be sent from Excel to the proper column on the Origin Worksheet template. The data should be written to column 2 (named "mono") for a mono pKa fit, and column 3 (named "bis") for a bis pKa fit.

```
Sub Send_Fitting_Data()
'Send appropriate pKa data to Origin
Dim ii As Integer, fittype As String
'Set Fit Type
If frm9OrigDDE!monoOptionButton.value = True Then
    fittype = "mono"
    Else
    fittype = "bis"
End If
For ii = 1 To 8
    Select Case Asc(frm9OrigDDE.TextBox_Array) - 64
      Case 2
        Call Send_Data("DDEfit", fittype, (ii),
        ActiveSheet.Cells(ii + 1, 2).value)

      Case 3
        Call Send_Data("DDEfit", fittype, (ii),
        ActiveSheet.Cells(ii + 1, 3).value)

      Case 4
        Call Send_Data("DDEfit", fittype, (ii),
        ActiveSheet.Cells(ii + 1, 4).value)
```

```
      Case 5
        Call Send_Data ("DDEfit", fittype, (ii),
        ActiveSheet.Cells(ii + 1, 5).value)
    End Select
Next ii
End Sub
```

The next subroutine transmits a discrete piece of data from Excel to a particular Worksheet, column, and row location within Origin.

```
Sub Send_Data (Worksheet$, column$, row&, data!)
'Send Data From Excel to a Specific Origin Worksheet/Cell
Dim SendString$
SendString$ = Worksheet$ & "_" & column$ & "[" &
Trim$(str(row&)) & "]=" & Trim$(str(data!)) & ";"
DDEExecute Channel, SendString$
End Sub
```

Convergence will not be possible without the selection of reasonable starting parameters. The next subroutine picks reasonable starting points for the various parameters based on the chosen data set and whether the data set is to be fit to a mono or bis pKa function. This subroutine invokes the SetParameter subroutine to actually assign the values for the starting points of the parameters.

```
Sub Set_Initial_Parameters()
'Define Initial Parameters Prior to Fit
Select Case frm9OrigDDE!monoOptionButton.value
  Case True
    Call SetParameter(1, ActiveSheet.Cells(2,
    (Asc(frm9OrigDDE.TextBox_Array) - 64)).value, True)

    Call SetParameter(2, ActiveSheet.Cells(9,
    (Asc(frm9OrigDDE.TextBox_Array) - 64)).value, True)

    Call SetParameter(3, 7, True)
  Case False
    Call SetParameter(1, ActiveSheet.Cells(2,
    (Asc(frm9OrigDDE.TextBox_Array) - 64)).value, True)

    Call SetParameter(2, ActiveSheet.Cells(6,
    (Asc(frm9OrigDDE.TextBox_Array) - 64)).value, True)

    Call SetParameter(3, ActiveSheet.Cells(9,
    (Asc(frm9OrigDDE.TextBox_Array) - 64)).value, True)

    Call SetParameter(4, 3, True)
    Call SetParameter (5, 7, True)
End Select
End Sub
```

This subroutine will set the nth parameter (or nlsf.pn) to a specific value. The subroutine also specifies if the parameter is allowed to vary during the regression process.

```
Sub SetParameter(nlsf_p_number, param_value, vary As Boolean)
'Set an NLSF Fitting Parameter
```

```
Dim SendString$, varyval As Integer
If vary = True Then varyval = 1 Else varyval = 0
'Set the Value for the Given Fitting Parameter
SendString$ = "nlsf.p" & Trim(str(nlsf_p_number)) & "=" &
Trim(str(param_value)) & ";"
DDEExecute Channel, SendString$
'Allow Parameter to Vary in Fitting Session?
SendString$ = "nlsf.v" & Trim(str(nlsf_p_number)) & "=" &
Trim(str(varyval)) & ";"

DDEExecute Channel, SendString$
End Sub
```

Once all the data have been written to the proper areas of the Origin Worksheet and the proper initial parameters have been set, all that is left to do is remotely "push" the button on the Origin Worksheet from Excel. The next subroutine accomplishes this.

```
Sub Click_Button()
Select Case frm9OrigDDE!monoOptionButton.value
  Case True
    '"PUSH" the mono pka fit button on the Worksheet using
    Labtalk Code

    DDEExecute Channel, "DDEfit!Text.run();"
  Case False
    '"PUSH" the bis pka fit button on the Worksheet using
    Labtalk Code
    DDEExecute Channel, "DDEfit!Text1.run();"
End Select
End Sub
```

Once the regression process has completed, the fitted parameters must be returned to the GUI in Excel. The following subroutine accomplishes this and utilizes the RequestData function to return the parameters calculated in Origin.

```
Sub Get_Parameters()
'Erase any pre-existing parameters
Workbooks(pKaDataBook).Worksheets(1).Cells(13,
(Asc(frm9OrigDDE.TextBox_Array) - 64)).value = ""
Workbooks(pKaDataBook).Worksheets(1).Cells(14,
(Asc(frm9OrigDDE.TextBox_Array) - 64)).value = ""
Workbooks(pKaDataBook).Worksheets(1).Cells(15,
(Asc(frm9OrigDDE.TextBox_Array) - 64)).value = ""

frm9OrigDDE!pKa1TextBox.Text = ""
frm9OrigDDE!pKa2TextBox.Text = ""
frm9OrigDDE!codTextBox.Text = ""
'Fetch Most Current Curve Fit Parameters - Write to Worksheet
& Form (GUI)
Select Case frm9OrigDDE!monoOptionButton.value
```

```
Case True
  Workbooks(pKaDataBook).Worksheets(1).Cells(13,
  (Asc(frm9OrigDDE.TextBox_Array) - 64)).value =
  RequestData(Channel, "nlsf.p3")

  Workbooks(pKaDataBook).Worksheets(1).Cells(15,
  (Asc(frm9OrigDDE.TextBox_Array) - 64)).value =
  RequestData(Channel, "nlsf.cod")

  frm9OrigDDE!pKa1TextBox.Text = RequestData(Channel,
  "nlsf.p3")

  frm9OrigDDE!codTextBox.Text = RequestData(Channel, "nlsf.cod")

Case False
  Workbooks(pKaDataBook).Worksheets(1).Cells(13,
  (Asc(frm9OrigDDE.TextBox_Array) - 64)).value =
  RequestData(Channel, "nlsf.p4")

  Workbooks(pKaDataBook).Worksheets(1).Cells(14,
  (Asc(frm9OrigDDE.TextBox_Array) - 64)).value =
  RequestData(Channel, "nlsf.p5")

  Workbooks(pKaDataBook).Worksheets(1).Cells(15,
  (Asc(frm9OrigDDE.TextBox_Array) - 64)).value = RequestData
  (Channel, "nlsf.cod")
  frm9OrigDDE!pKa1TextBox.Text = RequestData(Channel, "nlsf.p4")
  frm9OrigDDE!pKa2TextBox.Text = RequestData(Channel, "nlsf.p5")
  frm9OrigDDE!codTextBox.Text = RequestData(Channel, "nlsf.cod")
End Select
'Fetch Most Current Curve Fit Parameters - Write to Form (GUI)
End Sub
```

The next function will request the parameter specified. Notice that the requested parameter must be treated as an array even if only a single parameter has been requested. A stronger form of typecasting in VBA requires that all requested parameters be (1) treated as an array and (2) cast as type Variant.

```
Function RequestData(Channel, param$) As Single
Dim returnList As Variant, i As Integer
'The stronger typecasting requires the returnList in DDE to be
'(1) cast as Variant (2) treated as array EVEN IF ONLY
REQUESTING A SINGLE ELEMENT
returnList = Application.DDERequest(Channel, param$)
For i = LBound(returnList) To UBound(returnList)
    RequestData = Val(returnList(i))
Next i
End Function
```

Once Origin has finished the computations, and Excel has successfully requested the results from the Origin computational engine, the DDE link between Excel and Origin should be closed. The next subroutine terminates the DDE link at *both* ends. First, a DDE command is executed that sends a Labtalk command to Origin to shut down its DDE connections, and then a command is

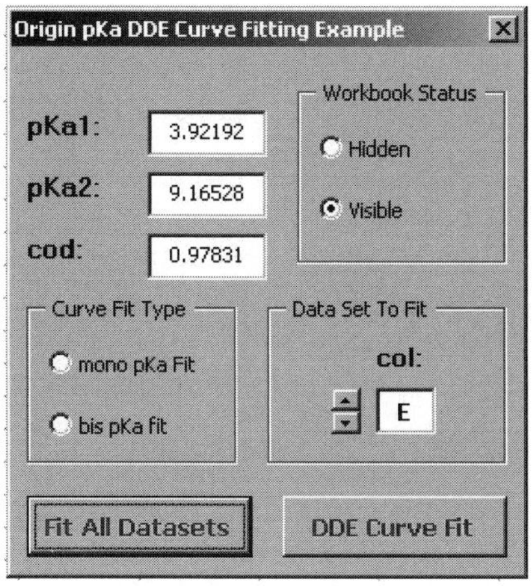

FIGURE 9.26 Origin DDE pKa curve-fitting example GUI.

sent to close Origin. A final VBA command is given to terminate the DDE Connection from the Excel side.

```
Sub Terminate_Conversation()
'Terminate the Excel-Origin DDE "Conversation" Link
'Exit Without Changing
DDEExecute Channel, "doc -s;dde -q 1;exit;"
DDETerminate Channel
End Sub
```

A working example of utilizing the Origin computational engine to compute pKa values can be run by selecting ADA->Chapter 9->Origin DDE Example from the menu bar. Doing so will bring up the GUI shown in Figure 9.26.

Pressing the DDE Curve Fit button when the bis pKa fit is selected upon column E will return the following results (pKa$_1$ ≈ 3.92, pKa$_2$ ≈ 9.16) shown in Figure 9.26. The sample is also capable of fitting all the data sets shown in Figure 9.24 by pressing the "Fit All Datasets" button. Doing so will cause Excel to utilize Origin to automatically curve fit all four data sets utilizing the proper functions. The results of the automated fitting sessions are shown in Figure 9.27.

The most important design component of controlling multiple curve-fitting sessions (or multiple calls to any third-party computational engine) is the use of the DoEvents function to allow each successive calculations to complete prior to proceeding to the next calculation. Use of the DoEvents function call is shown in the snippet of code called by the "Fit All Datasets" button.

```
Private Sub Button_All_Click()
'Fit all Datasets
Call Start_Origin(origin_path$, origin_sheet_path$,
vbMinimizedNoFocus)
Call Initiate_Conversation
For ii = 66 To 69
 'Set Dataset for Fitting
 TextBox_Array.Text = Chr(ii)
 'Set Fitting Type
```

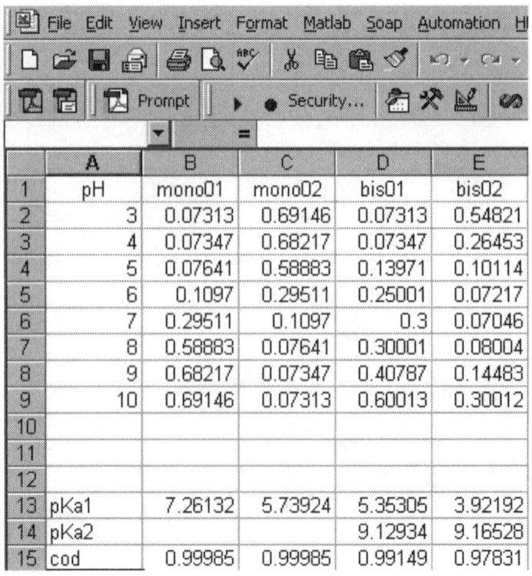

	A	B	C	D	E
1	pH	mono01	mono02	bis01	bis02
2	3	0.07313	0.69146	0.07313	0.54821
3	4	0.07347	0.68217	0.07347	0.26453
4	5	0.07641	0.58883	0.13971	0.10114
5	6	0.1097	0.29511	0.25001	0.07217
6	7	0.29511	0.1097	0.3	0.07046
7	8	0.58883	0.07641	0.30001	0.08004
8	9	0.68217	0.07347	0.40787	0.14483
9	10	0.69146	0.07313	0.60013	0.30012
10					
11					
12					
13	pKa1	7.26132	5.73924	5.35305	3.92192
14	pKa2			9.12934	9.16528
15	cod	0.99985	0.99985	0.99149	0.97831

FIGURE 9.27 Results of automated curve-fitting session using Origin.

```
    If ii > 67 Then
        monoOptionButton.value = False
        Else
        monoOptionButton.value = True
         End If
         Call Multi_Run
         DoEvents
   Next ii
   Call Terminate_Conversation
   End Sub
```

Although DDE is a somewhat dated technology, it continues to be utilized by many software applications. In the instance where an application has not been updated to support Microsoft COM, DDE may be the only method to enable external software components to communicate with Excel. DDE is a tried-and-true system that has nearly all of the bugs hashed out of it and continues to be widely utilized, which necessitates its coverage here.

9.9 INTERFACING EXCEL AND ORIGIN USING COM (COMPONENT OBJECT MODEL)

Microsoft COM technology in the Microsoft Windows family of operating systems enables differing software components to communicate with each other in a Windows environment. COM is used by developers to create reusable software components, link components together to build applications, and take advantage of Windows services. The family of COM technologies includes COM+, Distributed COM (DCOM), and ActiveX Controls. COM is the most current method that can be employed to link components from differing applications together to build a more robust application than either component could provide by itself working alone. In this instance, an example will be created to link Excel and Origin together using COM technologies to create a robust pKa curve-fitting application.

In building such an example, it would be desirable for the application to have certain built-in functionalities such as

1. The ability to launch a new instance of Origin or utilize a preexisting instance of Origin to perform the calculations.
2. The ability to select the Equation type to be fit (mono or bis) by means of an option button.
3. A means of selecting the data set to be fit. In this case, a drop-down box is utilized.
4. A button to trigger the fitting of the data set in Origin.
5. A plot template that provides a consistent look for all the analyzed data.
6. Columns that contain both the raw data set and the fitted data set.
7. A means of allowing the various fitting parameters to be fixed or to vary.

The difference in terms of functionality between the COM and DDE sample applications shown is that the COM application takes the analysis portion a step further. Not only are results calculated and returned, but the user can pick what parameters are allowed to vary during the fitting process. *In addition, a graph is returned to Excel that contains the raw data and the fitted curve.* Figure 9.28 shows the completed sample application.

Prior to utilizing COM commands, a COM object must be declared that refers to the Origin application. The easiest way to do this is to declare the COM application object at either the Module level or within the "ThisWorkbook" object. This sample uses the latter and places the following declaration in the "ThisWorkbook" object.

```
'Declare Origin Application as public object
Public ObjOrigin As Origin.Application
```

With the Origin application object declared, the application object variable can now be utilized to send specific commands to Origin via Excel. The "Connect to Origin" button's code utilizes the application object variable to start or reference the proper instance of Origin.

```
Private Sub Button_Start_Click()
    'Check if user wants to launch new instance or try connecting
    to an existing instance
    If (OptionButton_New.Value) Then
        On Error GoTo ErrMsg1
        'Launch a new instance of Origin and connect to it
        Set ThisWorkbook.ObjOrigin = GetObject("",
        "Origin.Application")

    Else
        On Error GoTo ErrMsg1
        'Connect to an existing Instance of Origin
        Set ThisWorkbook.ObjOrigin = GetObject("",
        "Origin.ApplicationSI")
    End If

    'Successfully connected to Origin
    'Show/hide appropriate controls
    CheckBox1.Visible = True
    'Button_Start.Visible = False
    Button_End.Visible = True
```

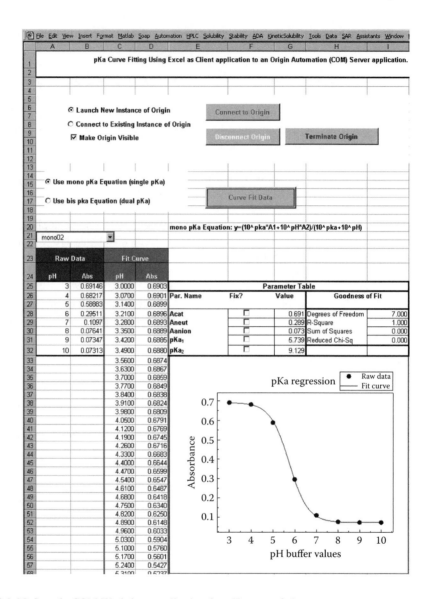

FIGURE 9.28 Sample COM Worksheet application for pKa curve fitting.

```
Button_Terminate.Visible = True
Button_Fit.Visible = True

   'Make Origin visible to begin with
   ThisWorkbook.ObjOrigin.Execute ("doc -mc 1;")
    CheckBox1.Value = True

   'Load a specific project template for performing pKa
   computation and graphing

    If (ThisWorkbook.ObjOrigin.IsModified) Then
       'Ask user to go to Origin, save the project, and
       come back to Excel
```

```
            MsgBox ("The project currently open in Origin needs
            to be saved" _

            & vbNewLine & _
            "Go to Origin and save the project if needed." _
            & vbNewLine & _
            "Click OK button below to continue.")
        'Reset the dirty flag
        ThisWorkbook.ObjOrigin.IsModified = False
    End If

    'Load specific template OPJ from same location as this Workbook
    Dim strFile As String
    strFile = ThisWorkbook.Path + "\pKaCurveFitting.opj"
    'Open the Origin project
    If Not (ThisWorkbook.ObjOrigin .Load(strFile)) Then
     Dim strTemp
     strTemp = "Could not open the Origin template project: " + strFile
     MsgBox (strTemp)
    End If

    'Return Focus to Excel
    AppActivate "Microsoft Excel"
     Exit Sub
    ErrMsg1:
     'Could not connect to Origin - report error
     MsgBox ("Could not connect to Origin!")
    End Sub
```

The Origin application object supports several properties and methods. The most commonly utilized are

PutWorksheet [Method] transfers two-dimensional data *to* the specified Origin Worksheet. Data can be placed starting at specified row and column location, or can be appended beneath the last existing row of the Worksheet.
Syntax:
Boolean PutWorksheet(BSTR name, VARIANT data, [optional] VARIANT r1, [optional] VARIANT c1)

GetWorksheet [Method] transfers data *from* the specified Origin Worksheet.
Syntax:
VARIANT GetWorksheet(BSTR name, [optional] VARIANT r1, [optional] VARIANT c1, [optional] VARIANT r2, [optional] VARIANT c2, [optional] VARIANT format)

LTVar [Property] allows access to LabTalk floating point variables.
Syntax:
Double LTVar(BSTR name)
Example:
'Query the r^2 value

```
ThisWorkbook.ObjOrigin.LTVar("nlsf.cod")
'Set the first fitting parameter to 0.663
ThisWorkbook.ObjOrigin.LTVar("nlsf.p1") = 0.663
```

LTStr [Property] provides access to LabTalk string variables.
Syntax:
BSTR LTStr(BSTR name)
Examples:

```
'Query the name of the active dataset (%C in Labtalk Script)
ThisWorkbook.ObjOrigin.LTStr("%C")
'Put "abc123" in temporary string storage (%Z in Labtalk Script)
ThisWorkbook.ObjOrigin.LTStr("%Z") = "abc123"
```

Run [Method] allows the Origin session to finish current operations. It can be used after
    calling a method that triggers lengthy auto update calculations in Origin. In many respects
    it is similar to the DoEvents Function in Visual Basic.
Syntax:
Boolean Run()
Example:
'Wait for auto update computation in Origin to complete

```
ThisWorkbook.ObjOrigin.Run
```

Execute [Method] executes a LabTalk script in specified context.
Syntax:

```
Boolean Execute(BSTR script, [optional] VARIANT name)
'Executes a series of Labtalk Commands
bb = ThisWorkbook.ObjOrigin.Execute("X1=0;X2=10," "Graph1")
'Executes the Script "Calcs" contained in the Worksheet named
  "Results"
bb = ThisWorkbook.ObjOrigin.Execute("Results!Calcs.run();")
bb = ThisWorkbook.ObjOrigin.Execute("X1=0;X2=10," "Graph1")
```

Will return True for success, and False for failure.

CopyPage [Method] replaces the clipboard with an image of the specified graph or layer.
Syntax:
BOOL CopyPage(BSTR name, [optional] VARIANT format, [optional] VARIANT dpi,
    [optional] VARIANT colordepth)

9.10 EXAMPLE: CREATING A COM TOOL TO PERFORM
CURVE FITTING USING ORIGIN FROM EXCEL

Utilizing the Origin application objects, properties, and methods covered in the previous section,
it is possible to construct a project such as that shown in Figure 9.28. To do so, however, will
require that Excel reference a specific project in Origin designed for this purpose. For this example,
a sample project named "pKaCurveFitting.opj" has been constructed in Origin for the sole purpose
of interfacing with Excel using COM-based methodologies. This project consists of (1) a graph
utilized as a uniform plotting template named "FitResult," (2) a Worksheet named COMFit that
contains two scripts attached to buttons that perform the curve fitting and the raw data utilized for
the curve fit, (3) a Worksheet named FitCurve that stores the fitted curve to the data consisting of
101 evenly spaced points.

Figure 9.29 shows the Origin pKa Curve Fitting Project that is included in the accompanying
CD-ROM under the folder \Chapter 9\Samples\Origin. If the user presses either the "mono fit" or

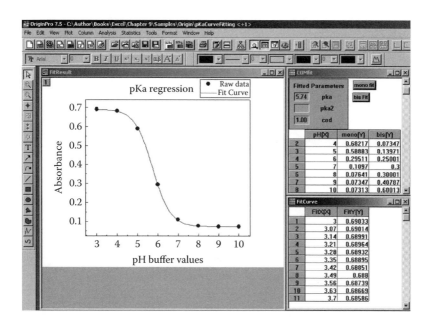

FIGURE 9.29 Origin pKa curve-fitting project.

"bis fit" button on the COMFit Worksheet, the appropriate curve fit will be executed, and the FitResult Graph and FitCurve Worksheet (which store the Fitted Curve of 101 plotted points) will both be updated. The trick here is to enable Excel to do the following tasks utilizing COM: (1) Write the appropriate data set into the Origin COMFit Worksheet, (2) set the curve-fitting parameters as necessary (initial guesses of parameters, parameters to remain fixed or vary, etc.), (3) trigger the appropriate script associated with a button on the COMFit Worksheet, (4) return a copy of the updated plot in Origin to Excel, (5) copy the curve-fitted data from Origin into Excel, (6) copy the curve-fitted parameters into Excel from Origin.

The scripts associated with the "mono fit" and "bis fit" buttons can easily be viewed by selecting Edit->Button Edit Mode from the Origin menu. If the user right-clicks on the text associated with the button and selects "Label Control," the contents of the script will appear in the Label Control form. It is beyond the scope of this text to explain Labtalk script in detail; however, a basic overview of the two scripts is in order. First the mono pKa script. Comments in Labtalk script are similar to "C" and preceded by "//" characters.

```
// COM Mono pKa Fitting Script
// Save the name of the current Worksheet
%W = %H;
// Choose the Dependant Dataset
dependataset=2;

// Set Plot Template as active Window
win -a FitResult;

// Erase any Data Present in the Plot Template
lay -c;
for (ii=1; ii<=count; ii++) {
```

```
  layer -e %[%Z,#ii];
};
// Initialize the nlsf engine
nlsf.init();
// Select the Custom Fitting Function
nlsf.func$=monopKa;
// initialize parameter values
A1=%(%W,dependataset,1); //.p1
A2=%(%W,dependataset,8); //.p2
pka = 5;          //.p3

// Dependant Data (Y-Axis) to be Fit
nlsf.y$ = %(%W,dependataset);
// Independent Data (X-Axis) to be Fit
nlsf.x$ = %(%W,1);

// Total Number of Points for Fitted Curve
nlsf.xpoints = 101;
// Improve accuracy of fit
nlsf.tolerance = 1e-5;
// Perform Curve Fitting
nlsf.fit(20);

// Add the original Data to the plot
layer -i %(%W,2);
// Rescale X & Y Axis to show All data
layer -at;

// Update "FitPams" Status Object
%W!FitPams.v1 = $(nlsf.p3,.2);
%W!FitPams.v2$ = "";
%W!FitPams.v3 = $(nlsf.cod,.2);

// Rename Axis Labels
label -xb pH Buffer Values;
label -yl Absorbance;

// Obtain Name of Generated NLSF Fitset
%J=%[%C,'_'];
// Copy Datasets
copy %J_A FitCurve_FitX;
copy %J_B FitCurve_FitY;
// Delete the Generated NLSF Fitset
win -c %J;

// Clean Up all internal Objects Used in nlsf
nlsf.cleanupfitdata();
```

Next follows the bis pKa script.

```
// COM Bis pKa Fitting Script
// Save the name of the current Worksheet
%W = %H;
// Set Dependant Dataset
dependataset=3;
// Set Plot Template as active Window
win -a FitResult;
// Erase any Data Present in the Plot Template
lay -c;
for (ii=1; ii<=count; ii++) {
  layer -e %[%Z,#ii];
};
// Initialize the nlsf engine
nlsf.init();
// Select the Custom Fitting Function
nlsf.func$=bispKa;
// initialize parameter values
Acat=%(%W,dependataset,1);        //.p1
Aneut=%(%W,dependataset,4);       //.p2
Aanion=%(%W,dependataset,8);      //.p3
pk1 = 5;           //.p4
pk2 = 8;           //.p5
// Do Not Allow Acat or Aanion to Vary!
nlsf.v1=0;
nlsf.v3=0;
// Dependant Data (Y-Axis) to be Fit
nlsf.y$ = %(%W,dependataset);
// Independent Data (X-Axis) to be Fit
nlsf.x$ = %(%W,1);
// Total Number of Points for Fitted Curve
nlsf.xpoints = 101;
// Improve Accuracy of Fit
nlsf.tolerance = 1e-5;
// Perform Curve Fitting
nlsf.fit(20);
// Add the original Data to the plot
layer -i %(%W,dependataset);
// Rescale X & Y Axis to show All data
rescale;

%W!FitPams.v1 = $(nlsf.p4,.2);
```

```
%W!FitPams.v2$ = $(nlsf.p5,.2);
%W!FitPams.v3 = $(nlsf.cod,.2);

// Rename Axis Labels
label -xb pH Buffer Values;
label -yl Absorbance;

// Obtain Name of Generated NLSF Fitset
%J=%[%C,'_'];
// Copy Datasets
copy %J_A FitCurve_FitX;
copy %J_B FitCurve_FitY;
// Delete the Generated NLSF Fitset
win -c %J;

// Clean Up all internal Objects Used in nlsf
nlsf.cleanupfitdata();
```

Both of these scripts operate in similar fashions. A summary of their operation follows. The data set of interest is identified. Any previously plotted material is erased, and the curve fitting engine is initialized. Initial parameters are set, and the independent and dependent data sets are identified to the nlsf fitting engine. The curve is fitted to the specified tolerance and the results plotted. The plot axes are rescaled to reflect the new data, the curve-fitted data is copied to the FitCurve Worksheet, and the original nlsf Worksheet is deleted. Finally, the nlsf engine is uninitialized (cleaned up).

The Origin_pKaCurveFitting.xls Workbook contains two Worksheets. The first, "pKaCurveFit," provides both a visual interface and access to the original and fitted data. The second Worksheet, "SampleData," holds pH vs. Absorbance data for various compounds.

Prior to executing any COM-based properties and methods, a connection must first be established to an existing version of Origin, or a new instance of Origin must be started and referenced.

```
Private Sub Button_Start_Click()
   'Launch a new instance or try connecting or connect to an
   existing instance
   If (OptionButton_New.Value) Then
   On Error GoTo ErrMsg1
   'Launch a new instance of Origin and connect to it
   Set ThisWorkbook.ObjOrigin = GetObject("",
"Origin.Application")

   Else
     On Error GoTo ErrMsg1
     'Connect to Existing Instance
     Set ThisWorkbook.ObjOrigin = GetObject("",
     "Origin.ApplicationSI")
   End If

     'Successfully connected to Origin
      'Show/hide appropriate controls
```

```
         CheckBox1.Visible = True
         'Button_Start.Visible = False
         Button_End.Visible = True
         Button_Terminate.Visible = True
         Button_Fit.Visible = True

         'Make Origin visible to begin with
         ThisWorkbook.ObjOrigin.Execute ("doc -mc 1;")
         CheckBox1.Value = True

'Now load a specific template OPJ for performing the computation
and graphing

'Before loading OPJ, give user chance to save current work if
dirty flag is set

If (ThisWorkbook.ObjOrigin.IsModified) Then
    'Ask user to go to Origin, save the project, and come back
    to Excel

    MsgBox ("The project currently open in Origin needs to be
    saved" _

          & vbNewLine & _
          "Go to Origin and save the project if needed." _
          & vbNewLine & _
          "Click OK button below to continue.")
    'Reset the dirty flag
    ThisWorkbook.ObjOrigin.IsModified = False
End If

'Load specific template "pKaCurveFitting.opj" OPJ from same
location as this Workbook
    Dim strFile As String
    strFile = ThisWorkbook.Path + "\pKaCurveFitting.opj"
    'Open the Origin project
    If Not (ThisWorkbook.ObjOrigin.Load(strFile)) Then
      Dim strTemp
      strTemp = "Could not open the Origin template project:
      " + strFile
      MsgBox (strTemp)
    End If
    'Return Focus to Excel
    AppActivate "Microsoft Excel"
    Exit Sub
ErrMsg1:
    'Could not connect to Origin - report error
      MsgBox ("Could not connect to Origin!")
End Sub
```

When the button labeled "Curve Fit Data" is pressed, the code in the Button_Fit_Click sub-routine is executed. It is the code that resides in this subroutine that accesses and executes appropriate COM calls from Excel to Origin.

```vba
Private Sub Button_Fit_Click()
  'Get check box values corresponding to fix/vary status of
  the four parameters
  Dim Params(1 To 10, 1 To 1) As Double
  Dim status
  Const OriginSheet As String = "COMFit"

  'Checkboxes To Fix or allow variance of Parameters
  Params(1, 1) = Worksheets(pKaSheet).CheckBox_Ac.Value
  Params(2, 1) = Worksheets(pKaSheet).CheckBox_An.Value
  Params(3, 1) = Worksheets(pKaSheet).CheckBox_Aa.Value
  Params(4, 1) = Worksheets(pKaSheet).CheckBox_pKa1.Value
  Params(5, 1) = Worksheets(pKaSheet).CheckBox_pKa2.Value
  'Parameter Values
  Params(6, 1) = Range("Acat").Value
  Params(7, 1) = Range("Aneut").Value
  Params(8, 1) = Range("Aanion").Value
  Params(9, 1) = Range("pKa1").Value
  Params(10, 1) = Range("pKa2").Value

  'Write in X Data
  Set rngDataAndParam = Range("XData")
  If Not ThisWorkbook.ObjOrigin.PutWorksheet(OriginSheet,
  rngDataAndParam.Value) Then Exit Sub
  'Write in Y Data
  Set rngDataAndParam = Range("YData")
  Select Case Sheet1.OptionButton_mono.Value
    Case True
    'Write to mono Column (2) in COMfit Origin Worksheet
    (Rows\Columns Zero Based!)
    If Not ThisWorkbook.ObjOrigin.PutWorksheet(OriginSheet,
    rngDataAndParam.Value, 0, 1) Then Exit Sub

    Case False
    'Write to bis Column (3) in COMfit Origin Worksheet
    (Rows\Columns Zero Based!)
    If Not ThisWorkbook.ObjOrigin.PutWorksheet(OriginSheet,
    rngDataAndParam.Value, 0, 2) Then Exit Sub

     End Select

    'Set Fixed Parameters
    Select Case Sheet1.OptionButton_mono.Value
      Case True
```

```
        'Mono Case
        ThisWorkbook.ObjOrigin.LTVar("nlsf.v1") = Params(1, 1)
        ThisWorkbook.ObjOrigin.LTVar("nlsf.v2") = Params(3, 1)
        ThisWorkbook.ObjOrigin.LTVar("nlsf.v3") = Params(4, 1)
      Case False
        'Bis Case
        For ii = 1 To 5
          Var$ = "nlsf.v" & Trim(Str(ii))
          ThisWorkbook.ObjOrigin.LTVar(Var$) = Params(ii, 1)
        Next ii
      End Select

  'If Not ThisWorkbook.ObjOrigin.PutWorksheet("RawData", Params,
  0, 3) Then Exit Sub

  'Push Button for desired fit
  Select Case Sheet1.OptionButton_mono.Value
      Case True
      'Push Mono Fit Button
      bRet = ThisWorkbook.ObjOrigin.Execute("COMfit!Mono.run();")
      Case False
      'Push bis Fit Button
      bRet = ThisWorkbook.ObjOrigin.Execute("COMfit!Bis.run();")
  End Select

  'Wait for auto update computation in Origin to complete
  ThisWorkbook.ObjOrigin.Run

  'Extract Fitted Parameters to Workbook
  Select Case Sheet1.OptionButton_mono.Value
      Case True
      'Mono Fitting Parameters
      Range("Acat").Value =
      ThisWorkbook.ObjOrigin.LTVar("nlsf.p1")

      Range("Aanion").Value =
      ThisWorkbook.ObjOrigin.LTVar("nlsf.p2")

      Range("pKa1").Value =
      ThisWorkbook.ObjOrigin.LTVar("nlsf.p3")

      Case False
      'Bis Fitting Parameters
      Range("Acat").Value =
      ThisWorkbook.ObjOrigin.LTVar("nlsf.p1")

      Range("Aneut").Value =
      ThisWorkbook.ObjOrigin.LTVar("nlsf.p2")

      Range("Aanion").Value =
      ThisWorkbook.ObjOrigin.LTVar("nlsf.p3")
```

```
    Range("pKa1").Value =
    ThisWorkbook.ObjOrigin.LTVar("nlsf.p4")

    Range("pKa2").Value =
    ThisWorkbook.ObjOrigin.LTVar("nlsf.p5")

End Select

    'Write Goodness of Fit Parameters to specified Excel
    Worksheet Range

Range("rcsq").Value =
ThisWorkbook.ObjOrigin.LTVar("nlsf.chiSqr")

    Range("ssq").Value =
    ThisWorkbook.ObjOrigin.LTVar("nlsf.ssr")

    Range("cod").Value =
    ThisWorkbook.ObjOrigin.LTVar("nlsf.cod")

    Range("dof").Value =
    ThisWorkbook.ObjOrigin.LTVar("nlsf.dof")

    'Get fit curve values from Origin
    Dim fit(1 To 101, 1 To 2) As Double
    Set rngFitCurve = Range("FitCurve")
    rngFitCurve.Value =
    ThisWorkbook.ObjOrigin.GetWorksheet("FitCurve")

    'Tell Origin to copy an image of the graph to the clipboard
    bRet = ThisWorkbook.ObjOrigin.CopyPage("FitResult", 4)

    'Now create a new chart and paste graph image from
    On Error Resume Next
    Worksheets(1).ChartObjects(1).Activate
    With ActiveChart
      .ChartArea.Clear
    End With

    With ActiveSheet.ChartObjects.Add _
      (Left:=207, Width:=368, Top:=457, Height:=325)
      .Chart.Paste
    End With
  End Sub
```

In order for the code in the `Button_Fit_Click` subroutine to function correctly, the user must first connect to an instance of Origin (using the push buttons on the sheet), pick the pKa fitting function (to fit the data to), and pick the data set of interest (from the combo box). Notice that the combo box contains the names of all the data sets contained in the SampleData Worksheet (Figure 9.30). This is accomplished using the following subroutine:

```
Sub HeadersToCombobox(Combo As Object, Headers() As String,
wkbook As String, _
wksheet As String, Optional ByVal startcol = 1, _
```

Origin_pKaCurveFitting.xls

	A	B	C	D	E
1	pH	mono01	mono02	bis01	bis02
2	3	0.07313	0.69146	0.07313	0.54821
3	4	0.07347	0.68217	0.07347	0.26453
4	5	0.07641	0.58883	0.13971	0.10114
5	6	0.1097	0.29511	0.25001	0.07217
6	7	0.29511	0.1097	0.3	0.07046
7	8	0.58883	0.07641	0.30001	0.08004
8	9	0.68217	0.07347	0.40787	0.14483
9	10	0.69146	0.07313	0.60013	0.30012

FIGURE 9.30 Sample data sets in SampleData Worksheet.

```
Optional ByVal endcol = 0, Optional ByVal header_row As Long = 1)
'Places the specified Headers in Combobox in wksheet in wkbook
'Dim Headers() As String
Dim col As Integer, arrayindex As Integer
Dim thissheet As String
Dim lastEntryCell As Range
'If ending column not specified use Last column in header row
If endcol = 0 Then endcol = LastColinRowX(wkbook, wksheet,
header_row)
'Clear any Pre-Existing Entries
Combo.Clear
'Debug.Print startcol, endcol
'Write All Headers into Array
For col = startcol To endcol
 arrayindex = arrayindex + 1
 ReDim Preserve Headers(1 To arrayindex)
 Headers(arrayindex) =
 Workbooks(wkbook).Worksheets(wksheet).Cells(header_row,
 col).Value
 'Write Headers into Combobox
 Combo.AddItem Headers(arrayindex)
Next col
End Sub
```

This subroutine is only called once, as soon as the Workbook is activated. The code to do this resides in the `Workbook_Activate` subroutine.

```
Private Sub Workbook_Activate()
'Only Execute Once at Startup
If Worksheets("pkaCurveFit").ComboBox_Sample.ListCount < 2 Then
   Call
   HeadersToCombobox(Worksheets("pkaCurveFit").ComboBox_Sample,
   Headers(), ActiveWorkbook.Name, "SampleData", 2)
   End If
   'Enable Start Button if Disabled
```

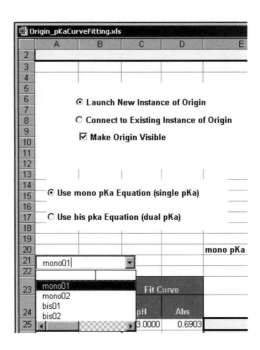

FIGURE 9.31 Combobox automatically populated with data set names.

```
Sheet1.Button_Start.Visible = True
Worksheets("pKaCurveFit").Activate
End Sub
```

Figure 9.31 shows the Combobox on the pKaCurveFit Worksheet which is automatically populated with the names of the datasets contained in the SampleData Worksheet.

Prior to running the sample pKa curve-fitting Workbook, fitting functions must be installed in Origin. This can be accomplished by copying the files monopKa.fdf and bispKa.fdf from the location \Chapter 9\Samples\Origin on the accompanying CD-ROM to the user's machine in the C:\Program Files\OriginLab\OriginPro75\FitFunc folder. Once the files are in this folder, the user must modify the nlsf.ini file located in C:\Program Files\OriginLab\OriginPro75\UserName, where UserName is the username of the person logged into Windows. The nlsf.ini file can easily be opened using notepad or any word processor. At the bottom of the nlsf.ini file, the following lines must be added for Origin to utilize the pKa fitting functions.

[pKa]
monopKa=monopKa
bispKa=bispKa
Default Function=monopKa

Once the nlsf.ini file has been modified, the user is free to push the "Curve Fit Data" button on the "pKaCurveFit" sheet and watch the sample application generate the fitted curves to the sample data sets. Figure 9.32 shows the results when the data sets are created for the bispKa system named "bis01." Figure 9.33 shows the results generated for the single pKa system named "mono01."

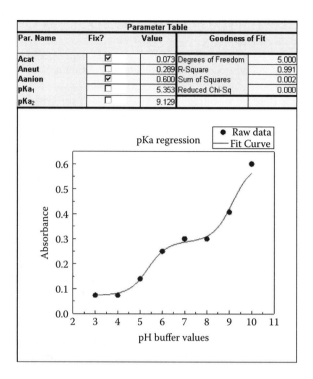

Parameter Table				
Par. Name	Fix?	Value	Goodness of Fit	
Acat	☑	0.073	Degrees of Freedom	5.000
Aneut	☐	0.289	R-Square	0.991
Aanion	☑	0.600	Sum of Squares	0.002
pKa$_1$	☐	5.353	Reduced Chi-Sq	0.000
pKa$_2$	☐	9.129		

FIGURE 9.32 Results for data set "bis01".

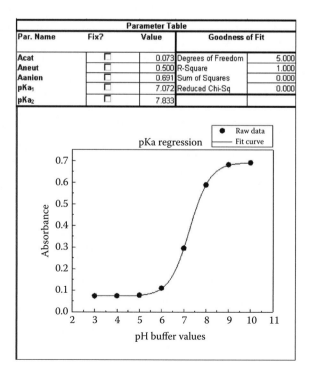

Parameter Table				
Par. Name	Fix?	Value	Goodness of Fit	
Acat	☐	0.073	Degrees of Freedom	5.000
Aneut	☐	0.500	R-Square	1.000
Aanion	☐	0.691	Sum of Squares	0.000
pKa$_1$	☐	7.072	Reduced Chi-Sq	0.000
pKa$_2$	☐	7.833		

FIGURE 9.33 Results for data set "mono01".

9.11 OPENING AND PLOTTING EXCEL WORKBOOKS IN ORIGIN FOR SUPERIOR GRAPHICS

Although Excel has the capability to store enormous amounts of data, it lacks the ability to display that which it holds in an eye-catching style. This is unfortunate because it is through visualization that the mind can process the true meaning of large amounts of information. What is truly disappointing is the fact that Excel is a mature product in every sense, yet no attempt has been made to upgrade the graphics capability of the program. The graphs, by and large, are boring, have limited customization, and often the data lines or plots are jagged, which is reminiscent of the DOS programs in the 1980s.

Origin, however, can import Worksheets created in Excel and produce presentation-quality graphics that would be expected in any type of business report, proposal, plan, etc. In this section, the reader will be shown how to utilize the graphics capabilities in Origin to create suitable plots for presentations.

Figure 9.34 shows *sample* pressure deformation information for HY-80 steel that is utilized for the hulls on fast attack submarines. The true performance characteristics of HY-80 steel as utilized in submarine hulls are classified, as is their estimated crush depth. The Worksheet in Figure 9.34 has three columns — pressure (given in pounds per square inch), defect size (radius, assumed circular defect in inches), and deformation (from normal) in inches.

The diagram in Figure 9.35 illustrates how the data in Figure 9.34 relate to the experiment in which it was taken. One side of a plate of HY-80 steel is subject to a certain uniform force (pressure). Each sampled plate has a defect within the steel measured (in inches) by X-Ray Computed Tomography, an effective way to look at internal defects and generate 3-D models of the sample being scanned. These defects weaken the steel's ability to maintain its integrity, as evidenced by an outward protrusion (puckering effect) on the opposing side of the steel plate. Five thousand plates of uniform thickness have been tested in an apparatus, and the data has been stored in the Worksheet "HY-80-deform.xls" that is contained in the Chapter 9\Samples directory of the included CD-ROM. It will now be shown how to import this information into Origin and construct an elegant 3-D plot.

From the Origin menu, the user should choose File->Open Excel and choose the file "Chapter 9\Samples\HY-80-deform.xls". Figure 9.36 shows the dialog box that will appear when an Excel Workbook is opened in Origin. Because Origin only supports one Worksheet per file (but one may import multiple files and thus have multiple Worksheets), the user must pick a Worksheet from the Excel Workbook that is to be imported as an Origin Worksheet. In this example, the Worksheet named "HY-80" is imported as an Origin Worksheet.

To create a 3-D graph, an Origin Worksheet must have an X, Y, and Z-column. After importation of the "HY-80" Worksheet, the X and Y columns are set (or mapped) correctly with $X \rightarrow A$ and $y \rightarrow B$. Column C, however, needs to be set as a Z-column. This can easily be accomplished by

	A	B	C
1	pressure (ppsi)	defect size (in)	deformation (in)
2	99854	0.004520526	0.028212037
3	107015	0.00630225	0.042152206
4	146953	0.006058773	0.055647175
5	89562	0.009207864	0.051542171
6	125724	0.005619815	0.044159099
7	98487	0.005202233	0.032022021
8	96087	0.002216031	0.013308235
9	122183	0.003142338	0.02399627
10	104808	0.003825213	0.025057055

FIGURE 9.34 Pressure deformation information for HY-80 steel.

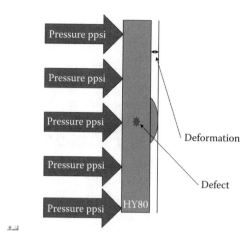

FIGURE 9.35 Free body diagram for data in Figure 9.34.

right-clicking on column C and choosing "Set As Z" as shown in Figure 9.37. With column C converted from type Y to type Z, the final step is to generate a matrix to create a 3-D plot.

Because the data in the new Origin Worksheet is not in order, the type of matrix that should be selected is "Random XYZ." To create the matrix, the user selects the Z column of the Worksheet (in this case column C), and selects Edit->Convert to Matrix->Random XYZ. When this happens, the dialog box shown in Figure 9.38 will come up.

Using the default number of columns and rows (70), and a smoothness of 0.5, the "Apply" button can be pressed. Note that smoothness is defined over an interval of 0.5–1.2, where 0.5 is the smoothest. The search radius is how many adjacent points are utilized on each side of a measurement for interpolating a spline between adjacent points. Once the Apply button is pressed, Origin begins generating the matrix and plotting the results. This can take anywhere from several seconds to several minutes, depending upon the number of data points and the degree of irregularities in the given data set. The progress bar in Figure 9.38 is shown while the matrix is being generated and plotted for the sample data set.

Figure 9.39 shows a rough plot generated for the given data set. Notice that this plot is done in "Speed Mode," which means only a subset of the given data points are fitted and plotted to save time. This can be a tremendous timesaver when working with a plot while adding Labels, formatting

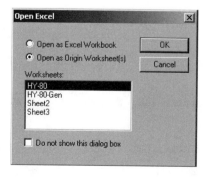

FIGURE 9.36 Opening an Excel Workbook in Origin.

HY80						
	A(X)	B(Y)	C(Y)			F(Y)	G(Y)
1	pressure (ppsi)	defect size (in)	deformation (in	Plot		
2	1000	0.00516	0.0051	Cut		
3	2000	0.00458	0.0091	Copy		
4	3000	0.00462	0.0138	Paste		
5	4000	0.00705	0.0282			
6	5000	0.00488	0.0244	Insert		
7	6000	0.00837	0.0502	Delete		
8	7000	0.00553	0.0386	Clear		
9	8000	0.00677	0.0541			
10	9000	0.00752	0.0677	Set As ▶ X X		
11	10000	0.00426	0.0425	Y Y		
12	11000	0.00828	0.091	Set Column Values... Z Z		
13	12000	0.00898	0.1077	Fill Column with ▶		
14	13000	0.00546	0.0710	Sort Column ▶ Label		
15	14000	0.00288	0.0403	Sort Worksheet ▶ Disregard		
16	15000	0.00359	0.0538			
17	16000	0.00881	0.1409	Normalize... X Error		
18	17000	0.0021	0.0356	Frequency Count Y Error		
19	18000	0.00289	0.052			
20	19000	0.00928	0.1763	Statistics on Columns		
21	20000	0.0053	0.1059			
22	21000	0.00629	0.1320	Mask ▶		
23	22000	0.00827	0.1818			
24	23000	0.00562	0.1291	Set as Categorical		
25	24000	0.00224	0.0537	Properties...		
26	25000	0.00266	0.06647			
27	26000	0.00744	0.19332			
28	27000	0.00422	0.11395			
29	28000	0.00357	0.10005			

FIGURE 9.37 Setting a column type in Origin.

Axis and legends, etc. The Speed Mode is easily turned off by right-clicking on the Layer indicator in the corner of the plot (in this case, a "1") and selecting "Layer Properties". A tabbed GUI will appear and, under the "Size/Speed" tab, uncheck the two checkboxes that enable the speed mode. The plot will now be plotted using all the generated data points. A formatted final plot in Origin is shown in Figure 9.40.

In the final plot, labels have been added for pressure (X-axis), defect size (Y-axis), and deformation (Z-axis). A color gradient has also been applied with blue shading denoting minimal deformity and red shading indicating severe deformity. A user looking at the plot can immediately discern four things. The deformity of HY-80 steel increases with increasing hull pressure. This deformation is aggravated by the size of defects present in the steel. The deformation due to particle size and pressure is more or less linear. If the defect size is kept to a minimum, the deformation in the steel will be minimized over the pressure range plotted.

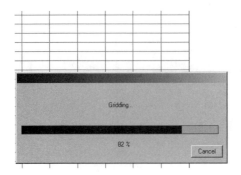

FIGURE 9.38 Matrix Creation of 3-D data set in Origin.

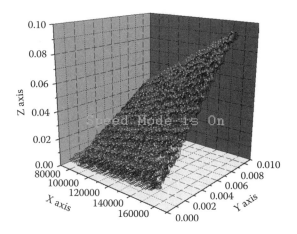

FIGURE 9.39 Plot generated in speed mode for sample data matrix.

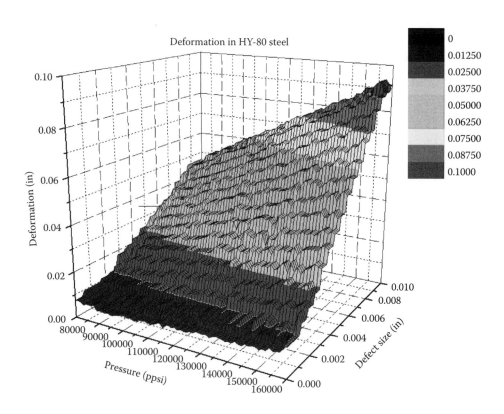

FIGURE 9.40 Formatted plot in Origin of Excel data.

9.12 SUMMARY

Although Excel offers great flexibility with its VBA macro language, both the calculation and graphics performance are somewhat limited. In this section, the reader is shown how to interface Matlab and Origin with Excel to make up for these deficiencies. Hopefully, at some point, Microsoft will upgrade the graphics performance of Excel to true presentation-quality graphics and enhance the computational engine in terms of its processing speed and functional abilities. Until such a time occurs, users will be forced to cross-integrate Excel with other applications to achieve optimum results.

10 An Example ADA Application That Generates a Report

10.1 INTRODUCTION AND PROBLEM DEFINITION

This is the final chapter in the text and where all of the information covered in the first nine chapters is put to use to produce a real-world application.

The final sample application automatically analyzes the results of a manufacturing process and prepares a report that describes the results of the various measured parameters related to the testing of the produced product. The automated report will determine if a particular item from the analyzed lot meets the design specifications and passes acceptance testing, or if an item is a reject that failed acceptance testing.

The application further tests the actual manufacturing process, making a determination if the process is in or out of control. Using six-sigma methodology, a determination can be made as to whether a manufactured product in a given lot is in control. Out-of-control processes are flagged in the reports as well as products that fail to meet the design specifications. (Note: It is possible for a process to be in control and produce some reject products. This can happen when a number of parameters go to their extremes but not outside their allowable quality limits. This often happens in semiconductor manufacturing, where if a number of electrical parameters choose to vary toward certain extremes, the resultant chip will not function correctly.)

This example focuses on testing strain gages that have been manufactured in lots. A strain gage is a circuit which produces a voltage that is proportional to the amount of strain that is applied to the gage. Typically, the gage will produce a linear output over some specified interval of strain.

Strain is a dimensionless quantity that is the amount of deformation of a body due to an applied force. Strain can be determined by measuring the fractional change in length of a particular body. The strain gage accomplishes this by changing its resistance in response to very small changes in length. When a voltage is applied to the strain gage, the voltage output by the gage will vary depending upon the amount of deformation the body it is attached to has undergone. Strain gages are typically produced using a Wheatstone bridge circuit as shown in Figure 10.1.

Looking at Figure 10.2, the voltage output between points B and D is simply the difference between the voltages at point B and point D. The voltages at points B and D are determined by the voltage divider circuits in the Wheatstone bridge. The principle here is that when all four resistors in the bridge are of equal value, there will be no voltage differential between points B and D because the voltage will be evenly divided in both voltage dividers. In such a state, the Wheatstone bridge is said to be balanced. However, if one of the resistors changes value, the bridge will become unbalanced, and a voltage differential will exist between points B and D.

The Wheatstone bridge configuration is utilized because even a small change in the resistance of the sensing grid will throw the bridge out of balance, resulting in an output voltage. This makes the Wheatstone bridge circuit ideal for the detection of strain using a strain gage whose elements change resistance under compression or tension. Because resistors can also change value with temperature, the Wheatstone configuration helps negate the effects of temperature on the output, as all four resistors should change resistance by the same amount, thus not affecting the output. In this application, strain gages are manufactured on a substrate. Each substrate can hold 96 resistors that are placed in 8 rows by 12 columns on the substrate. A block of four resistors (2 rows × 2 columns on a substrate) forms a single strain gage. Each manufactured substrate holds 24 strain gages.

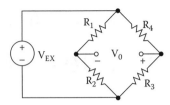

FIGURE 10.1 Typical Wheatstone bridge circuit.

The gages consist of three precision resistors (R1, R2, R3) that match a strain gage resistor (R4 or Rg) of the same value. The gages formed on the substrate are scored so that the substrate can be broken into 24 pieces or 24 individual gages. The substrates are manufactured in batches or lots of so many gages per lot. During each production run, measurements are taken of every resistor on each substrate. The measurements are stored in a text file that groups the measurements in sequential order of each manufactured substrate, with each substrates individual resistance measurements (96) in an 8-row by 12-column grid (the same layout as they are manufactured).

The desired task is to analyze each individual gage on a substrate and determine if it meets specifications. In other words, is the gage good, or is it defective? Is the gage within the quality tolerances to allow accurate measurement of strain? The process can be taken even further by analyzing *the entire process of a production run* of the gages to ensure that it is in control. It is important to look at the entire process of production because even processes that are in control can produce a defective product. Thus, a few defective products do not always indicate a defective process.

The sample application created in the text will accomplish the following:

1. Information will be obtained from a GUI.
2. Acceptable Tolerances will be calculated.
3. The text file of data will be imported.
4. Resistors out of tolerance will be identified.
5. Defective gages will be identified.
6. The manufacturing process will be analyzed.
7. A detailed report for the lot will be prepared.

To begin the process, a GUI will need to be designed to take in the pertinent information. Figure 10.3 shows the GUI that was designed for this sample application. The GUI is divided into

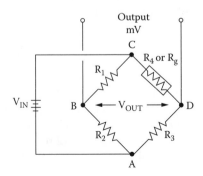

FIGURE 10.2 Strain gage substrate layout.

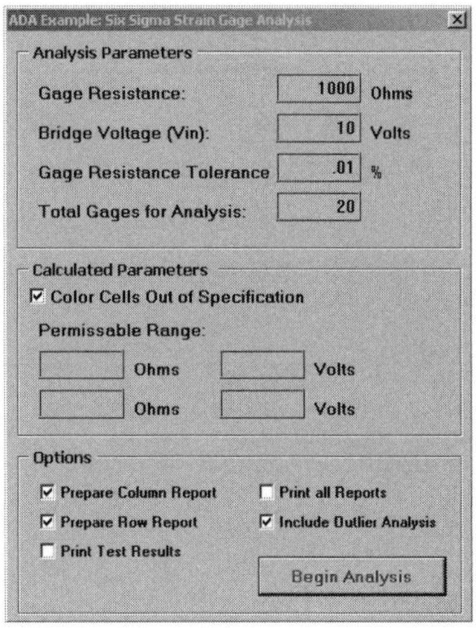

FIGURE 10.3 ADA example GUI for strain gage analysis.

three sections. The first contains the parameters that a gage in a specification should meet. In this example, the Wheatstone Bridge utilizes resistors that are 1000 Ω or 1.0K Ω. The applied voltage to the bridge for the application will be 10 V. Each resistor in the bridge may have a tolerance (or an expected variance of its nominal value) of 0.01%. The last parameter is the lot size, or number of substrates for analysis. In this instance, 20 substrates are produced per lot.

The second section of the GUI contains parameters calculated by the application. Each gage will have a permissible range of resistance for each resistor (in Ohms) based on the tolerance provided in the first section of the GUI. The voltage differential at the outputs of the gage can then be calculated using the specified tolerances.

The last section of the GUI contains the options for analysis. Here, the user can select which reports should be prepared, if the reports should be automatically printed at the conclusion of analysis, if an outlier analysis (more on this later in the chapter) should be performed, and if the test results should be printed at the end of the macro.

The component limits are calculated to the allowable extremes utilizing the CalculateLimits subroutine.

```
Sub CalculateLimits()
'Calculate Acceptable Limits of Voltage and Resistance
Dim R1 As Single, R2 As Single
Ohm = Val(frm10ADA.TextBox_GageR)
Vcc = Val(frm10ADA.TextBox_BridVolt)
OhmMin = Val(frm10ADA.TextBox_GageR) - Val(frm10ADA.TextBox_GageR)
* (Val(frm10ADA.TextBox_GageRT) / 100)
OhmMax = Val(frm10ADA.TextBox_GageR) + Val(frm10ADA.TextBox_GageR)
* (Val(frm10ADA.TextBox_GageRT) / 100)
R1 = OhmMax: R2 = OhmMin
VMin = Vcc * ((R2 / (R2 + R1)))
R1 = OhmMin: R2 = OhmMax
VMax = Vcc * ((R2 / (R2 + R1)))
'Write Values into GUI
```

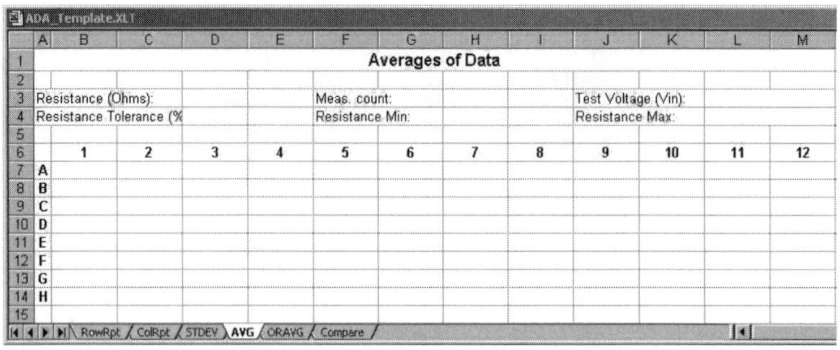

FIGURE 10.4 Workbook template utilized for example ADA macro.

```
frm10ADA.Textbox_Omax.Text = str(OhmMax)
frm10ADA.TextBox_Omin.Text = str(OhmMin)
frm10ADA.TextBox_Vmax.Text = str(VMax)
frm10ADA.TextBox_Vmin.Text = str(VMin)
End Sub
```

A Workbook is created to hold the Calculations the Application will make on the strain gage data. The Workbook is created from an existing template called "ADA_Template.xlt." The name of the Workbook to hold the data is stored in the Public variable CalcBook.

```
CalcBook = Open_Template ("C:\Program Files\Microsoft
Office\Templates\ADA_Template.xlt", "c:\temp\ADA_Example.xls")
```

Figure 10.4 shows the Workbook template utilized for the example macro. The Workbook template contains six Worksheets. Notice that some of the Worksheets have cells with predefined formulas in them. Developers may do this as a shortcut if they are certain that the reports they create will be *absolutely uniform each time*. If reports will always contain the same data and parameters in identical cells from run to run, this can be a big time-saver as it saves the developer from having to write VBA code to apply formulas to specific cells.

The template utilized for this example contains six Worksheets. The functions of each Worksheet are detailed in Table 10.1.

The challenge now is to create a macro that can run an analysis of the given data files and generate the report Worksheets detailed in Table 10.1.

TABLE 10.1
ADA Sample Application Worksheet Functionality

	Worksheet Name	Function
1	RowRpt	Analysis of rows
2	ColRpt	Analysis of columns
3	STDEV	STDEV of lot
4	AVG	Averages of lot
5	ORAVG	Averages of lot with outliers removed
6	Compare	Comparison of AVG and ORAVG Sheets

10.2 A QUICK WORD ON SIX SIGMA

Six Sigma is a quality management methodology. It allows only 3.4 defects per million opportunities per process. In Six Sigma, a defect is defined as anything that results in customer dissatisfaction.

Six Sigma relies heavily on statistical process control (SPC) methods to measure and reduce defects and increase quality. The central theme of Six Sigma is that if the number of defects present in a process can be measured, the defects can systematically be eliminated and the process can (in theory) approach a defect rate of zero.

Obviously, in any process there will be a value for production that is desired. In this example, the resistance of each element in the Wheatstone bridge should be 1,000 Ω. All processes, of course, have some degree of variation. Figure 10.5 shows the Normal distribution, which illustrates the likelihood of how close a produced part will be to its desired production value. Notice that, in a normally distributed process, 68% of the values lie within one Standard Deviation of the mean. A standard deviation is a computed measure of variability indicating the spread of the data set around the mean, whereas deviation is the difference or distance of an individual observation or data value from the center point (often the mean) in the set distribution.

Notice that two standard deviations from the mean captures 95% of all given occurrences for a process, whereas three Standard Deviations from the mean are nearly totally inclusive at 99.7%.

This gives rise to the data in Table 10.2, which are defects per million opportunities (DPMO). A Six Sigma process by definition encompasses all occurrences within ± 3 Standard Deviations of the mean. To meet Six Sigma specifications, a process must have no more than 3.4 defects per million products.

The ± 3 Standard Deviations from the mean has special significance in Six Sigma. These values are called the control limits of the given process. The +3 Standard Deviations from the mean are referred to as the UCL or Upper Control Limit, whereas –3 Standard Deviations from the mean are referred to as the LCL or Lower Control Limit. Very loosely, when a product occurrence falls between UCL and LCL, it is said to be in control. (In some instances this may not be true, if a trend is present, etc.)

Therefore, if a process produced 1 million units, and only 3 of the units were defective (above the UCL or below the LCL), such a process would be referred to as "in control." When a process is "in control," it should not be tampered with. Tampering is any action taken to compensate for variation within the control limits of a *stable system*. Tampering will increase rather than decrease process variation. A stable system is one in which special-cause variation has been removed and

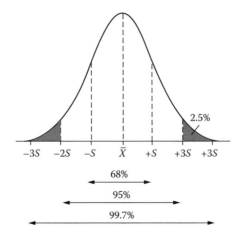

FIGURE 10.5 Normal distribution — occurrences from mean.

TABLE 10.2
Defects per Million Opportunities (DPMO)

Sigma Level	Defects per Million Opportunities (DPMO)
1	690,000
2	308,537
3	66,807
4	6,210
5	233
6	3.4

common-cause variation has been sufficiently reduced to make outputs consistent and predictable (i.e., ≤ 3.4 defects per million products).

What constitutes tampering as opposed to reducing variation? By definition, if an action taken does not reduce the variation in a process, it is tampering.

The sample application in this example needs to make two determinations. First, how many gages meet the specifications? (How many are good?) Second, is the *process* that produces the gages satisfactory or does it need to be refined?

10.3 DEALING WITH THE RAW DATA

Now the raw data from the manufacturing process must be imported, manipulated, and analyzed to provide the information desired.

Figure 10.6 shows the format for the raw data file that contains the measurements of the resistors fabricated on the substrates. The first line contains the substrate number. The next line contains the column numbers (1–12). The next eight lines contain the row number (labeled A–H for rows 1–8) in column 1, followed by the resistance measurements for every resistor produced on the substrate. Looking at Figure 10.6, an 8 × 12 matrix is formed for each substrate, where each row/column element is indicative of a resistance. The file repeats in this fashion for as many substrates that have been produced in the given lot.

	A	B	C	D	E	F	G	H	I	J	K	L	M
1	Substrate Number: 1												
2		1	2	3	4	5	6	7	8	9	10	11	12
3	A	600	600	600	600	1400	1400	1400	1400	999.9167	999.9678	999.9725	999.9104
4	B	999.8966	999.9723	999.9332	1000.03	999.9789	999.9418	999.876	999.904	999.9728	999.8739	1000.063	999.9311
5	C	1000.095	1000.012	1000.081	999.9646	999.9521	999.8962	1000.069	999.9646	999.8826	999.8774	999.949	999.998
6	D	999.9901	999.94	1000.096	999.9129	1000.046	999.8958	999.8876	999.8869	999.9902	1000.065	1000.077	1000.004
7	E	1000.018	1000.061	1000.046	999.9106	1000.026	999.9009	1000.062	1000.125	1000.068	999.8916	999.9247	999.9905
8	F	999.8899	999.8777	999.9101	1000.046	999.9361	1000.017	999.9233	999.9322	1000.042	999.9913	999.9237	999.9314
9	G	999.961	999.8957	1000.017	1000.115	1000.001	1000.089	1000.026	999.9951	1000.13	1000.039	999.8989	1000.034
10	H	999.9698	1000.049	999.9612	1000.109	1000.081	999.8719	999.9096	1000.002	1000.018	999.9568	999.9043	999.9449
11	Substrate Number: 2												
12		1	2	3	4	5	6	7	8	9	10	11	12
13	A	999.9076	999.9891	1000.098	999.9481	1000.022	1000.125	1000.025	999.9189	999.9167	999.9678	999.8988	999.9164
14	B	999.9146	1000.068	1000.025	1000.095	1000.016	999.902	1000.127	999.964	999.995	999.8908	999.9283	999.9778
15	C	1000.117	999.8943	999.9401	1000.071	999.984	999.9385	999.8896	999.8982	999.9835	1000.007	1000.128	999.9068
16	D	1000.065	999.9451	1000.059	1000.046	999.9516	1000.127	1000.09	999.8714	999.9865	999.966	999.9476	999.8992
17	E	1000.126	1000.028	999.996	1000.09	1000.123	999.8857	999.9391	999.9262	1000.111	1000.002	1000.062	999.9184
18	F	999.9788	1000.012	1000.121	999.9786	1000.042	1000.045	1000.06	1000.12	1000.044	999.9004	1000.04	999.882
19	G	1000.01	1000.029	999.9438	1000.117	1000.025	1000.034	999.8751	1000.091	999.9323	1000.028	999.9746	1000.053
20	H	1000.009	1000.066	1000.042	999.8952	1000.049	1000.016	999.9093	1000.038	1000.039	999.9844	1000.046	999.9608

FIGURE 10.6 Raw data format for strain gage manufacturing process.

The `ImportDataToWorkbook` subroutine created in Chapter 2 is utilized to bring this data into the existing Excel template. Notice how the `GetFileName` function is embedded in the call to allow the user to select the appropriate data file.

```
Call ImportDataToWorkbook(ActiveWorkbook.Name, SGDataSheet,
GetFileName ("txt", _
"Strain Gage Resistance Measurements"), False, True, False,
False, False)
```

One of the goals of this exercise is to readily identify which gages are defective. One way to do this would be to highlight the good and bad gages contained on the substrate. However, it would be bad form to highlight the data contained on the original sheet as it will make it difficult to look at later. To preserve the original data sheet, a copy will be created of it that will be used to display the good and the bad strain gages on the substrate. The `CopyWorksheet` subroutine created in Chapter 6 is utilized for this purpose.

```
Call CopyWorksheet(ActiveWorkbook.Name, SGDataSheet,
ActiveWorkbook.Name, "RowRpt", True, "SGDefects")
```

On each of the sheets in the calculation Workbook, it is desired to have a footer that displays the Worksheet name at the bottom center of every printed Worksheet. The `UniversalFooter` subroutine created in Chapter 6 can do this quite easily.

```
Call UniversalFooter(CalcBook, ''", OrigWkstName$, ''")
```

Once the data has been loaded into a Worksheet, the analysis of the data can begin. The first step is to perform some basic statistics on the lot of substrates that house the gages. A calculation is made to determine the average and standard deviations of the resistors located in identical regions of all the substrates within the given lot. The sheets "AVG" and "STDEV" are utilized to hold this information.

When performing calculations of this nature, two variations are possible. One is to simply perform the calculations in code and write the value of the calculation into a cell. This is really an inferior method for several reasons. First, the person looking at the spreadsheet will have just a number, with no idea how it was calculated or what data was utilized to calculate it. Another limitation of such an approach is that it prevents users from using the created Worksheets to perform "what if"-types of analyses. In other words, changing a value in one area of the report will not propagate through the entire Worksheet and change the results that depend on the altered value in other sections of the Worksheet.

Using the `AddFormula` subroutine developed in Chapter 4, it is possible to utilize the Worksheets to do the calculations by writing the appropriate formula to a particular cell. The only caveat here is that the formulas to be written must be constructed specifically for each cell to reflect the values they must reference. This is easily done in the `OriginalCalcs` subroutine by means of a looping structure that builds a string representation of the formula desired, which can then be assigned to the cell.

```
Sub OriginalCalcs()
'Here we extract the original data and compute the mean and
stdev using all of the data
'Place the Average on hidden Worksheet AVG and the STDEV on
hidden W orksheet STDEV
Dim ii As Integer, msg$
Dim row As Integer, col As Integer, cellstr$, funcstr$
Sheets(SGDataSheet).Activate
TRows = TotalRows(CalcBook, SGDataSheet)
ColOrigin = FindCol(FindPos2(CalcBook, SGDataSheet, "A")) + 1
```

```
RowOrigin = FindRow(FindPos2(CalcBook, SGDataSheet, "A"))
RowPos = RowOrigin
For ii = 1 To 100
  If RowPos > TRows Then Exit For
  NoScans = NoScans + 1
  RowPos = RowPos + 10
Next ii
If UtilScans > NoScans Then
    msg$ = "Fatal Error: " & "Total Scans in File = " &
    str(NoScans) & " Scans to Utilize = " & str(UtilScans) & "!"
    MsgBox "You wish to Utilize More Scans than the Data File
    has Stored!", vbCritical + vbOKOnly, msg$
     End
End If
  For col = 0 To 11
    For row = 0 To 7
      For CellPos = 1 To UtilScans
        Select Case CellPos
        Case UtilScans 'Omit ","
              cellstr$ = cellstr$ & SGDataSheet & "!" &
              Chr(ColOrigin + 64 + col) & _
                       Trim(str((RowOrigin + row + (CellPos
                       - 1) * 10)))
        Case Else
              cellstr$ = cellstr$ & SGDataSheet & "!" &
              Chr(ColOrigin + 64 + col) & _
                       Trim(str((RowOrigin + row + (CellPos - 1)
                       * 10))) & ","
        End Select
        'Color Out of Spec Cells in Original Data
        If WithinSpec
        (Workbooks(CalcBook).Worksheets(SGDataSheet).range(Chr(
        ColOrigin + 64 + col) & _ Trim(str((RowOrigin + row +
        (CellPos - 1) * 10)))).Value) = False Then
        'Out of Spec - Color Cell
        Call ColorCell (CalcBook, SGDataSheet, (Chr(ColOrigin +
        64 + col) & _
                       Trim(str((RowOrigin + row + (CellPos -
                       1) * 10)))))
        End If
Next CellPos
funcstr$ = "=AVERAGE(" & cellstr$ & ")"
Call AddFormula (CalcBook, "AVG", ((row + 7)), ((col + 2)), funcstr$)
CellAvg(row + 1, col + 1) = Worksheets("AVG").Cells(row + 7,
col + 2).Value
```

```
funcstr$ = "=STDEV(" & cellstr$ & ")"
Call AddFormula (CalcBook, "STDEV", ((row + 7)), ((col + 2)),
funcstr$)
CellDev(row + 1, col + 1) = Worksheets("STDEV").Cells(row +
7, col + 2).Value
'Debug.Print funcstr$
'If ALL Values in LOT Not within 6 sigma limits color cell
For CellPos = 1 To UtilScans
  If WithinSixSigma ("AVG", row + 1, col + 1, (RowOrigin +
  row + (CellPos - 1) * 10), ColOrigin + col, CellPos) _
  = False Then
  Call ColorCell (CalcBook, "AVG", Chr(64 + col + 2) &
  Trim(str(row + 7)))

  Exit For
 End If
Next CellPos
'Clean Up built strings for next pass
 cellstr$ = "": funcstr$ = ""
 Next row
Next col
'Add HEADINGS to AVG & STDEV Sheet
Workbooks(CalcBook).Worksheets("AVG").range("H3").Value =
Trim(str(UtilScans)) & "/" & Trim(str(NoScans))
Workbooks(CalcBook).Worksheets("STDEV").range("H3").Value =
Trim(str(UtilScans)) & "/" & Trim(str(NoScans))
Workbooks(CalcBook).Worksheets("AVG").range("H4").Value =
str(OhmMin)
Workbooks(CalcBook).Worksheets("STDEV").range("H4").Value =
str(OhmMin)
Workbooks(CalcBook).Worksheets("AVG").range("D3").Value =
Val(frm10ADA.TextBox_GageR.Text)
Workbooks(CalcBook).Worksheets("STDEV").range("D3").Value =
Val(frm10ADA.TextBox_GageR.Text)
Workbooks(CalcBook).Worksheets("AVG").range("D4").Value =
Val(frm10ADA.TextBox_GageRT.Text)
Workbooks(CalcBook).Worksheets("STDEV").range("D4").Value =
Val(frm10ADA.TextBox_GageRT.Text)
Workbooks(CalcBook).Worksheets("AVG").range("L3").Value =
Val(frm10ADA.TextBox_BridVolt.Text)
Workbooks(CalcBook).Worksheets("STDEV").range("L3").Value =
Val(frm10ADA.TextBox_BridVolt.Text)
Workbooks(CalcBook).Worksheets("AVG").range("L4").Value =
str(OhmMax)
Workbooks(CalcBook).Worksheets("STDEV").range("L4").Value =
str(OhmMax)
End Sub
```

10.4 PROCESS ANALYSIS

Once the calculations have been done on the manufactured substrates an analysis can be run on the process to determine if the manufactured lot is "in control" and falls within the Six Sigma quality standards. The WithinSixSigma function determines on a point-by-point basis which locations on the substrate (resistors) meet the Six Sigma manufacturing standards (Figure 10.7).

```
Function WithinSixSigma(ByVal wksheet, ByVal row As Integer,
ByVal col As Integer, ByVal dprow As Integer, ByVal dpcol As
Integer, ByVal point As Integer) As Boolean
'Does point X satisfy six sigma criteria?
If wksheet = "ORAVG" And SixSigmaPt(row, col, point) = False Then

    'DO NOT ANALYZE THIS POINT
    WithinSixSigma = True
    Exit Function
End If
If Worksheets(SGDataSheet).Cells(dprow, dpcol).Value <=
(CellAvg(row, col) + (3 * CellDev(row, col))) And _
    Worksheets(SGDataSheet).Cells(dprow, dpcol).Value >=
    (CellAvg(row, col) - (3 * CellDev(row, col))) Then
            WithinSixSigma = True
    Else
            WithinSixSigma = False
            Worksheets(SGDataSheet).Cells(dprow, dpcol).Value,
            (CellAvg(row, col) + (3 * CellDev(row, col)))
End If
End Function
```

Those points that do not conform to Six Sigma are then highlighted and colored to indicate which positions on the substrate are not meeting manufacturing standards (Figure 10.8). In this example, locations A1–A8 on the substrate were substandard.

The next analysis will focus only on *individual* resistor elements within each substrate. The goal here is to identify out-of-tolerance resistors and identify gages that contain such resistors. The WithinSpec function makes a determination as to whether an individual resistor on a substrate is within allowable specifications (± 0.01%).

	A	B	C	D	E	F	G	H	I	J	K	L	M
1	Substrate Number: 1												
2		1	2	3	4	5	6	7	8	9	10	11	12
3	A	600	600	600	600	1400	1400	1400	1400	999.9167258	999.9678466	999.9725319	999.9103573
4	B	999.8966111	999.9723489	999.9331509	1000.029669	999.9788623	999.9417922	999.8759545	999.9040238	999.9726458	999.8738792	1000.062836	999.9310925
5	C	1000.094878	1000.011598	1000.081239	999.9646204	999.9521455	999.8961700	1000.068664	999.9645869	999.8826497	999.8773087	999.948979	999.9980343
6	D	999.9900871	999.9399879	1000.096245	999.9128736	1000.046445	999.8957553	999.8876013	999.886897	999.9902322	1000.065419	1000.077431	1000.003888
7	E	1000.017871	1000.061392	1000.045838	999.9106477	1000.026246	999.9009355	1000.062327	1000.125302	1000.067783	999.8915762	999.9247017	999.9904526
8	F	999.8898771	999.8776759	999.9100673	1000.045973	999.9360617	1000.01685	999.9233456	999.9321726	1000.04214	999.9913297	999.9237162	999.9313701
9	G	999.9610142	999.8957155	1000.017267	1000.114547	1000.000665	1000.089452	1000.026362	999.9950835	1000.129509	1000.038504	999.8968941	1000.033534
10	H	999.9698191	1000.048729	999.9611683	1000.108757	1000.080938	999.871931	999.9095863	1000.002035	1000.018231	999.9667793	999.9043122	999.9449463
11	Substrate Number: 2												
12		1	2	3	4	5	6	7	8	9	10	11	12
13	A	999.90759	999.9891042	1000.097867	999.9480626	1000.022174	1000.124715	1000.024575	999.9188809	999.9167258	999.9678466	999.9725319	999.9164325
14	B	999.9146076	1000.068368	1000.025496	1000.094774	1000.015518	999.9020426	1000.127059	999.9639764	999.9949607	999.8907891	999.9262655	999.9778285
15	C	1000.116742	999.8942961	999.9400732	1000.071282	999.9840114	999.9385469	999.889645	999.8981911	999.9835173	1000.006902	1000.12789	999.9068458
16	D	1000.064502	999.9451206	1000.058681	1000.045959	999.9516048	1000.127091	1000.089939	999.8713934	999.9864532	999.9660006	999.9475727	999.89915
17	E	1000.12637	1000.027858	999.9959732	1000.090217	1000.123345	999.8856663	999.9391193	999.926232	1000.111098	1000.002362	1000.061917	999.9184082
18	F	999.978754	1000.01172	1000.121299	999.9785894	1000.042112	1000.044793	1000.059909	1000.120165	1000.044218	999.9003512	1000.039933	999.8820044
19	G	1000.009804	1000.028958	999.9437933	1000.116739	1000.024945	1000.033725	999.8750676	1000.090946	999.9323099	1000.027653	999.9746088	1000.053231
20	H	1000.00871	1000.066445	1000.041803	999.895215	1000.04851	1000.015831	999.9093494	1000.037737	1000.039047	999.9844317	1000.046481	999.9607913
21	Substrate Number: 3												

SGData / SGDefects / RowRpt / ColRpt / STDEV / AVG / ORAVG / Compare /

FIGURE 10.7 Individual resistor elements out of spec are identified.

	A	B	C	D	E	F	G	H	I	J	K	L	M
1							Averages of Data						
2													
3	Resistance (Ohms):		1000			Meas. count:		20/20		Test Voltage (Vin):			10
4	Resistance Tolerance (%		0.01			Resistance Min:		999.900		Resistance Max:			1000.100
5													
6		1	2	3	4	5	6	7	8	9	10	11	12
7	A	980.017	979.984	980.008	979.985	1020.002	1019.968	1020.007	1019.983	1000.014	1000.001	999.987	1000.015
8	B	1000.012	1000.012	999.997	999.995	999.959	999.986	1000.024	999.994	999.992	999.974	1000.030	999.994
9	C	1000.023	999.984	999.997	1000.024	1000.027	999.992	999.981	999.993	999.998	1000.023	999.998	999.982
10	D	999.978	1000.001	1000.015	999.977	1000.001	1000.008	999.981	999.960	999.991	1000.004	999.980	1000.009
11	E	999.981	1000.004	999.982	1000.029	1000.028	1000.008	999.988	1000.012	999.996	999.986	1000.038	999.992
12	F	1000.031	1000.018	999.975	999.997	1000.002	1000.007	1000.015	1000.023	1000.021	999.983	999.985	999.959
13	G	1000.017	999.977	999.967	1000.014	1000.007	1000.004	1000.003	999.969	999.995	999.981	1000.003	999.985
14	H	1000.012	999.995	1000.028	999.991	1000.004	999.996	999.989	1000.004	1000.004	999.965	1000.006	999.960
15													

FIGURE 10.8 Nonconforming substrate locations highlighted.

Recall that each gage is made up of four resistors. If a single resistor on a gage is out of allowable tolerances, then the entire gage is deemed a failure. A looping structure is set up in the ShadeGages subroutine to loop through each gage one at a time, and then through each resistor in a given gage (starting in the upper left-hand corner and proceeding clockwise). Figure 10.9 shows the pattern that the algorithm utilizes to search through each gage. Note that all 24 gages are searched, starting with the gage in the upper left-hand corner. Each gage is then examined proceeding downward until the end of the matrix is encountered, at which point the search is shifted two columns to the right and starts again at the top of the column. The RowCoord and ColCoord functions provide the proper (x,y) coordinates to rotate the WithinSpec function through to check the value of each resistor within each gage.

Once a determination has been made as to whether a gage is acceptable or defective, the ColorRangeBackground subroutine can be called to color each gage an appropriate color. In this instance, defective gages have their background set to red, whereas acceptable gages have their background set to green. The coloring scheme is applied to the "SGDefects" sheet that is simply a copy of the raw data sheet created earlier in the macro.

```
Sub ShadeGages()
'Shade Out of Spec Gages Light Red, Good Gages Light Green
Dim ColOrigin As Integer, RowOrigin As Integer, RowPos As Integer
Dim gage As Integer, defect As Boolean, SQRange As String
Sheets("SGDefects").Activate
TRows = TotalRows(CalcBook, "SGDefects")
ColOrigin = FindCol(FindPos2(CalcBook, "SGDefects", "A")) + 1
RowOrigin = FindRow(FindPos2(CalcBook, "SGDefects", "A"))
RowPos = RowOrigin
For ii = 1 To 100
```

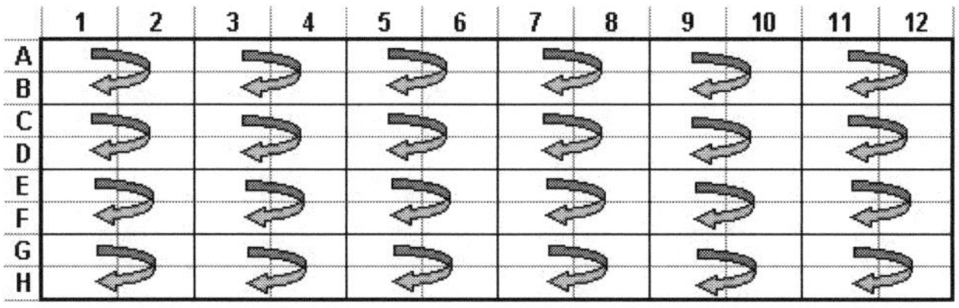

FIGURE 10.9 Strain gage examination algorithm.

```vbnet
      If RowPos > TRows Then Exit For
      NoScans = NoScans + 1
      RowPos = RowPos + 10
   Next ii
   For col = 1 To 12 Step 2
      For row = 1 To 8 Step 2
         For CellPos = 1 To UtilScans
            'Rotate through Each Cell in Gage
            defect = False 'Reset
            For gage = 1 To 4
                  Debug.Print gage, Chr(ColOrigin + 64 + ColCoord(gage,
                  col + 1) - 2) & _

                  Trim(str((RowOrigin - 1 + RowCoord(gage, row) +
                  (CellPos - 1) * 10)))
                  If WithinSpec
                  (Workbooks(CalcBook).Worksheets("SGDefects").range(
                  Chr(ColOrigin + 64 + ColCoord(gage, col + 1) - 2)
                  & Trim(str((RowOrigin - 1 + RowCoord(gage, row) +
                  (CellPos - 1) * 10)))).Value) = False Then
                     defect = True
                  End If
            Next gage
            'If ANY Defects present in block of 4 - gage is bad
            'Define Square Range
            SQRange = Chr(ColOrigin + 64 + ColCoord(1, col + 1)
            - 2) & _

            Trim(str((RowOrigin - 1 + RowCoord(1, row) + (CellPos
            - 1) * 10))) & ":" & _

            Chr(ColOrigin + 64 + ColCoord(3, col + 1) - 1) & _
            Trim(str((RowOrigin - 1 + RowCoord(3, row) + (CellPos
            - 1) * 10)))
            Debug.Print "SQRange: ", SQRange
            If defect = True Then
               Call ColorRangeBackground(CalcBook, "SGDefects",
               SQRange, 9)

               Else
               Call ColorRangeBackground(CalcBook, "SGDefects",
               SQRange, 35)
            End If
         Next CellPos
      Next row
   Next col
End Sub
```

```
Function WithinSpec(RVal As Single) As Boolean
'Determines if Resistance Measurement is Within Spec
If RVal < OhmMin Or RVal > OhmMax Then
  'Out of Specification
  WithinSpec = False
  Else
  WithinSpec = True
End If
End Function
```

```
Sub ColorRangeBackground(wkbook, wksheet, cellrange, color)
'Colors the Background of a Range of Cells
'This subroutine WILL NOT WORK if the Sheet with the Selection
Range is not Active!
'Color Index Values: (Light Green = 35: Light Red = 9)
Workbooks(wkbook).Worksheets(wksheet).Activate
Workbooks(wkbook).Worksheets(wksheet).range(cellrange).Select
With Selection.Interior
  .ColorIndex = color
  .Pattern = xlSolid
End With
End Sub
```

The next two functions create coordinates that rotate clockwise around each gage.

```
Function RowCoord(ByVal index As Integer, ByVal row As Integer)
As Integer
If index > 2 Then
   RowCoord = row + 1
   Else
   RowCoord = row
End If
End Function
```

```
Function ColCoord(ByVal index As Integer, ByVal col As Integer)
As Integer
If IsEven (index) = True Then
   ColCoord = col + 1
   Else
   ColCoord = col
End If
End Function
```

The IsEven and IsOdd functions are very handy and can be utilized in a variety of scenarios.

```
Function IsEven(ByVal lngNumber As Long) As Boolean
 IsEven = (lngNumber \ 2 = lngNumber / 2)
End Function
```

```
Function IsOdd(ByVal lngNumber As Long) As Boolean
 IsOdd = (lngNumber \ 2 <> lngNumber / 2)
End Function
```

Often, it is worthwhile to do an analysis of the effect outliers have on a series of data. Outliers are data points that are located in excess of ± 3 Standard Deviations from the mean. If outliers make up only a small number of data points within the data set, eliminating them from consideration can yield quite different results. Outlier analysis can often provide insight on what is causing the quality control problems in a process.

If the user chooses to conduct an outlier analysis, the `EliminOutliers` subroutine is invoked, and the Averages and Standard Deviations are recalculated omitting any points that were greater than ± 3 Standard Deviations from the originally calculated mean and standard deviation. This is accomplished by using the `IncludePoint` function, which makes a determination if a particular point should be included in an analysis or omitted from the analysis. Notice the callout in Figure 10.10 that shows an instance of where only 19 of 20 points were utilized in the revised calculation (the omission of 1 outlier).

```
Sub EliminOutliers()
'Here we extract the original data and compute the mean and
stdev using all of the data
```

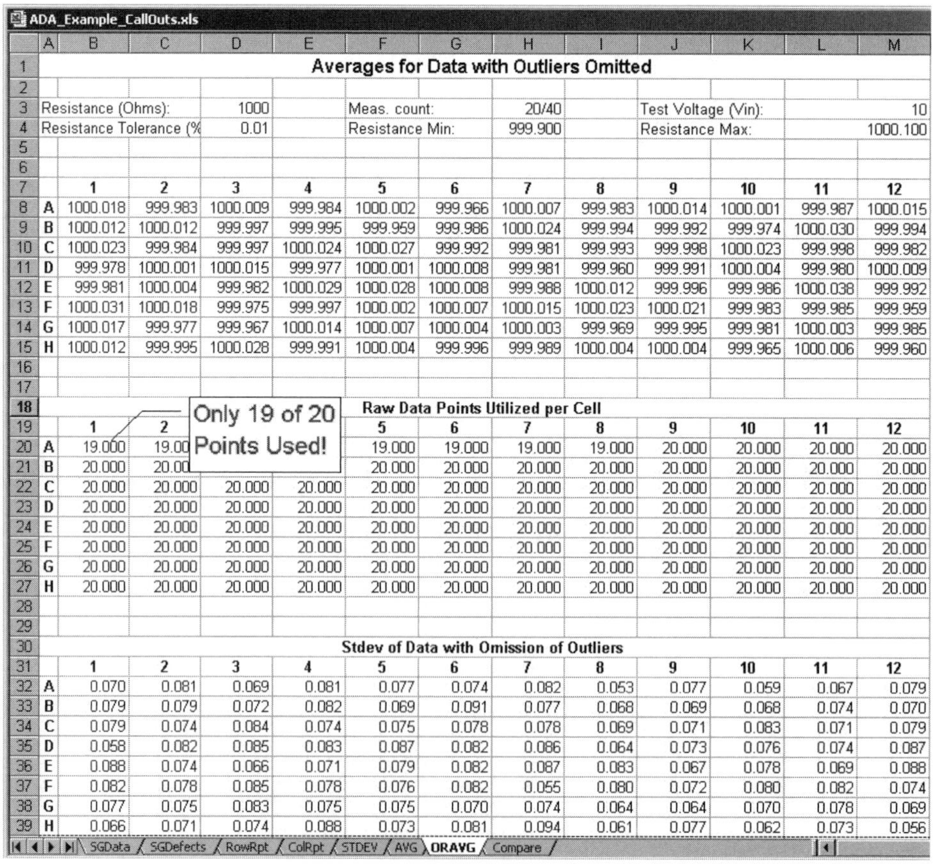

ADA_Example_CallOuts.xls

Averages for Data with Outliers Omitted

Resistance (Ohms):		1000		Meas. count:		20/40		Test Voltage (Vin):			10
Resistance Tolerance (%		0.01		Resistance Min:		999.900		Resistance Max:			1000.100

	1	2	3	4	5	6	7	8	9	10	11	12
A	1000.018	999.983	1000.009	999.984	1000.002	999.966	1000.007	999.983	1000.014	1000.001	999.987	1000.015
B	1000.012	1000.012	999.997	999.995	999.959	999.986	1000.024	999.994	999.992	999.974	1000.030	999.994
C	1000.023	999.984	999.997	1000.024	1000.027	999.992	999.981	999.993	999.998	1000.023	999.998	999.982
D	999.978	1000.001	1000.015	999.977	1000.001	1000.008	999.981	999.960	999.991	1000.004	999.980	1000.009
E	999.981	1000.004	999.982	1000.029	1000.028	1000.008	999.988	1000.012	999.996	999.986	1000.038	999.992
F	1000.031	1000.018	999.975	999.997	1000.002	1000.007	1000.015	1000.023	1000.021	999.983	999.985	999.959
G	1000.017	999.977	999.967	1000.014	1000.007	1000.004	1000.003	999.969	999.995	999.981	1000.003	999.985
H	1000.012	999.995	1000.028	999.991	1000.004	999.996	999.989	1000.004	1000.004	999.965	1000.006	999.960

Raw Data Points Utilized per Cell

	1	2			5	6	7	8	9	10	11	12
A	19.000	19.00			19.000	19.000	19.000	19.000	20.000	20.000	20.000	20.000
B	20.000	20.00			20.000	20.000	20.000	20.000	20.000	20.000	20.000	20.000
C	20.000	20.000	20.000	20.000	20.000	20.000	20.000	20.000	20.000	20.000	20.000	20.000
D	20.000	20.000	20.000	20.000	20.000	20.000	20.000	20.000	20.000	20.000	20.000	20.000
E	20.000	20.000	20.000	20.000	20.000	20.000	20.000	20.000	20.000	20.000	20.000	20.000
F	20.000	20.000	20.000	20.000	20.000	20.000	20.000	20.000	20.000	20.000	20.000	20.000
G	20.000	20.000	20.000	20.000	20.000	20.000	20.000	20.000	20.000	20.000	20.000	20.000
H	20.000	20.000	20.000	20.000	20.000	20.000	20.000	20.000	20.000	20.000	20.000	20.000

Only 19 of 20 Points Used!

Stdev of Data with Omission of Outliers

	1	2	3	4	5	6	7	8	9	10	11	12
A	0.070	0.081	0.069	0.081	0.077	0.074	0.082	0.053	0.077	0.059	0.067	0.079
B	0.079	0.079	0.072	0.082	0.069	0.091	0.077	0.068	0.069	0.068	0.074	0.070
C	0.079	0.074	0.084	0.074	0.075	0.078	0.078	0.069	0.071	0.083	0.071	0.079
D	0.058	0.082	0.085	0.083	0.087	0.082	0.086	0.064	0.073	0.076	0.074	0.087
E	0.088	0.074	0.066	0.071	0.079	0.082	0.087	0.083	0.067	0.078	0.069	0.088
F	0.082	0.078	0.085	0.078	0.076	0.082	0.055	0.080	0.072	0.080	0.082	0.074
G	0.077	0.075	0.083	0.075	0.075	0.070	0.074	0.064	0.064	0.070	0.078	0.069
H	0.066	0.071	0.074	0.088	0.073	0.081	0.094	0.061	0.077	0.062	0.073	0.056

SGData / SGDefects / RowRpt / ColRpt / STDEV / AVG / ORAVG / Compare

FIGURE 10.10 Parameter report for lot with outliers omitted from calculations.

```
'Place the Average on hidden Worksheet AVG and the STDEV on
hidden Worksheet STDEV
Dim ii As Integer
Dim row As Integer, col As Integer, cellstr$, funcstr$
Sheets(SGDataSheet).Activate
For col = 0 To 11
 For row = 0 To 7
   For CellPos = 1 To UtilScans
    If IncludePoint (CellPos, row + 1, col + 1, ((RowOrigin +
    row + (CellPos - 1) * 10)), (ColOrigin + col)) = True _
Then
    Select Case CellPos
    Case UtilScans 'Omit ","
      cellstr$ = cellstr$ & SGDataSheet & "!" & Chr(ColOrigin
      + 64 + col) & _
       Trim(str((RowOrigin + row + (CellPos - 1) * 10)))
    Case Else
      cellstr$ = cellstr$ & SGDataSheet & "!" & Chr(ColOrigin
      + 64 + col) & _
          Trim(str((RowOrigin + row + (CellPos - 1) * 10))) & ","
    End Select
    End If
Next CellPos
funcstr$ = "=AVERAGE(" & cellstr$ & ")"
Call AddFormula(CalcBook, "ORAVG", ((row + 8)), ((col + 2)),
funcstr$)
'Report Number of Cells (Datapoints) used in revised average
Worksheets("ORAVG").Cells(row + 20, col + 2).Value =
TotalPts(row + 1, col + 1)
funcstr$ = "=STDEV(" & cellstr$ & ")"
Call AddFormula(CalcBook, "ORAVG", ((row + 32)), ((col + 2)),
funcstr$)
'If ALL Values in LOT Not within 6 sigma limits color cell
For CellPos = 1 To UtilScans
  If WithinSixSigma("ORAVG", row + 1, col + 1, (RowOrigin +
  row + (CellPos - 1) * 10), ColOrigin + col, CellPos)   =
  False Then
    Call ColorCell(CalcBook, "ORAVG", Chr(64 + col + 2) &
    Trim(str(row + 8)))
    Exit For
  End If
Next CellPos
    'Clean Up built strings for next pass
    cellstr = "": funcstr$ = ""
  Next row
```

```
Next col
'Add Heading Parameters to ORAVG Sheet
Workbooks(CalcBook).Worksheets("ORAVG").range("H3").Value =
Trim(str(UtilScans)) & "/" & Trim(str(NoScans))
Workbooks(CalcBook).Worksheets("ORAVG").range("H4").Value =
str(OhmMin)
Workbooks(CalcBook).Worksheets("ORAVG").range("D3").Value =
Val(frm10ADA.TextBox_GageR.Text)
Workbooks(CalcBook).Worksheets("ORAVG").range("D4").Value =
Val(frm10ADA.TextBox_GageRT.Text)
Workbooks(CalcBook).Worksheets("ORAVG").range("L3").Value =
Val(frm10ADA.TextBox_BridVolt.Text)

Workbooks(CalcBook).Worksheets("ORAVG").range("L4").Value =
str(OhmMax)
End Sub

Function IncludePoint(ByVal point As Single, ByVal row As
Integer, ByVal col As Integer, ByVal dprow As Integer, ByVal
dpcol As Integer) As Boolean
'Should point X be included in the averaging process
If Worksheets(SGDataSheet).Cells(dprow, dpcol).Value <=
(CellAvg(row, col) + (3 * CellDev(row, col))) And
Worksheets(SGDataSheet).Cells(dprow, dpcol).Value >=
(CellAvg(row, col) - (3 * CellDev(row, col))) Then

     IncludePoint = True
     TotalPts(row, col) = TotalPts(row, col) + 1
     SixSigmaPt(row, col, point) = True
     Else
     IncludePoint = False
     SixSigmaPt(row, col, point) = False
End If
End Function
```

If a second analysis is to be done on the data, there is little point in completing both analyses unless there is a mechanism by which they can be compared. What is really needed is a sheet that presents the analyzed data for both cases, where all the data is utilized in the analysis and where the outliers are omitted from the analysis. Then a comparison of the deviation between the two analyzed data sets can follow on the same sheet. The ComparisonAnalysis subroutine accomplishes this task.

```
Sub ComparisonAnalysis()
'Create a Comparison Report between Data Calculated with all
points vs
'Data Calculated with Outliers Removed
Dim ii As Integer
Dim row As Integer, col As Integer, ROrigin As Integer,
cellstr$, funcstr$, wsheet$
```

```
Sheets("Compare").Activate
'Fill in Data Calculated utilizing all points
wsheet$ = "AVG!": ROrigin = 0
For col = 1 To 12
    For row = 0 To 7
        'Assemble string with cell information
        cellstr$ = wsheet$ & Chr(col + 65) & Trim(str(7 +
        ROrigin + row))
        funcstr$ = "=" & cellstr$
        Call AddFormula(CalcBook, "Compare", ((row + 8)),
        ((col + 1)), funcstr$)

    Call CopyCellFormat(7 + ROrigin + row, col + 1, 7 +
    ROrigin + row, col + 1, "AVG", CalcBook, (row + 8), (col
    + 1), (row + 8), (col + 1), "Compare", CalcBook)
        'Clean Up built strings for next pass
        cellstr$ = "": funcstr$ = ""
    Next row
Next col

'Fill in Data Calculated eliminating Outliers
wsheet$ = "ORAVG!": ROrigin = 1
For col = 1 To 12
  For row = 0 To 7
    'Assemble string with cell information
    cellstr$ = wsheet$ & Chr(col + 65) & Trim(str(7 + ROrigin + row))
    funcstr$ = "=" & cellstr$
    Call AddFormula(CalcBook, "Compare", ((row + 20)), ((col +
    1)), funcstr$)

    Call CopyCellFormat(7 + ROrigin + row, col + 1, 7 + ROrigin
    + row, col + 1, "AVG", CalcBook, (row + 20), (col + 1),
    (row + 20), (col + 1), "Compare", CalcBook)

    'Clean Up built strings for next pass
    cellstr$ = "": funcstr$ = ""
  Next row
Next col
'Add Heading Parameters to Compare Sheet
Workbooks(CalcBook).Worksheets("Compare").range("H3").Value =
Trim(str(UtilScans)) & "/" & Trim(str(NoScans))

Workbooks(CalcBook).Worksheets("Compare").range("H4").Value =
str(OhmMin)

Workbooks(CalcBook).Worksheets("Compare").range("D3").Value =
Val(frm10ADA.TextBox_GageR.Text)

Workbooks(CalcBook).Worksheets("Compare").range("D4").Value =
Val(frm10ADA.TextBox_GageRT.Text)
```

ADA_Example_CallOuts.xls

Averages for Data (No Points Excluded)

Resistance (Ohms):	1000	Meas. count:	20/40	Test Voltage (Vin):	10
Resistance Tolerance (%	0.01	Resistance Min:	999.900	Resistance Max:	1000.100

	1	2	3	4	5	6	7	8	9	10	11	12
A	**980.017**	**979.984**	**980.008**	**979.985**	**1020.002**	**1019.968**	**1020.007**	**1019.983**	1000.014	1000.001	999.987	1000.015
B	1000.012	1000.012	999.997	999.995	999.959	999.986	1000.024	999.994	999.992	999.974	1000.030	999.994
C	1000.023	999.984	999.997	1000.024	1000.027	999.992	999.981	999.993	999.998	1000.023	999.998	999.982
D	999.978	1000.001	1000.015	999.977	1000.001	1000.008	999.981	999.960	999.991	1000.004	999.980	1000.009
E	999.981	1000.004	999.982	1000.029	1000.028	1000.008	999.988	1000.012	999.996	999.986	1000.038	999.992
F	1000.031	1000.018	999.975	999.997	1000.002	1000.007	1000.015	1000.023	1000.021	999.983	999.985	999.959
G	1000.017	999.977	999.967	1000.014	1000.007	1000.004	1000.003	999.969	999.995	999.981	1000.003	999.985
H	1000.012	999.995	1000.028	999.991	1000.004	999.996	999.989	1000.004	1000.004	999.965	1000.006	999.960

Averages for Data with Outliers Omitted

	1	2	3	4	5	6	7	8	9	10	11	12
A	1000.018	999.983	1000.009	999.984	1000.002	999.966	1000.007	999.983	1000.014	1000.001	999.987	1000.015
B	1000.012	1000.012	999.997	999.995	999.959	999.986	1000.024	999.994	999.992	999.974	1000.030	999.994
C	1000.023	999.984	999.997	1000.024	1000.027	999.992	999.981	999.993	999.998	1000.023	999.998	999.982
D	999.978	1000.001	1000.015	999.977	1000.001	1000.008	999.981	999.960	999.991	1000.004	999.980	1000.009
E	999.981	1000.004	999.982	1000.029	1000.028	1000.008	999.988	1000.012	999.996	999.986	1000.038	999.992
F	1000.031	1000.018	999.975	999.997	1000.002	1000.007	1000.015	1000.023	1000.021	999.983	999.985	999.959
G	1000.017	999.977	999.967	1000.014	1000.007	1000.004	1000.003	999.969	999.995	999.981	1000.003	999.985
H	1000.012	999.9955	1000.028	999.9908	1000.004	999.996	999.9892	1000.004	1000.004	999.965	1000.006	999.96

% Difference Between of Data with/without Omission of Outliers

	1	2	3	4	5	6	7	8	9	10	11	12
A	2.041	2.041	2.041	2.041	1.961	1.961	1.961	1.961	0.000	0.000	0.000	0.000
B	0.000	0.000	0.000	0.000	0.000	0.000	0.000	0.000	0.000	0.000	0.000	0.000
C	0.000	0.000	0.000	0.000	0.000	0.000	0.000	0.000	0.000	0.000	0.000	0.000
D	0.000	0.000	0.000	0.000	0.000	0.000	0.000	0.000	0.000	0.000	0.000	0.000
E	0.000	0.000	0.000	0.000	0.000	0.000	0.000	0.000	0.000	0.000	0.000	0.000
F	0.000	0.000	0.000	0.000	0.000	0.000	0.000	0.000	0.000	0.000	0.000	0.000
G	0.000	0.000	0.000	0.000	0.000	0.000	0.000	0.000	0.000	0.000	0.000	0.000
H	0.000	0.000	0.000	0.000	0.000	0.000	0.000	0.000	0.000	0.000	0.000	0.000

SGData / SGDefects / RowRpt / ColRpt / STDEV / AVG / ORAVG / Compare

FIGURE 10.11 A comparison report on the analyzed data sets.

```
Workbooks(CalcBook).Worksheets("Compare").range("L3").Value =
Val(frm10ADA.TextBox_BridVolt.Text)

Workbooks(CalcBook).Worksheets("Compare").range("L4").Value =
str(OhmMax)

End Sub
```

Notice that the compare Worksheet in Figure 10.11 shows both sets of data. In the first data series (inclusive of all data points), notice that resistors A1–A8 are highlighted, indicating that the fabrication of the chip at these locations does not conform to Six Sigma manufacturing standards. On the second data series (which omits outliers), notice that the remaining (19) substrates do meet Six Sigma standards. Such an analysis can help track down quality problems. Is the failure really due to a manufacturing error, or is there a quality problem with 1 of the 20 substrates (more likely)? The last matrix in this report shows the percentage differential between the two data sets. Notice how the elimination of a single outlier can greatly improve the variability of the data set.

Another analysis that might be of use is to look at trends in the columns and rows of resistors on the substrate (Figure 10.12 and Figure 10.13). A manufacturing defect could manifest itself during the generation of a particular row or column of resistors on the substrate. The following two subroutines will perform the row and column calculations should the user elect to have them done via the GUI.

ADA_Example_CallOuts.xls

Row Averages for Data

Resistance (Ohms):			1000		Meas. count:		20/40		Test Voltage (Vin):			10	
Resistance Tolerance (%):			0.01		Resistance Min:		999.900		Resistance Max:			1000.100	
	1	**2**	**3**	**4**	**5**	**6**	**7**	**8**	**9**	**10**	**11**	**12**	**Avg**
A	1000.018	999.983	1000.009	999.984	1000.002	999.966	1000.007	999.983	1000.014	1000.001	999.987	1000.015	999.997
B	1000.012	1000.012	999.997	999.995	999.959	999.986	1000.024	999.994	999.992	999.974	1000.030	999.994	999.997
C	1000.023	999.984	999.997	1000.024	1000.027	999.992	999.981	999.993	999.998	1000.023	999.998	999.982	1000.002
D	999.978	1000.001	1000.015	999.977	1000.001	1000.008	999.981	999.960	999.991	1000.004	999.980	1000.009	999.992
E	999.981	1000.004	999.982	1000.029	1000.028	1000.008	999.988	1000.012	999.996	999.986	1000.038	999.992	1000.004
F	1000.031	1000.018	999.975	999.997	1000.002	1000.007	1000.015	1000.023	1000.021	999.983	999.985	999.959	1000.001
G	1000.017	999.977	999.967	1000.014	1000.007	1000.004	1000.003	999.969	999.995	999.981	1000.003	999.985	999.994
H	1000.012	999.995	1000.028	999.991	1000.004	999.996	999.989	1000.004	1000.004	999.965	1000.006	999.960	999.996

SGData / SGDefects / **RowRpt** / ColRpt / STDEV / AVG / ORAVG / Compare

FIGURE 10.12 Analysis of substrate rows.

```
Sub DoRowCalcs()
'Create a Worksheet with Row Wise Averages if desired
Dim ii As Integer
Dim row As Integer, col As Integer, ROrigin As Integer,
cellstr$, funcstr$, wsheet$
Sheets("RowRpt").Activate
'Determine Which Sheet Data is to be drawn from
If CalcOutliers = True Then
   wsheet$ = "ORAVG!": ROrigin = 1
   Else
   wsheet$ = "AVG!": ROrigin = 0
End If

For col = 1 To 12
  For row = 0 To 7
    'Assemble string with cell information
    cellstr$ = wsheet$ & Chr(col + 65) & Trim(str(7 + ROrigin + row))
    funcstr$ = "=" & cellstr$
    Call AddFormula(CalcBook, "RowRpt", ((row + 7)), ((col +
    1)), funcstr$)
    'If Average > Threshold Value in GUI Color iff option set
    'If Workbooks(CalcBook).Worksheets("RowRpt").Cells(row +
    7, col + 1).Value > Val(frm10ADA!ThreshBox.Text) And _
```

ADA_Example_CallOuts.xls

Column Averages for Data

Resistance (Ohms):			1000		Meas. count:		20/40		Test Voltage (Vin):			10
Resistance Tolerance (%):			0.01		Resistance Min:		999.900		Resistance Max:			1000.100
	1	**2**	**3**	**4**	**5**	**6**	**7**	**8**	**9**	**10**	**11**	**12**
A	1000.018	999.983	1000.009	999.984	1000.002	999.966	1000.007	999.983	1000.014	1000.001	999.987	1000.015
B	1000.012	1000.012	999.997	999.995	999.959	999.986	1000.024	999.994	999.992	999.974	1000.030	999.994
C	1000.023	999.984	999.997	1000.024	1000.027	999.992	999.981	999.993	999.998	1000.023	999.998	999.982
D	999.978	1000.001	1000.015	999.977	1000.001	1000.008	999.981	999.960	999.991	1000.004	999.980	1000.009
E	999.981	1000.004	999.982	1000.029	1000.028	1000.008	999.988	1000.012	999.996	999.986	1000.038	999.992
F	1000.031	1000.018	999.975	999.997	1000.002	1000.007	1000.015	1000.023	1000.021	999.983	999.985	999.959
G	1000.017	999.977	999.967	1000.014	1000.007	1000.004	1000.003	999.969	999.995	999.981	1000.003	999.985
H	1000.012	999.995	1000.028	999.991	1000.004	999.996	999.989	1000.004	1000.004	999.965	1000.006	999.960
Avg:	1000.009	999.997	999.996	1000.001	1000.004	999.996	999.999	999.992	1000.001	999.990	1000.003	999.987

SGData / SGDefects / RowRpt / **ColRpt** / STDEV / AVG / ORAVG / Compare

FIGURE 10.13 Analysis of substrate columns.

```
            'frm10ADA!SolCheck.Value = True Then
               'Call ColorCell(row + 7, col + 1)
            'End If
            'Clean Up built strings for next pass
            cellstr$ = "": funcstr$ = ""
             Next row
        Next col
        'Add Heading Parameters to RowRpt Sheet
        Workbooks(CalcBook).Worksheets("RowRpt").range("H3").Value =
        Trim(str(UtilScans)) & "/" & Trim(str(NoScans))
        Workbooks(CalcBook).Worksheets("RowRpt").range("H4").Value =
        str(OhmMin)
        Workbooks(CalcBook).Worksheets("RowRpt").range("D3").Value =
        Val(frm10ADA.TextBox_GageR.Text)
        Workbooks(CalcBook).Worksheets("RowRpt").range("D4").Value =
        Val(frm10ADA.TextBox_GageRT.Text)
        Workbooks(CalcBook).Worksheets("RowRpt").range("L3").Value =
        Val(frm10ADA.TextBox_BridVolt.Text)
        Workbooks(CalcBook).Worksheets("RowRpt").range("L4").Value =
        str(OhmMax)
        End Sub
        Sub DoColCalcs()
        'Create a Worksheet with Col Wise Averages if desired
        Dim ii As Integer
        Dim row As Integer, col As Integer, ROrigin As Integer,
        cellstr$, funcstr$, wsheet$
        Sheets("ColRpt").Activate
        'Determine Which Sheet Data is to be drawn from
        If CalcOutliers = True Then
            wsheet$ = "ORAVG!": ROrigin = 1
            Else
            wsheet$ = "AVG!": ROrigin = 0
        End If

        For col = 1 To 12
          For row = 0 To 7
            'Assemble string with cell information
            cellstr$ = wsheet$ & Chr(col + 65) & Trim(str(7 + ROrigin + row))
            funcstr$ = "=" & cellstr$
            Call AddFormula(CalcBook, "ColRpt", ((row + 7)), ((col + 1)),
            funcstr$)
            'If Average > Threshold Value in GUI Color iff option set
            'If Workbooks(CalcBook).Worksheets("ColRpt").Cells(row + 7,
            col + 1).Value > Val(frm10ADA!ThreshBox.Text) And _
            'frm10ADA!SolCheck.Value = True Then
              'Call ColorCell(row + 7, col + 1)
```

```
    'End If
    'Clean Up built strings for next pass
    cellstr$ = "": funcstr$ = ""
    Next row
 Next col
 'Add Heading Parameters to ColRpt Sheet
 Workbooks(CalcBook).Worksheets("ColRpt").range("H3").Value =
 Trim(str(UtilScans)) & "/" & Trim(str(NoScans))
 Workbooks(CalcBook).Worksheets("ColRpt").range("H4").Value =
 str(OhmMin)

 Workbooks(CalcBook).Worksheets("ColRpt").range("D3").Value =
 Val(frm10ADA.TextBox_GageR.Text)
 Workbooks(CalcBook).Worksheets("ColRpt").range("D4").Value =
 Val(frm10ADA.TextBox_GageRT.Text)
 Workbooks(CalcBook).Worksheets("ColRpt").range("L3").Value =
 Val(frm10ADA.TextBox_BridVolt.Text)
 Workbooks(CalcBook).Worksheets("ColRpt").range("L4").Value =
 str(OhmMax)
 End Sub
```

10.5 THE FINAL REPORT

What started as a raw data file of text has been transformed into an Excel Workbook that contains eight Worksheets. Each generated sheet does a specific analysis and serves a selected purpose. The name of each generated sheet and the function it serves are listed below.

SGData — The original strain gage data file is imported into this Worksheet.

SGDefects — This is a copy of the SGData sheet. Defective strain gages on this sheet are highlighted in red, and good strain gages are highlighted in green.

RowRpt — A row-by-row analysis of each resistor on the substrate.

ColRpt — A column-by-column analysis of each resistor on the substrate.

STDEV — This sheet contains the calculated standard deviations using the data in all the data sets.

AVG — This sheet contains the calculated means (or averages) using the data in all the data sets. Elements in the process that do not meet Six Sigma quality standards are highlighted.

ORAVG — This sheet contains the results of a new analysis that omits outliers in the original data set. The means, standard deviations, and data points utilized in the new analysis are all listed in matrix format.

Compare — A sheet that compares the results of the AVG sheet with the results from the ORAVG sheet. Note that the results will be identical in the case where no outliers are present in the first data set.

10.6 SAVING THE FINAL REPORT

Saving the final report always poses some vexing questions. First, should the user have the ability to choose the name the final report is saved under? Although this may seem like a good idea, the reality is that letting users choose the file names for automated reports usually proves to be problematic. Often, the names they choose do not lend themselves well to sorting or even indicating exactly what the file contains and under what conditions the analysis was run.

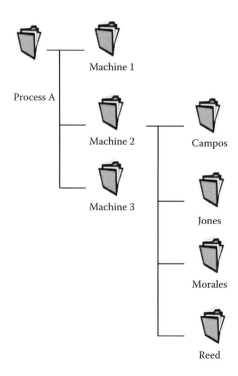

FIGURE 10.14 Logical file structure creation.

Automated file saving eliminates these problems at the cost of some degree of flexibility to the user. Recall that in Chapter 2 a date–time stamp generator was created that allowed the user to generate any type of date–time stamp they wished.

The author has a personal preference when it comes to naming and saving reports. All filenames should have a consistent date–time stamp embedded in them at some location in the filename. The date–time stamp can then have a prefix, or a suffix, or both added to it. For example, the filename might use a machine name for a prefix, and the operator's last name for suffix: MachineA_ 20051117-153322_Bissett.xls. In this example, the date–time stamp is of the form YYYYMMDD "-" HHMMSS (in military time format (24 h)).

The developer should go to even greater extremes to maintain a semblance of organization over the files their macros generate. This would include utilizing the techniques covered in Chapter 2. Whenever possible, a macro should create an "intelligent" directory structure to store its associated data files. For example, suppose a user is running Process A, and Process A can be run using one of three machines; Machine1, Machine2, or Machine3. It is also known that the process will only be run by one of four technicians whose last names are Jones, Morales, Reed, and Campos. An intelligent file structure to store analyzed data files for this process is shown in Figure 10.14.

Here, everything related to Process A is stored under a common directory, in this case "Process A." Under the Process A directory, are three additional directories named "Machine 1," "Machine 2," and "Machine 3," each of which stores information acquired from each machine capable of performing Process A. Within *each* machine directory resides a series of subdirectories that are last names of *all* the possible operators for the machines that perform Process A: Campos, Jones, Morales, and Reed.

What such a file structure does for the end user is that it allows files that contain certain information to be segregated without the necessity of doing a query on the data. The directory structure can be set up in any order that suits the user's application. In this instance, data is grouped by machines before operators. The grouping scheme could be reversed if presorting data based on the operator took preference for a particular application. Such would be the case if a certain individual was suspected of sloppy work, perhaps running the process much less carefully than their peers.

10.7 SUMMARY OF THE FINAL APPLICATION

The final sample application can be run by selecting from the menu ADA->Chapter 10->Sample ADA Application. Doing so will bring up the GUI shown in Figure 10.3. When the "Begin Analysis" button is pressed, the user will be prompted for a file. The user may test the macro by selecting the sample data file on the CD-ROM under the directory Chapter10\Samples\ADA_Data.txt. Doing so will generate the report shown in all figures in this chapter.

This sample is not the end but the beginning for the reader. Throughout the text the reader has been prepared toward the final eventuality of having the ability to construct applications specific to a current need. The first three chapters of the text were dedicated toward teaching the reader two basic principles: (1) how to get data into the Excel environment and (2) how to manipulate data around within the Excel environment. The next two chapters illustrated how to apply functions on the data that exist within the Excel environment and how to identify that information that is of interest to the user within the Excel environment. The reader is then shown how to take the information that has been identified, extracted, and calculated on, and create a standardized report within Excel to place this information. Notice that the first six chapters in the text have one common theme — they all focus on conducting operations *within the Excel environment*.

However, the Excel environment is not the end-all and be-all of the entire universe. The next three chapters were dedicated toward allowing developers to accomplish two tasks: (1) to extract information from any type of source (notably databases) and (2) to harness the computational engines of other tools to supplement Excel's functionality where it is deficient. Chapters 7 and 8 gave the reader the tools to access data contained in any Microsoft Access compatible format as well as discussing how to go about connecting to any ODBC-compliant data sources such as Oracle. Chapter 9 explored using DDE, OLE, and COM to allow Excel to control other applications, pass data to those applications, utilize their computational engines, and return any calculated results to Excel. Excel is often utilized as a "carrier vehicle" for data, meaning data is delivered to individuals by means of an Excel spreadsheet, but often, the computational work is done elsewhere, in a program such as SAS, Mathematica, Matlab, MathCad, Origin, and others. It is regrettable that Microsoft has not exploited the fact that Excel is so widely utilized in industry as a carrier vehicle. If the company added more functionality to the product, it is very likely that users would not seek out these third-party computational products but simply perform their calculations within the carrier vehicle (Excel itself). Using the examples and the information contained within the text, the user can construct Automated Data Analysis applications to perform a variety of tasks. The included sample could be readily modified to accomplish such tasks as extracting information from databases as well as files, doing custom calculations utilizing third-party programs, autoloading data files, and autosaving data files using a specific file structures to any storage medium.

Appendix A: Option Explicit

NOTE: The brilliant AddrOf function herein contained is the work of Ken Getz and Michael Kaplan. Published in the May 1998 issue of 'Microsoft Office & Visual Basic for Applications Developer (page 46).

Office 97 does not support the "AddressOf" operator which is needed to tell Windows where our "call back" function is. Getz and Kaplan figured out a workaround.

The rest of this module is entirely their work.

Declarations

These function names were puzzled out by using DUMPBIN /exports with VBA332.DLL and then puzzling out parameter names and types through a lot of trial and error and over 100 IPFs in MSACCESS.EXE and VBA332.DLL.

These parameters may not be named properly but seem to be correct in light of the function names and what each parameter does.

EbGetExecutingProj: Gives you a handle to the current VBA project
TipGetFunctionId: Gives you a function ID given a function name
TipGetLpfnOfFunctionId: Gives you a pointer a function given its function ID

```
Private Declare Function GetCurrentVbaProject _
 Lib "vba332.dll" Alias "EbGetExecutingProj" _
 (hProject As Long) As Long
Private Declare Function GetFuncID _
 Lib "vba332.dll" Alias "TipGetFunctionId" _
 (ByVal hProject As Long, ByVal strFunctionName As String, _
 ByRef strFunctionId As String) As Long
Private Declare Function GetAddr _
 Lib "vba332.dll" Alias "TipGetLpfnOfFunctionId" _
 (ByVal hProject As Long, ByVal strFunctionId As String, _
 ByRef lpfn As Long) As Long

'Main Browse for directory function
Declare Function SHBrowseForFolder Lib "shell32.dll" _
 Alias "SHBrowseForFolderA" (lpBrowseInfo As BROWSEINFO) As Long
 'Gets path from pidl
Declare Function SHGetPathFromIDList Lib "shell32.dll" _
 Alias "SHGetPathFromIDListA" (ByVal pidl As Long, ByVal _
 pszPath As String) As Long
 'Used by callback function to communicate with the browser
```

```
Declare Function SendMessage Lib "user32" _
  Alias "SendMessageA" (ByVal hwnd As Long, _
  ByVal wMsg As Long, ByVal wParam As Long, ByVal lParam As
  Any) As Long

Declare Sub CopyMemory Lib "kernel32" Alias "RtlMoveMemory"
(hpvDest As Any, _
    hpvSource As Any, ByVal cbCopy As Long)
```

Public Type BROWSEINFO

hOwner As Long	Handle to the owner window for the dialog box.
pidlRoot As Long	Address of an ITEMIDLIST structure specifying the location of the root folder from which to browse. Only the specified folder and its subfolders appear in the dialog box. This member can be NULL; in that case, the namespace root (the desktop folder) is used.
pszDisplayName As String	Address of a buffer to receive the display name of the folder selected by the user. The size of this buffer is assumed to be MAX_PATH bytes.
lpszTitle As String	Address of a null-terminated string that is displayed above the tree view control in the dialog box. This string can be used to specify instructions to the user.
ulFlags As Long	Flags specifying the options for the dialog box. See constants below.
lpfn As Long	Address of an application-defined function that the dialog box calls when an event occurs. For more information, see the BrowseCallbackProc function. This member can be NULL.
lParam As Long	Application-defined value that the dialog box passes to the callback function (in pData), if one is specified.
iImage As Long	Variable to receive the image associated with the selected folder. The image is specified as an index to the system image list.
End Type	

Public Const WM_USER = &H400
Public Const MAX_PATH = 260

'ulFlag constants

Public Const BIF_RETURNONLYFSDIRS = &H1 Only return file system directories. If the user selects folders that are not part of the file system, the OK button is grayed.
Public Const BIF_DONTGOBELOWDOMAIN = &H2 Do not include network folders below the domain level in the tree view control.
Public Const BIF_STATUSTEXT = &H4 Include a status area in the dialog box. The callback function can set the status text by sending messages to the dialog box.
Public Const BIF_RETURNFSANCESTORS = &H8 Only return file system ancestors. If the user selects anything other than a file system ancestor, the OK button is grayed.
Public Const BIF_EDITBOX = &H10 Version 4.71. The browse dialog includes an edit control in which the user can type the name of an item.

Public Const BIF_VALIDATE = &H20Version 4.71. If the user types an invalid name into the edit box, the browse dialog will call the application's BrowseCallbackProc with the BFFM_VALIDATEFAILED message. This flag is ignored if BIF_EDITBOX is not specified.

Public Const BIF_NEWDIALOGSTYLE = &H40Version 5.0. New dialog style with context menu and resizability.

Public Const BIF_BROWSEINCLUDEURLS = &H80Version 5.0. Allow URLs to be displayed or entered. Requires BIF_USENEWUI.

Public Const BIF_BROWSEFORCOMPUTER = &H1000Only return computers. If the user selects anything other than a computer, the OK button is grayed.

Public Const BIF_BROWSEFORPRINTER = &H2000Only return printers. If the user selects anything other than a printer, the OK button is grayed.

Public Const BIF_BROWSEINCLUDEFILES = &H4000The browse dialog will display files as well as folders.

Public Const BIF_SHAREABLE = &H8000Version 5.0. Allow display of remote shareable resources. Requires BIF_USENEWUI.

Message from browser to callback function constants.

Public Const BFFM_INITIALIZED = 1Indicates the browse dialog box has finished initializing. The lParam parameter is NULL.

Public Const BFFM_SELCHANGED = 2Indicates the selection has changed. The lParam parameter contains the address of the item identifier list for the newly selected folder.

Public Const BFFM_VALIDATEFAILED = 3Version 4.71. Indicates the user typed an invalid name into the edit box of the browse dialog. The lParam parameter is the address of a character buffer that contains the invalid name. An application can use this message to inform the user that the name entered was not valid. Return zero to allow the dialog to be dismissed or nonzero to keep the dialog displayed.

messages to browser from callback function

```
Public Const BFFM_SETSTATUSTEXTA = WM_USER + 100
Public Const BFFM_ENABLEOK = WM_USER + 101
Public Const BFFM_SETSELECTIONA = WM_USER + 102
Public Const BFFM_SETSELECTIONW = WM_USER + 103
Public Const BFFM_SETSTATUSTEXTW = WM_USER + 104

Public Const LMEM_FIXED = &H0
Public Const LMEM_ZEROINIT = &H40
Public Const LPTR = (LMEM_FIXED Or LMEM_ZEROINIT)

Public Declare Sub CoTaskMemFree Lib "ole32.dll" (ByVal pv As Long)

Public Declare Function LocalAlloc Lib "kernel32" _
  (ByVal uFlags As Long, _
  ByVal uBytes As Long) As Long

Public Declare Function LocalFree Lib "kernel32" _
  (ByVal hMem As Long) As Long
```

"The following declarations for the option to center the dialog in the user's screen

```
Public Declare Function GetWindowRect Lib "user32" (ByVal hwnd
As Long, lpRect As RECT) As Long
Public Declare Function GetSystemMetrics Lib "user32" (ByVal
nIndex As Long) As Long
```

```
Public Const SM_CXFULLSCREEN = 16
Public Const SM_CYFULLSCREEN = 17
Public Type RECT
     Left As Long
     Top As Long
     Right As Long
     Bottom As Long
End Type
Public Declare Function MoveWindow Lib "user32" (ByVal hwnd
As Long, _
  ByVal X As Long, _
    ByVal y As Long, _
    ByVal nWidth As Long, _
      ByVal nHeight As Long, ByVal bRepaint As Long) As Long
''End of dialog centering declarations
Dim CntrDialog As Boolean
```


AddrOf

Returns a function pointer of a VBA public function given its name. This function gives similar functionality to VBA as VB5 has with the AddressOf param type.

NOTE: This function only seems to work if the proc you are trying to get a pointer to is in the current project. This makes sense, since we are using a function named EbGetExecutingProj.


```
  Public Function AddrOf(strFuncName As String) As Long
    Dim hProject As Long
    Dim lngResult As Long
    Dim strID As String
    Dim lpfn As Long
    Dim strFuncNameUnicode As String

Const NO_ERROR = 0

    ' The function name must be in Unicode, so convert it.
    strFuncNameUnicode = StrConv(strFuncName, vbUnicode)

    ' Get the current VBA project
    ' The results of GetCurrentVBAProject seemed inconsistent,
    in our tests,

    ' so now we just check the project handle when the function
    returns.
    Call GetCurrentVbaProject(hProject)

  ' Make sure we got a project handle... we always should, but
  you never know!
  If hProject <> 0 Then
```

```
    ' Get the VBA function ID (whatever that is!)
    lngResult = GetFuncID( _
    hProject, strFuncNameUnicode, strID)

    ' We have to check this because we GPF if we try to get a
    function pointer

    ' of a non-existent function.
    If lngResult = NO_ERROR Then
        ' Get the function pointer.
        lngResult = GetAddr(hProject, strID, lpfn)

        If lngResult = NO_ERROR Then
            AddrOf = lpfn
        End If
    End If
 End If
End Function

Function GetDirectory(InitDir As String, Flags As Long, CntrDlg
As Boolean, msg) As String

    Dim bInfo As BROWSEINFO
    Dim pidl As Long, lpInitDir As Long

CntrDialog = CntrDlg ''Copy dialog centering setting to module
level variable so callback function can see it

    With bInfo
        .pidlRoot = 0 'Root folder = Desktop
        .lpszTitle = msg
        .ulFlags = Flags

        lpInitDir = LocalAlloc(LPTR, Len(InitDir) + 1)
        CopyMemory ByVal lpInitDir, ByVal InitDir, Len(InitDir) + 1
        .lParam = lpInitDir

        If Val(Application.Version) > 8 Then 'Establish the
        callback function
            .lpfn = BrowseCallBackFuncAddress
        Else
            .lpfn = AddrOf("BrowseCallBackFunc")
        End If
    End With
        'Display the dialog
        pidl = SHBrowseForFolder(bInfo)
        'Get path string from pidl
        GetDirectory = GetPathFromID(pidl)
        CoTaskMemFree pidl
        LocalFree lpInitDir
End Function
```

Windows calls this function when the dialog events occur

```
Function BrowseCallBackFunc(ByVal hwnd As Long, ByVal msg As
Long, ByVal lParam As Long, ByVal pData As Long) As Long
  Select Case msg
    Case BFFM_INITIALIZED
      'Dialog is being initialized. I use this to set the
      initial directory and to center the dialog if the requested
      SendMessage hwnd, BFFM_SETSELECTIONA, 1, pData 'Send
      message to dialog
      If CntrDialog Then CenterDialog hwnd
    Case BFFM_SELCHANGED
      'User selected a folder - change status text ("show
      status text" option must be set to see this)
    SendMessage hwnd, BFFM_SETSTATUSTEXTA, 0,
    GetPathFromID(lParam)

    Case BFFM_VALIDATEFAILED
    'This message is sent  to the callback function only if
    "Allow direct entry" and

    '"Validate direct entry" have been be set on the Demo
    worksheet

    'and the user's direct entry is not valid.
    '"Show status text" must be set on to see error message
    we send back to the dialog

    Beep
    SendMessage hwnd, BFFM_SETSTATUSTEXTA, 0, "Bad Directory"

    BrowseCallBackFunc = 1 'Block dialog closing
    Exit Function
  End Select
  BrowseCallBackFunc = 0 'Allow dialog to close
End Function

'Converts a PIDL to a string
Function GetPathFromID(ID As Long) As String
  Dim Result As Boolean, Path As String * MAX_PATH
  Result = SHGetPathFromIDList(ID, Path)
  If Result Then
    GetPathFromID = Left(Path, InStr(Path, Chr$(0)) - 1)
  Else
    GetPathFromID = ""
  End If
End Function
```

XL8 is very unhappy about using Excel 9s AddressOf operator, but as long as it is in a
function that is not called when run on XL8, it seems to allow it to exist.

```
Function BrowseCallBackFuncAddress() As Long
  BrowseCallBackFuncAddress = Long2Long(AddressOf
  BrowseCallBackFunc)
End Function
```

It is not possible to assign the result of AddressOf (which is a Long) directly to a member of a user defined data type. This explicitly "converts" it to a Long and allows the assignment

```
Function Long2Long(X As Long) As Long
  Long2Long = X
End Function

'Centers dialog on desktop
Sub CenterDialog(hwnd As Long)
  Dim WinRect As RECT, ScrWidth As Integer, ScrHeight As Integer
  Dim DlgWidth As Integer, DlgHeight As Integer
  GetWindowRect hwnd, WinRect
  DlgWidth = WinRect.Right - WinRect.Left
  DlgHeight = WinRect.Bottom - WinRect.Top
  ScrWidth = GetSystemMetrics(SM_CXFULLSCREEN)
  ScrHeight = GetSystemMetrics(SM_CYFULLSCREEN)
  MoveWindow hwnd, (ScrWidth - DlgWidth) / 2, (ScrHeight -
  DlgHeight) / 2, DlgWidth, DlgHeight, 1
End Sub
```

To include Browse for Folder in your project, copy the above code into a single module in your project.

The following function allows the Browse for Directory parameters to be easily set

```
Function SteroidBrowse(ByVal initialpath$, ByVal
AllowDirectEntry As Boolean, ByVal ValidateEntry As Boolean, _
ByVal ShowStatusText As Boolean, ByVal ShowFilesToo As Boolean,
ByVal UseNewDialogStyle As Boolean, _
ByVal CenterOnScreen As Boolean) As String
'Set up Browse Options via Boolean Flags
  Dim RetStr As String, Flags As Long, DoCenter As Boolean
  Flags = BIF_RETURNONLYFSDIRS
  If AllowDirectEntry = True Then
     Flags = Flags + BIF_EDITBOX
  End If
  If ValidateEntry = True Then
     Flags = Flags + BIF_VALIDATE
  End If
  If ShowStatusText = True Then
      Flags = Flags + BIF_STATUSTEXT
  End If
  If ShowFilesToo = True Then
      Flags = Flags + BIF_BROWSEINCLUDEFILES
  End If
```

```
If UseNewDialogStyle = True Then
    Flags = Flags + BIF_NEWDIALOGSTYLE
End If
If DirectoryExists(initialpath$) = False Then
    'An Invalid initial path was specified - Change to Root
    initialpath$ = "c:\"
    End If
    SteroidBrowse = GetDirectory(initialpath$, Flags,
    CenterOnScreen, "Please select a location to store data files")
End Function
```

Appendix B: Excel Developer Tip

A very special thank you to John Walkenbach for allowing reprint and use of his work. Please see: http://j-walk.com/ss and http://j-walk.com/ss/excel/tips/tip53.htm.

B.1 CREATING CUSTOM MENUS

If you've moved up to Excel 97 (or later) from a previous version, you may have noticed that the former menu editor is no longer available. The menu editor made it easy to create a custom menu that was stored with a particular workbook. Opening the workbook added the menu, and closing the workbook removed the menu.

The menu editor was removed because Excel 97 and later versions use CommandBars — a new (and much move superior) method of dealing with menus. Workbook-specific menus must be created programmatically using VBA code. This tip describes a relatively simple way to create a custom menu (on the Worksheet Menu Bar) that appears when a particular workbook is opened, and is deleted when the workbook is closed.

To create a custom menu, simply modify the data in the table which consists of five columns.

B.2 DOWNLOAD AN EXAMPLE

You can download an example workbook that demonstrates this technique.

- Download menumakr.xls (61K)

The example file contains all of the VBA code that you need to create your own custom menus. In most cases, you will not need to make any changes to the VBA code — simply customize the MenuSheet worksheet.

Please note that this technique will *not* work if you need to add a menu item to an existing menu.

B.3 HOW IT WORKS

This technique uses a table, which is stored in a worksheet. The figure below shows such a table. To create a custom menu, simply modify the data in the table. This table contains five columns:

- **Level:** The "level" of the particular item. Valid values are 1, 2, and 3. A level of 1 is for a menu; 2 is for a menu item; and 3 is for a submenu item. Normally, you'll have one level 1 item, with level 2 items below it. A level 2 item may or may not have level 3 (submenus) items.
- **Caption:** The text that appears in the menu, menu item, or submenu. Use an ampersand (&) to specify a character that will be underlined.
- **Position/Macro:** For level 1 items, this should be an integer that represents the position in the menubar. For level 2 or level 3 items, this will be the macro that is executed when the item is selected. If a level 2 item has one or more level 3 items, the level 2 item may not have a macro associated with it.

Budgeting.xls

	A	B	C	D	E
1	Level	Caption	Position/Macro	Divider	FaceID
2	1	&Budget Tools	10		
3	2	&Activate a sheet			
4	3	&Assumptions	GotoAssumptions		
5	3	&Model	GotoModel		
6	3	&Scenarios	GotoScenarios		
7	3	&Notes	GotoNotes		
8	2	&View scenarios...	ViewScenarios		
9	2	&Data entry...	DataEntry	TRUE	387
10	2	&Printing...	Printing	TRUE	4
11	2	&Charts			
12	3	&Budget vs. Actual	Chart1		433
13	3	&Year-To-Date	Chart2		436
14	3	&Quarterly Summary	Chart3		427
15	2	&Help		TRUE	
16	3	&Help Contents	Help1		
17	3	&Terminology	Help2		
18	3	&About	Help3	TRUE	
19					

MenuSheet

- **Divider:** True if a "divider" should be placed before the menu item or submenu item.
- **FaceID:** Optional. A code number that represents the built -in graphic images that are displayed next to an item. Follow this link for more information about determining FaceID code numbers.

B.4 AN EXAMPLE MENU

The figure below shows the menu that is created using the table above.

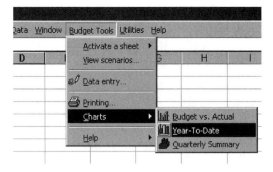

B.5 USING THIS TECHNIQUE

To use this technique in your workbook or add-in, follow these general steps:

1. Download menumakr.xls. This file contains the VBA code, plus a worksheet name MenuSheet.
2. Copy all of the code in Module1 to a module in your project.
3. Add subroutines like these to the code module for the ThisWorkbook object:

```
Private Sub Workbook_Open() Call CreateMenu
End Sub

Private Sub Workbook_BeforeClose(Cancel As Boolean) Call
DeleteMenu

End Sub
```

4. The Workbook_Open subroutine is executed when the workbook is opened; the Workbook_BeforeClose subroutine is executed before the workbook is closed.
5. Insert a new worksheet and name it MenuSheet. Better yet, copy the MenuSheet from the menumakr.xls file.
6. Customize the MenuSheet to correspond to your custom menu.
7. There is no error handling, so it's up to you to make sure that everything works.

Appendix C: Regression Analysis and Best Fit Lines (XE0124)

I am sorry to report that Microsoft © did not grant us permission to reproduce this document for this text. In most instances Microsoft's © reluctance seemed to stem from the fact that the documents requested were written for older versions of Excel. Considering that Microsoft © did not have immediate plans to update any of the requested documents, and that the said documents were freely available on the knowledgebase of their website, the denial of permission seems silly if not absurd. By keeping the material available on Microsoft's © website the company seems to feel the material still has some value for its consumers, but not enough value to update or grant permission for a third party to reprint. The last known link to the material is placed below, along with keywords that should enable the user to find the material using a Google search.

If the reader finds this policy somewhat frustrating or it has caused them to waste time hunting down documents on the internet, I would encourage them to write Microsoft and politely ask them to review their policy. My point of contact was Ceal Howell. Typical Microsoft © email addresses are First Initial + Last Name @ microsoft.com (Ex. John Doe = jdoe@microsoft.com).

Last known locations for "Regression Analysis and Best Fit Lines.":

http://support.microsoft.com/kb/103839

http://www.microsoft.com/downloads/details.aspx?FamilyID=5e0bff27-b103-49bf-8eb1-fa507709157b&displaylang=en

Keywords: Regression, "Best Fit Lines", XE0124.

Appendix D: Built-in Constants in Visual Basic for Applications (WC0993)

I am sorry to report that Microsoft © did not grant us permission to reproduce this document for this text. In most instances Microsoft's © reluctance seemed to stem from the fact that the documents requested were written for older versions of Excel. Considering that Microsoft © did not have immediate plans to update any of the requested documents, and that the said documents were freely available on the knowledgebase of their website, the denial of permission seems silly if not absurd. By keeping the material available on Microsoft's © website the company seems to feel the material still has some value for its consumers, but not enough value to update or grant permission for a third party to reprint. The last known link to the material is placed below, along with keywords that should enable the user to find the material using a Google search.

If the reader finds this policy somewhat frustrating or it has caused them to waste time hunting down documents on the internet, I would encourage them to write Microsoft and politely ask them to review their policy. My point of contact was Ceal Howell. Typical Microsoft © email addresses are First Initial + Last Name @ microsoft.com (Ex. John Doe = jdoe@microsoft.com).

Last known locations for "Built-in Constants in Visual Basic for Applications.":

http://support.microsoft.com/kb/112671

Keywords: "Built-in Constants", "Visual Basic for Applications", WC0993, 112671.

Appendix E: xlodbc.exe Add-In

I am sorry to report that Microsoft © did not grant us permission to distribute the xlodbc.exe file with this text. In most instances Microsoft's © reluctance seemed to stem from the fact that the xlodbc.exe file was written for older versions of Excel (Earlier than 2003.) Considering that Microsoft © does not have an immediate plan to update this file, and that the file is freely available on the knowledgebase of their website, the denial of permission seems silly if not absurd. While the XLODBC add-in will not install or function in Excel 2003 without some tweaking, a fix for this problem is described in this Appendix (see "Installing the XLODBC.XLA Add-In in Excel 2003".) By keeping the file available on Microsoft's © website, the company seems to feel the material still has some value for its consumers, but not enough value to update or grant permission for a third party to distribute. The last known link to the material is placed below, along with keywords that should enable the user to find the material using a Google search.

If the reader finds this policy somewhat frustrating or it has caused them to waste time hunting down documents on the internet, I would encourage them to write Microsoft and politely ask them to review their policy. My point of contact was Ceal Howell. Typical Microsoft © email addresses are First Initial + Last Name @ microsoft.com (Ex. John Doe = jdoe@microsoft.com).

Note to Reader: Microsoft © will not allow anyone to download this file unless they can prove to Microsoft's satisfaction that they are running an unpirated version of Windows™ and/or Microsoft Office™. When trying to download the software you will see a message like this: "This download is available to customers running genuine Microsoft Office. Please click the Continue button to begin Office validation. As described in our privacy statement, Microsoft will not use the information collected during validation to identify or contact you." If Microsoft does not determine that the software you are running is legitimate, there is an unsubstantiated rumor that if you offer your first born child to serve as a slave in the Gates mansion you will be allowed to download anything you want from Microsoft © for free.

Last known location for the xlodbc.exe file:

http://www.microsoft.com/downloads/details.aspx?familyid=57E79367-13A0-4895-9942-5B177846AB8A&displaylang=en

Keywords: xlodbc.exe, "Open Database Connectivity", ODBC, Excel

SUNGARD DENIED PERMISSION TO REPRINT THIS MANUAL

Sungard has put together a handy little manual on using ODBC Functions within Excel. I had hoped to reprint the 20 page document here in the appendix titled "EXCEL VBA ODBC FUNCTIONS." The document outlines in a simple understandable manner how to use ODBC Functions within Microsoft Excel. At the time of editing this document was freely available on the web at these locations:

http://www.tradeline.com/doclib/library/TLOS_Excel_ODBC_Functions.doc

http://www.tradeline.com/doclib/library/TLOS_Excel_ODBC_Functions.pdf

Sadly, after deliberating on the merits of allowing me to reprint the manual which they freely post on the internet, Sungard Data Management Solutions declined to give me permission to do so. I made the request in October and after three months they phoned me and declined to give me permission. The domain www.tradeline.com listed above has moved to http://www.marketdata.sungard.com/, but at the time this document was edited the "EXCEL VBA ODBC FUNCTIONS" manual was still available on the above URLs, and the "EXCEL VBA ODBC FUNCTIONS" manual could be found using the Google Search Engine at the above URLs using the search terms "Excel ODBC VBA Functions" or "TLOS_Excel_ODBC_Functions", which is how I found it in the first place.

A very nice young lady named Michelle tried to obtain permission for me, and ironically *Sungard was not even aware the document existed* when I inquired about permission to reproduce it. (see note below.) In any event, it is always handy to have a hard copy available when working on something, and the author regrets that he was unable to provide it for the readers here.

Hello Brian,

I apologize for the delay getting back to you. What I had expected was a simple process has had to pass through several departments at several levels (further delayed as the mega-conference "SunGard World" was last week).

In short, the issue is that neither product management, corporate marketing, nor the copyright counsel know of this document. The FAME products were acquired by SunGard in 2004, and as the copyright for your document is 2002, it is likely some other SunGard affiliate. I'm sorry to ask, but could you please send me a copy of the programmers manual? I might then be able to track down what exactly is the copyrighted material.

Sorry again for the delay.

Michelle

Appendix F: XL: Scope of Variables in Visual Basic for Applications

This article was previously published under Q141693

Article ID: 141693

Last Review : August 17, 2005

Revision: 2.3

F.1 SUMMARY

The scope of a variable is determined at the time the variable is declared. In Microsoft Visual Basic for Applications, the three scopes available for variables are procedure, module, and public. The "More Information" section of this article describes each scope in detail.

F.2 MORE INFORMATION

F.2.1 PROCEDURE (LOCAL) SCOPE

A local variable with procedure scope is recognized only within the procedure in which it is declared. A local variable can be declared with a Dim or Static statement.

```
Dim:
```

When a local variable is declared with the Dim statement, the variable remains in existence only as long as the procedure in which it is declared is running. Usually, when the procedure is finished running, the values of the procedure's local variables are not preserved, and the memory allocated to those variables is released. The next time the procedure is executed, all of its local variables are reinitialized.

For example, in the following sample macros, "Example1" and "Example2," the variable X is declared in each of the modules. Each variable X is independent of the other — the variable is only recognized within its respective procedure.

```
Sub Example1()
   Dim X As Integer
   ' Local variable, not the same as X in Example2.
   X = 100
   MsgBox "The value of X is " & X

End Sub

Sub Example2()
   Dim X As String
   ' Local variable, not the same as X in Example1.
```

```
    X = "Yes"
    MsgBox "The answer is " &X
End Sub
Static:
```

A local variable declared with the Static statement remains in existence the entire time Visual Basic is running. The variable is reset when any of the following occur:

The macro generates an untrapped run-time error.
Visual Basic is halted.
You quit Microsoft Excel.
You change the module.

For example, in the RunningTotal example, the Accumulate variable retains its value every time it is executed. The first time the module is run, if you enter the number **2**, the message box will display the value "2." The next time the module is run, if the value 3 is entered, the message box will display the running total value to be 5.

```
Sub RunningTotal()
    Static Accumulate
    ' Local variable that will retain its value after the module
    ' has finished executing.
    num = Application.InputBox(prompt:="Enter a number: ",
    Type:=1)
    Accumulate = Accumulate + num
    MsgBox "The running total is " & Accumulate
End Sub
```

F.2.2 Module Scope

A variable that is recognized among all of the procedures on a module sheet is called a "module-level" variable. A module-level variable is available to all of the procedures in that module, but it is not available to procedures in other modules. A module-level variable remains in existence while Visual Basic is running until the module in which it is declared is edited. Module-level variables can be declared with a Dim or Private statement at the top of the module above the first procedure definition.

At the module level, there is no difference between Dim and Private. Note that module-level variables cannot be declared within a procedure.

NOTE: If you use Private instead of Dim for module-level variables, your code may be easier to read (that is, if you use Dim for local variables only, and Private for module-level variables, the scope of a particular variable will be more clear).

In the following example, two variables, A and B, are declared at the module level. These two variables are available to any of the procedures on the module sheet. The third variable, C, which is declared in the Example3 macro, is a local variable and is only available to that procedure.
Note that in Example4, when the macro tries to use the variable C, the message box is empty. The message box is empty because C is a local variable and is not available to Example4, whereas variables A and B are.

```
 Dim A As Integer' Module-level variable.
 Private B As Integer' Module-level variable.
```

```
Sub Example1()
    A = 100
    B = A + 1
End Sub

Sub Example2()
    MsgBox "The value of A is " & A
    MsgBox "The value of B is " & B
End Sub
Sub Example3()
    Dim C As Integer' Local variable.
    C = A + B
    MsgBox "The value of C is " & C

End Sub
Sub Example4()
    MsgBox A
    ' The message box displays the value of A. MsgBox B
    ' The message box displays the value of B. MsgBox C
    ' The message box displays nothing because C was a local
    variable.

End Sub
```

F.2.3 PUBLIC SCOPE

Public variables have the broadest scope of all variables. A public variable is recognized by every module in the active workbook. To make a public variable available to other workbooks, from a new workbook select the workbook containing the public variable in the Available References box of the References dialog box (from a module sheet, click **References** on the **Tools** menu). A public variable, like a module-level variable, is declared at the top of the module, above the first procedure definition. A public variable cannot be declared within a procedure. A public variable is always declared with a "Public" statement. A public variable may be declared in any module sheet.

It is possible for multiple module sheets to have public variables with the same name. To avoid confusion and possible errors, it is a good idea to use unique names or to precede each variable name with a module qualifier (for example, in a module named "Feb_Sales" you may want to precede all public variables with the letters "FS"). To create the macros:

1. Create a new workbook and name it CDSales.xls.
2. In the **CDSales.xls** workbook, insert a module sheet. Name the module sheet **CDSales**. In Microsoft Excel 97 or later follow these steps to insert a new module sheet and name the module sheet:
   a. In the **CDSales.xls** workbook, point to **Macro** on the **Tools** menu, and then click **Visual Basic Editor**.
   b. On the **Insert** menu, click **Module**.
   c. In Microsoft Excel for Windows (version 97 and later), you can rename a module by activating the module, clicking to the right of "(Name)" in the **Properties** window of the Visual Basic Editor, type a new module name, and then press ENTER.

3. In the CDSales module sheet, type the following code:

```
Public SalesPrice As Integer
Public UnitsSold As Integer
Public CostPerUnit As Integer
Private Markup As Long

Sub CDSales()
    Dim X as String
    SalesPrice = 12
    UnitsSold = 1000
    CostPerUnit = 5
    Markup = 1.05
    X = "yes"
    MsgBox "The Gross Profit for CD Sales is $" & (SalesPrice _
    * UnitsSold) -(UnitsSold * CostPerUnit * Markup)
    ' Displays the value of 7000 as the gross profit.
End Sub
```

4. Create a new workbook and name it **VideoSales.xls**.
5. In the **VideoSales.xls** workbook, insert a module sheet. Name the module sheet **VideoSales**. In Microsoft Excel 97 or later follow these steps to insert a new module sheet and name the module sheet:
   a. In the CDSales.xls workbook, point to **Macro** on the **Tools** menu, and then click **Visual Basic Editor**.
   b. On the **Insert** menu, click **Module**.
   c. In Microsoft Excel for Windows (version 97 and later), you can rename a module by activating the module, clicking to the right of "(Name)" in the **Properties** window of the Visual Basic Editor, type a new module name, and then press ENTER.
6. In the **VideoSales** module sheet, type the following code:

```
Public SalesPrice As Integer
Public UnitsSold As Integer
Public CostPerUnit As Integer

Sub VideoSales()
    SalesPrice = CDSales.SalesPrice * 1.05
    UnitsSold = CDSales.UnitsSold * 1.456
    CostPerUnit = CDSales.CostPerUnit * 1.75
    MsgBox "The Projected Gross Profit for video sales is $" & _
        (SalesPrice * UnitsSold) - (UnitsSold * CostPerUnit)
    ' Displays the value of 5824 as the projected gross profit.
End Sub
```

F.3   TO RUN THE SAMPLE MACROS IN MICROSOFT EXCEL 5.0, 7.0

1. To create a reference from the workbook VideoSales.xls to CDSales.xls, select the **Video Sales** module sheet in **VideoSales.xls**, and then click **References** on the **Tools** menu.
2. In the **References** dialog box, select the **CDSales.xls** check box, and then click OK.
3. Run the CDSales macro and then run the VideoSales macro.

F.4 TO RUN THE SAMPLE MACROS IN MICROSOFT EXCEL 97 AND LATER

1. Rename the project name of the two workbooks so that they are unique by following these steps:

   a. In the Project Explorer pane in the Visual Basic Editor, look at the projects that are listed. You should see entries similar to the following:

   ```
   <VBAProject>  (VideoSales.xls)
   <VBAProject>  (CDSales.xls)
   <VBAProject>  (Personal.xls)
   ```

   where <VBAProject> is the name of the project.

   b. Click the entry for VideoSales.xls.
   c. In the **Properties** pane, in the box to the right of "(Name)," type a new, unique project name, and then press **Enter**.

   NOTE: Do not use a project name that you use in any other workbook.

   d. On the **File** menu, click **Save <bookname>**, where <bookname> is the name of the workbook you modified.

2. To Create a reference from the workbook VideoSales.xls to CDSales.xls, select the **VideoSales** module sheet in VideoSales.xls, and then click **References** on the **Tools** menu.
3. In the **References** dialog box, select the check box of the project name specified in step 1c, and then click **OK**.
4. Run the CDSales macro and then run the VideoSales macro.

Note that the VideoSales macro uses the public variables declared in the CDSales module of CDSales.xls.

F.5 EXAMPLE OF MACRO FAILURE WHEN YOU TRY TO ACCESS LOCAL VARIABLE

The following example tries to use the module-level variable, CDSales.Markup or the local variable CDSales.X in the VideoSales module sheet:

```
Sub VideoSales2()
MsgBox CDSales.Markup
End Sub
Sub VideoSales3()
CDSales.X
End Sub
```

In Microsoft Excel 5.0 or 7.0, the following error message appears when you run either of these procedures:

Member not defined

In Microsoft Excel 97 or later, the following error message appears when you run either of these procedures: Compile error:

Method or data member not found

REFERENCES

For more information about scope, click the Index tab in Microsoft Excel 7.0 Help, type the following text:

scope

and then double-click the selected text to go to the "Understanding scope" topic.

For more information about how long the value of a variable is retained, click the Index tab in Microsoft Excel 7.0

Help, type the following text

```
variables, lifetime
```

and then double-click the selected text to go to the "Lifetime of variables" topic. "Visual Basic User's Guide," version 5.0, Chapter 6, "Making Your Variables Available Within Procedures, Modules, or Publicly."

F.6   APPLIES TO

- Microsoft Excel 2000 Standard Edition
- Microsoft Visual Basic for Applications 1.0
- Microsoft Excel 95 Standard Edition
- Microsoft Excel 5.0 Standard Edition
- Microsoft Excel 5.0c
- Microsoft Excel 5.0 for Macintosh
- Microsoft Excel 5.0a for Macintosh
- Microsoft Excel 97 Standard Edition
- Microsoft Excel 98 for Macintosh
- Microsoft Excel 2002 Standard Edition

Keywords: kbcode kbinfo kbprogramming KB141693

Appendix G: How to Convert Hexadecimal Numbers to Long Integer

This article was previously published under Q161304

Article ID: 161304

Last Review : July 13, 2004

Revision: 3.2

G.1 SUMMARY

Microsoft Visual Basic has a Hex$() function to convert a number into a string of Hexadecimal digits. However, it has no obvious function to perform the inverse conversion. This article details how to use the Val() function to perform the Hexadecimal to Long Integer conversion, plus a trap to avoid.

G.2 MORE INFORMATION

The Val() function can be used to convert a string of decimal digits into a number. It can also be used on a string of hexadecimal digits, but you have to append the characters "&H" to the string of digits so that the Val() function will use the correct number base in the conversion process.

For example:

```
A = Val("1234")' performs decimal conversion
A = Val("7FFF")' results in 7 - the F's are ignored
A = Val("&H7FFF")' performs hexadecimal conversion
```

G.3 INTEGER TRAP

If you are assigning the results of the conversion to a Long integer variable, you will probably only want numbers in the range 80000000 to FFFFFFFF to be treated as negative numbers. However, without taking special precautions, numbers in the range 8000 to FFFF will also be treated as negative.

This occurs because numbers with four or fewer digits are converted to signed 2-byte integers. When the signed integer intermediate value is converted to a Long (or 4-byte) integer, the sign is propagated.

G.4 VISUAL BASIC 4.0/ACCESS 95 AND LATER

By appending the "&" Long suffix to the hexadecimal string, even small numbers are treated as Long. The following function performs the conversion correctly:

```
Function HexToLong(ByVal sHex As String) As Long
  HexToLong = Val("&H" & sHex & "&")
End Function
```

G.5 VISUAL BASIC 3.0/ACCESS 2.0 AND EARLIER

The method illustrated above will not work in Microsoft Visual Basic 3.0 or earlier for hexadecimal numbers greater or equal to 80000000. Please see the REFERENCES section of this article for more information.

Because this problem does not occur with numbers that fall in the Integer Trap range, the alternate function given below adds error trapping to try again without the appended "&" Long suffix:

```
Function HexToLong(ByVal sHex As String) As Long
    On Error Resume Next
    HexToLong = Val("&H" & sHex & "&")
    If Err Then
        On Error Goto 0
        HexToLong = Val("&H" & sHex) End If
End Function
```

Usage of either version of the function is as follows:

```
A = HexToLong("8000")
```

REFERENCES

Microsoft Visual Basic Help topic: Val Function
Microsoft Visual Basic 4.0 ReadMe topic: Coercion of Hexadecimal Values
For more information, please see the following article in the Microsoft Knowledge Base: 95431 (http://support.microsoft.com/kb/95431/EN-US/) FIX: Type Mismatch Error If Use VAL Function on Big Value

G.6 APPLIES TO

- Microsoft Visual Basic 6.0 Learning Edition
- Microsoft Visual Basic 6.0 Professional Edition
- Microsoft Visual Basic 6.0 Enterprise Edition
- Microsoft Visual Basic 5.0 Control Creation Edition
- Microsoft Visual Basic 5.0 Learning Edition
- Microsoft Visual Basic 5.0 Professional Edition
- Microsoft Visual Basic 5.0 Enterprise Edition
- Microsoft Visual Basic 4.0 Standard Edition
- Microsoft Visual Basic 4.0 Professional Edition
- Microsoft Visual Basic 4.0 Enterprise Edition
- Microsoft Visual Basic 2.0 Standard Edition
- Microsoft Visual Basic 3.0 Professional Edition
- Microsoft Visual Basic 2.0 Professional Edition
- Microsoft Visual Basic 3.0 Professional Edition

Keywords: kbhowto KB161304
    &H8000 to &HFFFF Hex = −32,768 to −1, Affects LONG Bit Masking

This article was previously published under Q38888

Article ID: 38888

Last Review : January 9, 2003

Revision: 1.0

G.7   SUMMARY

This article concerns assigning a LONG integer to a hexadecimal constant in the range &H8000 through &HFFFF hex. This information is especially important if you intend to do any bit manipulation with the logical operators (AND, OR, NOT, XOR, EQV, or IMP) with any LONG integer larger than &HFFFF& hex (or 65,535 decimal) that has at least one bit from 16 through 32 on.

Many programmers may not realize that the constants &H8000 through &HFFFF default to a type of short integer, representing decimal values −32,768% through −1%, respectively. Also note that the LONG-integer constants &H8000& through &HFFFF& represent decimal values +32,768 to +65,535. Basic must follow these integer-type notation standards and behaviors since it doesn't have an unsigned data type.

This information applies to Microsoft QuickBasic versions 4.00, 4.00b, and 4.50, to Microsoft Basic Compiler versions 6.00 and 6.00b for MS-DOS and MS OS/2, and to Microsoft Basic Professional Development System (PDS) versions 7.00 and 7.10 for MS-DOS and MS OS/2.

G.8   MORE INFORMATION

Assigning a LONG integer variable to a short integer hexadecimal constant in the range &H8000 through &HFFFF adds &HFFFF0000 to the number, resulting in the LONG integer being stored as &HFFFF8000 to &HFFFFFFFF (that is, −32,768& to −1& decimal).

This behavior occurs because constants &H8000 through &HFFFF default to a type of short integer (%). In short integer notation, constants &H8000 through &HFFFF have decimal values −32,768% through −1%, respectively. For example:

```
PRINT VAL("&HFFFF"); &HFFFF' Prints:-1 -1
PRINT VAL("&H8000"); &H8000' Prints:-32768 -32768
```

To assign hex constants &H8000 through &HFFFF to a LONG integer without turning on bits 16 through 32 (&HFFFF0000), you must change them to a type of LONG by appending an ampersand (&) character as follows:

```
LongInt& = &H8000&' Use &H8000& (or
+32768) instead of &H8000
LongInt& = &H8001&' Use &H8001& (or
+32769) instead of &H8001

  . . .                   . . .
LongInt& = &HFFFF&' Use &HFFFF& (or
+65535) instead of &HFFFF
```

Appending "&" to the constant is not necessary for hex constants outside the range &H8000& through &HFFFF&.

The LONG integer &HFFFF& hex is equal to 65,535 decimal. The short integer &HFFFF hex is equal to −1 decimal (according to the signed, two's complement, integer format standard). A −1 in decimal notation is &HFFFFFFFF in hex LONG-integer notation.

The hexadecimal constants &H8000 through &HFFFF default to short integers &H8000% through &HFFFF%, which represent the decimal numbers −32,768% to −1% in the two's complement, signed binary integer format. You must append an ampersand (&) character to the end of the constant to make it a LONG integer, as follows: &H8000& through &HFFFF& (which represent the decimal numbers 32,768 through 65,535).

The hexadecimal constants &HFFFF8000 to &HFFFFFFFF are LONG integers that represent the decimal numbers −32,768& to −1& in the two's complement, signed binary integer format.

Note: Bit masking (manipulation) is normally NOT done with floating-point numbers, because of the nature of the floating-point format. Bit masking normally is useful only with integers.

G.9 CODE EXAMPLE 1

```
'a& is a variable, and b& will serve as a bit mask:
a& = &HFFFF0000 ' &HFFFF0000 is a constant of type LONG
b& = &HFFFF&' ATTENTION: Assign &HFFFF& instead of &HFFFF
'This masks out bits 16 through 32, and keeps bits 1 through 15:
a& = a& AND b&
PRINT "a& AND b& = "; a&, HEX$(a&); " prints zero (all bits off)"
' Assigning b& to &HFFFF instead of &HFFFF& changes the result:
b& = &HFFFF
' Now, b& contains &HFFFFFFFF, or -1 (all bits on). ANDing
with b&
' does not change a&:
a& = a& AND b&
PRINT "a& AND b& = "; a&, HEX$(a&); " prints -65536, hex
FFFF0000"
```

G.10 CODE EXAMPLE 2

```
' The following assigns short constant &HFFFF (-1) to a LONG
' integer. &HFFFF is converted to &HFFFFFFFF; the decimal
' value (-1) stays the same:
longint& = &HFFFF
shortint% = &HFFFF
PRINT "longint& =", longint&, HEX$(longint&) PRINT "shortint%
=", shortint%, HEX$(shortint%)

' The following assigns short constant &H8000 (-32,768) to a LONG
' integer. &H8000 is converted to &HFFFF8000; the decimal
' value (-32,768) stays the same:
longint& = &H8000
shortint% = &H8000
PRINT "longint& =", longint&, HEX$(longint&)
PRINT "shortint% =", shortint%, HEX$(shortint%)
```

Keywords: KB38888

From: Randy ***** (PSS TEXAS) [*******@microsoft.com]
Sent: Tuesday, May 25, 2004 11:30 AM
To: brian_d_bissett@*************
Subject: Support Case #*********************
Hi Brian,
I put both issues in the same case, and changed the type to advisory.
Included is the file I made from your code without changing the code and just adding "&" to the end of each hex number. You could test for "&" on the end of the hex input and add it if it isn't there. I will be happy to write the code for you if needed but it looks like you are an accomplished programmer.
As we discussed, I would not use the Hex2Dec function name as it conflicts with the name used in the Analysis tool pack. I changed your function name to Hx2Dx.
Here is the text of the article I found on why your code didn't work....we both got to learn something...<grin>....
I'll start on the Conditional formatting issue now and also drop the severity to a Sec C (Important).
Thanks
-randy
&H8000 to &HFFFF Hex = −32,768 to −1, Affects LONG Bit Masking WGID:188
ID: 38888.KB.EN-US   CREATED: 1988-12-07   MODIFIED: 2003-01-09
Public |
* Security   : Public

==

G.11 SUMMARY

This article concerns assigning a LONG integer to a hexadecimal constant in the range &H8000 through &HFFFF hex. This information is especially important if you intend to do any bit manipulation with the logical operators (AND, OR, NOT, XOR, EQV, or IMP) with any LONG integer larger than &HFFFF& hex (or 65,535 decimal) that has at least one bit from 16 through 32 on.

Many programmers may not realize that the constants &H8000 through &HFFFF default to a type of short integer, representing decimal values −32,768% through −1%, respectively. Also note that the LONG-integer constants &H8000& through &HFFFF& represent decimal values +32,768 to +65,535. Basic must follow these integer-type notation standards and behaviors since it doesn't have an unsigned data type.

This information applies to Microsoft QuickBasic versions 4.00, 4.00b, and 4.50, to Microsoft Basic Compiler versions 6.00 and 6.00b for MS-DOS and MS OS/2, and to Microsoft Basic Professional Development System (PDS) versions 7.00 and 7.10 for MS-DOS and MS OS/2.

G.12 MORE INFORMATION

Assigning a LONG integer variable to a short integer hexadecimal constant in the range &H8000 through &HFFFF adds &HFFFF0000 to the number, resulting in the LONG integer being stored as &HFFFF8000 to &HFFFFFFFF (that is, −32,768& to −1& decimal).

This behavior occurs because constants &H8000 through &HFFFF default to a type of short integer (%). In short integer notation, constants &H8000 through &HFFFF have decimal values −32,768% through −1%, respectively. For example:

```
PRINT VAL("&HFFFF"); &HFFFF    ' Prints:   -1 -1
PRINT VAL("&H8000"); &H8000    ' Prints:   -32768 -32768
```

To assign hex constants &H8000 through &HFFFF to a LONG integer without turning on bits 16 through 32 (&HFFFF0000), you must change them to a type of LONG by appending an ampersand (&) character as follows:

```
LongInt& = &H8000&     ' Use &H8000& (or
+32768) instead of &H8000

LongInt& = &H8001&     ' Use &H8001& (or
+32769) instead of &H8001

  . . .          . . .

LongInt& = &HFFFF&     ' Use &HFFFF& (or
+65535) instead of &HFFFF
```

Appending "&" to the constant is not necessary for hex constants outside the range &H8000& through &HFFFF&.

The LONG integer &HFFFF& hex is equal to 65,535 decimal. The short integer &HFFFF hex is equal to −1 decimal (according to the signed, two's complement, integer format standard). A −1 in decimal notation is &HFFFFFFFF in hex LONG-integer notation.

The hexadecimal constants &H8000 through &HFFFF default to short integers &H8000% through &HFFFF%, which represent the decimal numbers −32,768% to −1% in the two's complement, signed binary integer format. You must append an ampersand (&) character to the end of the constant to make it a LONG integer, as follows: &H8000& through &HFFFF& (which represent the decimal numbers 32,768 through 65,535).

The hexadecimal constants &HFFFF8000 to &HFFFFFFFF are LONG integers that represent the decimal numbers −32,768& to −1& in the two's complement, signed binary integer format.

Note: Bit masking (manipulation) is normally NOT done with floating-point numbers because of the nature of the floating-point format. Bit masking normally is useful only with integers.

CODE EXAMPLE 1

```
'a& is a variable, and b& will serve as a bit mask:
a& = &HFFFF0000 ' &HFFFF0000 is a constant of type LONG
b& = &HFFFF&   ' ATTENTION: Assign &HFFFF& instead of &HFFFF

'This masks out bits 16 through 32, and keeps bits 1 through 15:
a& = a& AND b&
PRINT "a& AND b& = "; a&, HEX$(a&); " prints zero (all bits off)"

' Assigning b& to &HFFFF instead of &HFFFF& changes the result:
b& = &HFFFF
' Now, b& contains &HFFFFFFFF, or -1 (all bits on). ANDing with b&
' does not change a&:
a& = a& AND b&
PRINT "a& AND b& = "; a&, HEX$(a&); " prints -65536, hex FFFF0000"
```

CODE EXAMPLE 2

```
' The following assigns short constant &HFFFF (-1) to a LONG
' integer. &HFFFF is converted to &HFFFFFFFF; the decimal
```

```
' value (-1) stays the same:
longint& = &HFFFF
shortint% = &HFFFF
PRINT "longint& =", longint&, HEX$(longint&)
PRINT "shortint% =", shortint%, HEX$(shortint%)
' The following assigns short constant &H8000 (-32,768) to a LONG
' integer. &H8000 is converted to &HFFFF8000; the decimal
' value (-32,768) stays the same:
longint& = &H8000
shortint% = &H8000
PRINT "longint& =", longint&, HEX$(longint&)
PRINT "shortint% =", shortint%, HEX$(shortint%)
```

Appendix H: SendKeys Statement

Description

Sends one or more keystrokes to the active window as if typed at the keyboard. Syntax

SendKeys string[, wait]

The SendKeys statement syntax has these named arguments:

Part	Description
string	Required. String expression specifying the keystrokes to send.
wait	Optional. Boolean value specifying the wait mode. If False (default), control is returned to the procedure immediately after the keys are sent. If True, keystrokes must be processed before control is returned to the procedure.

Remarks

Each key is represented by one or more characters. To specify a single keyboard character, use the character itself. For example, to represent the letter A, use "A" for string. To represent more than one character, append each additional character to the one preceding it. To represent the letters A, B, and C, use "ABC" for string.

The plus sign (+), caret (^), percent sign (%), tilde (~), and parentheses ( ) have special meanings to SendKeys. To specify one of these characters, enclose it within braces({}). For example, to specify the plus sign, use {+}. Brackets ([ ]) have no special meaning to SendKeys, but you must enclose them in braces. In other applications, brackets do have a special meaning that may be significant when dynamic data exchange (DDE) occurs. To specify brace characters, use {{} and {}}.

To specify characters that aren't displayed when you press a key, such as ENTER or TAB, and keys that represent actions rather than characters, use the codes shown below:

Key	Code
BACKSPACE	{BACKSPACE}, {BS}, or {BKSP}
BREAK	{BREAK}
CAPS LOCK	{CAPSLOCK}
DEL or	DELETE{DELETE} or {DEL}
DOWN ARROW	{DOWN}
END	{END}
ENTER	{ENTER} or ~
ESC	{ESC}
HELP	{HELP}
HOME	{HOME}
INS or INSERT	{INSERT} or {INS}

LEFT ARROW	{LEFT}
NUM LOCK	{NUMLOCK}
PAGEDOWN	{PGDN}
PAGE UP	{PGUP}
PRINT SCREEN	{PRTSC}
RIGHT ARROW	{RIGHT}
SCROLL LOCK	{SCROLLLOCK}
TAB	{TAB}
UPARROW	{UP}
F1	{F1}
F2	{F2}
F3	{F3}
F4	{F4}
F5	{F5}
F6	{F6}
F7	{F7}
F8	{F8}
F9	{F9}
F10	{F10}
F11	{F11}
F12	{F12}
F13	{F13}
F14	{F14}
F15	{F15}
F16	{F16}

To specify keys combined with any combination of the SHIFT, CTRL, and ALT keys, precede the key code with one or more of the following codes:

Key	Code
SHIFT	+
CTRL	^
ALT	%

To specify that any combination of SHIFT, CTRL, and ALT should be held down while several other keys are pressed, enclose the code for those keys in parentheses. For example, to specify to hold down SHIFT while E and C are pressed, use "+(EC)." To specify to hold down SHIFT while E is pressed, followed by C without SHIFT, use "+EC."

To specify repeating keys, use the form {key number}. You must put a space between key and number. For example, {LEFT 42} means press the LEFT ARROW key 42 times; {h 10} means press H 10 times.

NoteSendKeys to send keystrokes to an application that is not designed to run in Microsoft Windows. SendKeys also can't send the PRINT SCREEN key {PRTSC} to any application.

See Also

AppActivate statement, DoEvents function.

Example

This example uses the Shell function to run the Calculator application included with Microsoft Windows. It uses the SendKeys statement to send keystrokes to add some numbers, and then quit the Calculator. The SendKeys statement is not available on the Macintosh. (To see the example, paste it into a procedure, then run the procedure. Because AppActivate changes the focus to the Calculator application, you can't single step through the code.)

```
Dim ReturnValue, I
ReturnValue = Shell("Calc.exe", 1)  ' Run Calculator.
AppActivate ReturnValue                 ' Activate the Calculator.
For I = 1 To 100                        ' Set up counting loop.
    SendKeys I & "{+}", True            ' Send keystrokes to
                                        Calculator

Next I' to                              add each value of I.

SendKeys "="                            , True' Get grand total.

SendKeys "%{F4}", True      ' Send ALT+F4 to close Calculator.
```

Index